W0113511

Coal Surface Mining:
Impacts of Reclamation

Other Titles in This Series

Westview Special Studies in Natural Resources and Energy Management

Coal Surface Mining: Impacts of Reclamation
edited by James E. Rowe

Since coal is seen by many as the logical solution to the nation's energy crisis, strip mining operations will continue. But they will continue amid intense public debate, much of it centering on the standards that will govern reclamation. In this book leading authorities address the economic, environmental, and legal ramifications of land reclamation following coal surface mining, review the status of the coal industry, and look at possible future developments.

James E. Rowe is executive director of the Sauk County Development Corporation, Wisconsin. He previously served as chief of planning research and development for the Tri-County Council for western Maryland and as community development planner for the Mount Rogers Planning District Commission, Virginia.

Coal Surface Mining:
Impacts of Reclamation
edited by James E. Rowe

Routledge
Taylor & Francis Group

LONDON AND NEW YORK

First published 1979 by Westview Press

Published 2018 by Routledge
52 Vanderbilt Avenue, New York, NY 10017
2 Park Square, Milton Park, Abingdon, Oxon OX14 4RN

Routledge is an imprint of the Taylor & Francis Group, an informa business

Copyright © 1979 by Taylor & Francis

All rights reserved. No part of this book may be reprinted or reproduced or utilised in
any form or by any electronic, mechanical, or other means, now known or hereafter
invented, including photocopying and recording, or in any information storage or
retrieval system, without p ermission in writing fromthe publishers.

Notice:
Product or corporate names may be trademarks or registered trademarks, and are used
only for identification and explanation without intent to infringe.

Library of Congress Cataloging in Publication Data
Main entry under tide:
Coal surface mining: Impacts of reclamation.
   (Westview special studies in natural resources and energy management)
   1. Coal mines and mining-Environmental aspects.   2. Strip mining-Environ-
mental aspects.   3. Reclamation of land. I. Rowe, James E.
TD195.CS8C62   333.7'6   79-761
ISBN D-89158-475-7

ISBN 13: 978-0-367-02126-9 (hbk)
ISBN 13: 978-0-367-17113-1 (pbk)

To Clara, Lisa, and Heather
They light up my life

Contents

Coal Surface Mining:
Impacts of Reclamation

Introduction

The impacts of land reclamation following coal surface mining are three-fold: environmental, economic, and legal. Environmental impact has received the most notice but the real issue has been economic. Can the coal industry, the coal consumer, and the nation afford the cost of addressing the environmental issue raised by decades of weak reclamation laws and enforcement? Although the fact remains that these costs must be absorbed, the question is by whom—the coal industry or the consumer? The third type of reclamation impact is the legal problem associated with upgrading reclamation standards. This collection of articles is designed to review the various aspects of the reclamation issue and to provide an overview of the status of the coal industry and a look into the future. The need for such a compilation is self-evident by the unusual dispersion of the literature throughout many different academic disciplines and obscure government research documents.

Many Americans look to coal for the logical solution to the nation's energy crisis. Since the late 1960s, however, the destruction of the landscape by surface mining and resulting adverse environmental effects have received much publicity, particularly the environmental impact of surface mining on steep slopes. The amount of land disturbed by strip mining as of January 1, 1972, exceeded 4 million acres nationally.

In response to the increased demand for surface-mined coal and to the nation's growing environmental awareness, the federal government has sought to establish minimum standards for surface mine reclamation. After several attempts the Surface Mining Control and Reclamation Act of 1977 was signed into law on April 3, 1977, by President Carter (HR 2, now P.L. 95-87).

Adapted and expanded from James E. Rowe, "Surface Mine Reclamation: The 'Back-to-Contour' Constraint," *Journal of Soil and Water Conservation* 32, no. 2 (March-April 1977):74-75. Copyrighted by the Soil Conservation Society of America. Reprinted by permission.

Much of the debate surrounding enactment of a federal surface mine reclamation law has centered on the economic impact of requiring "back-to-contour" reclamation. The fear is that such a strict reclamation standard would cripple the coal industry and greatly hinder America's chances of achieving some measure of energy independence. President Ford used this argument in vetoing the 1975 version of the bill. The House Rules Committee tabled a similar bill in 1976. However, that bill provided that in special circumstances the back-to-contour constraint would not apply. Without that provision, many postmining land uses could not even be considered.

The act marks the first successful federal attempt to regulate surface mining since Everett Dirksen introduced the idea in 1937, and culminates five years of an intensive federal legislative effort. It establishes the first minimum federal standards for environmental protection that must be met by strip-mining operations. States will have primary responsibility for enforcing the law, with the secretary of interior having backup enforcement authority. The act also creates an Abandoned Mine Reclamation Fund to be used to restore lands adversely affected by surface-mining operations in the past.

According to Title V of the act, states that wish to assume jurisdiction over the regulation of surface coal-mining and reclamation operations must prepare a state regulatory program. Under the act the secretary of interior has authority to make grants to the states to develop and implement these regulatory programs. Grants can provide as much as 80 percent of costs during the first year, 60 percent the second year, and 50 percent each year thereafter. These programs must be submitted to the secretary of interior within eighteen months after enactment of this bill.

State programs must include laws that define violations of state laws concerning strip-mining and reclamation operations; establish a process for designating areas unsuitable for strip mining; and provide for the implementation, maintenance, and enforcement of a permit system to regulate strip-mining and reclamation operations within the state. Under the state programs, states will be able to issue permits for mining sites where the following requirements of the act are met: reclamation can be accomplished, damage to water sources is prevented, and the area has not been specifically designated as unsuitable for mining. The secretary of interior may plan and implement a federal strip-mining regulation program in any state that fails to submit its own program.

The Abandoned Mine Reclamation Fund will be established as a Department of the Treasury trust fund. Money for reclamation will come mainly from a reclamation fee of $.35 per ton of coal produced by surface coal mining and $.15 per ton of coal produced by underground methods, paid by mine owners to the secretary of interior. Fifty percent of the annual funds will be returned to the states where they were collected for use in mine reclamation programs. The balance of the funds may be expended at the discretion of the secretary of interior. Of this 50 percent, 10 percent or not more than $10 million may pay the cost of test borings and core samplings for operators of small coal mines (coal mines that produce a total of less than 100,000 tons a year) who request assistance. As much as 20 percent of the 50 percent may be transferred from the Interior Department to the Agriculture Department's Soil Conservation Service to help to reclaim rural lands.

Some of the other important highlights of the Surface Mining Control and Reclamation Act of 1977 are as follows:

• Authorization of $10 million each year for FY 1978–FY 1979 for development of initial regulatory procedures and administration of the program; $10 million each year for fifteen years beginning in FY 1978 for hydrologic studies and test borings for small mine operations; and $20 million in FY 1978 and $30 million each in FY 1979 and FY 1980 for grants to states to prepare their regulatory plans. Authorization of $2 million in FY 1977 for the secretary of interior to begin implementing the act.

• Establishment of an Office of Surface Mining Reclamation and Enforcement in the Interior Department. The new office has responsibility for administering the act's regulatory and reclamation programs, approving state programs, and providing grant and technical assistance to the states.

• Authorization of each state to establish, or continue to support, a state mining and mineral resources research institute at a public or private college. Appropriations for each state are $200,000 in FY 1978, $300,000 in FY 1979, and $400,000 for each of the following five fiscal years. These funds are to be matched dollar-for-dollar with nonfederal funds. An additional $15 million in FY 1978, which will be increased by $2 million in each fiscal year for six years, will fund specific mineral research and demonstration projects.

• Authorization of the administrator of the Energy Research and Development Administration (ERDA), now Department of Energy (DOE), to designate ten institutions of higher education to establish university coal research laboratories. Grants of as much as $6 million may be made for initial costs and $1.5 million annually for operating expenses for each laboratory.

• Authorization of the administrator of ERDA to award 1,000 graduate fellowships annually from FY 1979 through FY 1984 for research in applied science and engineering related to the production, conservation, and use of fossil fuels for energy. Eleven million dollars for each of the six fiscal years was appropriated for these purposes.

P.L. 95-87 requires that operators restore the land to the approximate original contour, except where mining removed an entire coal seam or seams running through the upper fraction of a mountain by stripping the overburden and creating a level plateau capable of supporting postmining uses. In situations where an industrial, commercial, residential, or public development is proposed for postmining use, the state may grant a variance from the back-to-contour requirement.

To obtain a variance, specific postmining land-use plans must be prepared and assurances made that guarantee the site will be (a) compatible with adjacent land uses, (b) realistic with regard to expected needs and markets, (c) assured of investment in necessary public facilities, (d) supported by commitments from public agencies where needed, (e) practical with respect to private financing of the proposed development, (f) planned in advance and integrated with mining operations and reclamation, and (g) designed by a registered engineer to assure the configuration stability and drainage needed for intended use.

The coal and electric power industries opposed the federal surface-mine legislation. The steam electric coal industry consumes over 85 percent of all coal extracted. This rate of consumption is increasing at a rate of 6.4 percent annually, and most of the coal consumed is obtained by surface mining. They argued that existing state statutes were adequate to protect the environment and that the federal law would prohibit rapid expansion of coal production. Largely undocumented estimates have been quoted widely indicating that a back-to-contour reclamation requirement would result in lost output of 50 million tons of coal (about one-sixth of current annual surface mine output), preclude contour mountain mining, cost 50,000 jobs, and increase the nation's energy bill by millions of dollars.

The controversy over a back-to-contour requirement is particularly heated in Appalachia. Coal mining in this region often takes place on slopes exceeding twenty-five degrees where back-to-contour constraints could make surface mining completely uneconomic. It is with regard to the economic impact, rather than the environmental impact, that the debate about back-to-contour reclamation has continued.

Lin et al. found that in 1972 full reclamation requirements would have resulted in an increase of $.35 per ton in production costs. This would also have meant a 10 million ton decrease in production and a $.40 per ton increase in the delivered price of coal. They also estimated that the implementation of a strict law at that time would have resulted in a direct loss of 885 jobs with another indirect loss of a total of 1,593 jobs if the 1.8 national employment multiplier were used. In 1972, 47 percent of all Appalachian coal was strip mined and it was estimated that the economic impacts would be temporary and not long term.[1]

In an earlier study by Goldstein and Smith, the 1971 projection of a back-to-contour requirement would have increased the price of coal by $.10 per ton. This in turn would have caused coal production and employment in strip mining to fall by 2.3 percent. However, they found that underground output and employment would have risen by 1.75 percent for almost no net change.[2]

Therefore, the key question is how to mine coal on steep slopes and not destroy the environment. Many excellent examples of environmentally compatible reclamation have been documented. One such example is at Moraine State Park in Butler, Pennsylvania.[3] At Moraine, back-to-contour reclamation was found to cost approximately $1,100 per acre. The cost per ton would be $.15 if one foot of coal per acre yields 2,500 tons of coal and the seams were three feet thick. If the coal sold for $25 per ton, the gross revenue would be $187,500 per acre or $.15 per ton. A realistic national figure would be approximately $3,000 per acre.

The cost per ton should decrease because of the rapidly accelerating cost of coal. The *Wall Street Journal* reported that the Tennessee Valley Authority (TVA) has just signed (summer 1978) a ten-year contract with Island Creek Sales for delivery of 14 million tons of West Virginia and Kentucky coal at a price of $44.42 per ton. Another TVA contract with the Pittston Coal Sales Company calls for 15 million tons of low sulfur southwestern Virginia coal at $50.89 a ton.

Several methods of extraction that restore the land to its approximate original contour are being developed. The TVA is mining coal on Massengale Mountain in Campbell County, Tennessee, in a manner that meets the federal regulations. The method is basically truck haulback. Zwick concluded that at Massengale the incremented overburden handling cost per ton of coal was $2.50.[4] Another method known as Pennsylvania block-cut mining uses bulldozers and loaders to move spoil over relatively short distances to the point of final placement.

Federal legislation to control surface mining can be construed as a response to the adverse environmental impacts of coal surface mining. The environmental degradation of strip mining is exemplified by the U.S. Geological Survey study describing the sedimentation rate at Fishtrap Lake near Pikeville, Kentucky. It found that due to strip mining, the reservoir was filling with sediment seven times faster than anticipated.

Yet the current controversy over the legislation concerns all the possible economic impacts: increased coal costs; increased prices; increased, and in some cases decreased, output; as well as employment and income in coal mining regions. LaFevers found that reclamation costs vary widely. At Black Mesa, Arizona, the Peabody Coal Company found that complete

reclamation costs were $5,000 per acre. However, this produced more pro-
ductive land after mining because the original land was so overgrazed by
the livestock of the Navajo and Hopi Indian tribes. He also found that in
the North Dakota lignite mining areas the cost of mining increased from
$2,400 per acre under the 1976 state law to $6,613 under the new proposals.[5]

The controversy is furthered by some of the taxation policies now
being proposed in various coal producing states. Many of these states tax
coal at a rate far below the costs of direct industry damage to public roads.
Maryland is a prime example. Coal is taxed at a rate of only $.15 per ton.
A recent proposal called for $.30 per ton; however, this caused an uproar
and industry claimed that they were being taxed out of business. At the
other extreme is the Montana severance tax of 30 percent of value on all
surface mined coal.

The economics and the geography of the two mining areas are different
but the extraction of coal is increasing in both states. The tax has not
forced anybody out of business in Montana. Most of Appalachia is not
even recovering enough taxes to pay for road and bridge damages let alone
any environmental damage. The new federal laws will force increased
taxation in the form of a $.35 surcharge per ton but this is not enough to
offset all of the peripheral costs.

There are three main areas of coal production in the United States.
The coal fields of Appalachia stretch from Pennsylvania to Alabama. Most
underground coal production also occurs in this region. Here, surface
mining generally occurs on steep slopes over fifteen degrees. The environmental problems associated with contour mining on such slopes are particularly severe.

The major midwestern coal fields—Illinois, Indiana, and western Kentucky—are dominated by surface mining on gently rolling terrain. The
relatively recent development of western coal fields has mainly involved
surface mining because of the relatively thick coal seams and the shallow
overlying strata.

The explicit purpose of some sponsors of back-to-contour legislation
is to stimulate underground production. In our opinion, the regional
impacts of a back-to-contour constraint include the following:

• It costs more to mine under a back-to-contour method than under
conventional mining.

• Present technology allows for efficient back-to-contour techniques.

• Equipment requirements will increase with passage of the new
regulations.

• Increased costs may put small mine operators out of business.

• Total production will remain stable.

The economic impacts of a back-to-contour reclamation requirement will not ruin the coal industry or the nation. The impact of P.L. 95-87 today should be significantly less than it would have been in the early 1970s. Since then, state laws have become more rigid. Many now require back-to-contour reclamation. Also, markets have been given ample time to anticipate and adjust to the changes that will arise in conjunction with back-to-contour reclamation.

While a national energy plan has not been delineated, there seems to be an emerging consensus that Appalachian coal will provide an increasingly important share of U.S. energy supplies. The energy crisis will in all likelihood precipitate further strip mining on the Appalachian region's already scarred landscape. Therefore, effective reclamation depends upon the quality and enforcement of mining and reclamation laws.

The regional impact of the back-to-contour constraint will not upset Appalachian coal production as much as people fear. Overall, the regulation should have a beneficial impact on the region both environmentally and, in the long term, economically.

Organization of This Book

Coal Surface Mining: Impacts of Reclamation is divided into seven main sections. The Introduction sets the stage for the following sections.

Part 1, "Surface Mining in Perspective," answers the questions of what, when, where, and how much. Clements's article describes the geography (spatial distribution) of coal. The two articles from *Coal Age* describe the industry's view of the problem and the techniques being used today. Paone et al. list the statistics needed for an understanding of the scope of the problem and the numbers involved.

Part 2, "Environmental Impacts," contains articles favoring both the coal industry and the environmentalist. They describe the massive problems associated with surface mining. All the articles point out that environmental damage can be overcome, but results take time and are expensive.

Part 3, "Economic Impacts," contains five articles written by economists. These articles, except Dials and Moore's "The Cost of Coal," are highly technical and detailed. All were written before the 1977 national law was signed; however, they merit careful reading and analysis because of the approaches taken toward an objective understanding of the economic problems associated with surface mining.

Part 4, "Legal Impacts," offers five articles on the legal problems of surface mining. They are to be read with the understanding that they were written in anticipation of the enactment of a federal law. Some of the arguments presented in the articles were based upon P.L. 95-87 and some were not. The Surface Mining Control and Reclamation Act of 1977 is also described in detail. However, the final regulations have not been presented at the time of publication of this volume. We would urge readers to investigate the final regulations carefully when they are available.

Part 5, "The Future of Surface Mining," describes the unique Appalachian and German experiences. Schlottmann's article on the steam electric coal market is pertinent since such a large portion of the nation's strip mined coal is destined for utilities. Our article on land use alternatives and Lafevers and Imhoff's article on land use planning offer some solutions. The article on the allocation of coal offers a look into the future demands for coal.

The Glossary of Terms is designed to give a clear definition for the many unexplained terms used by most of the authors including the editor.

Intended Use

This collection of articles is designed to make available in one volume the many scattered research reports describing the impacts of coal surface-mining reclamation. It should provide an important reference for anyone involved with reclamation.

The book should also prove useful as a text for college courses on reclamation, resource management, or conservation. Lastly, it should be an excellent supplemental reader for courses in natural resources, ecology, energy, economics, and geography.

Acknowledgments

I am indebted to the many authors of the various scholarly articles included in this reader, as well as to their editors for granting reprint permission. I also appreciate the comments from Jon Loff, Allegany Community College, and Alan Schlottmann, University of Tennessee, on the Introduction to this volume. I want to thank my editor, Lynne Rienner, for her valuable suggestions, and my typist, Peggy Lane.

Notes

1. William Lin, Robert L. Spore, and Edmund A. Nephew, "Land Reclamation and Strip-mined Coal Production in Appalachia," *Journal of Environmental Economics and Management* 3 (1976):251.

2. Morris Goldstein and Robert S. Smith, "Land Reclamation Requirements and their Estimated Effect on the Coal Industry," *Journal of*

Environmental Economics and Management 2 (1975):145.

3. L. M. McNay, *Surface Mine Reclamation, Moraine State Park, Pennsylvania* PB 225 165 (Washington, D.C.: U.S. Government Printing Office, 1970), pp. 1-27.

4. Burton Zwick, *The Cost of Back-to-Contour Reclamation of Steep Slope Surface Coal Mines: The TVA Massengale Mountain Experience* UTEC-ARP-TM-1 (Knoxville, Tenn.: University of Tennessee Environment Center, May 1975), p. 21.

5. James R. LaFevers, "A Regional Perspective on Land Reclamation Economics," *Proceedings*, 3rd Conference on High Altitude Revegetation at Fort Collins, Colorado, March 1978.

Part 1
Surface Mining in Perspective

1
Recent Trends
in the Geography of Coal

Donald W. Clements

Conditions for the location of economic activity frequently are stated in physical and economic terms. Institutional factors must also be considered. A pervasive institutional element is legal restraint which derives from legislative and administrative fiat. Governmental regulations imposed since the 1960s have caused profound changes in the coal industry of the United States. This study is concerned with these events which have coincided with rapidly escalating demands and prices for energy.

Coal is this nation's most copious energy source. The current demonstrated coal reserve base (economically and legally available) is 434,000,000,000 tons (393,718,000,000 metric tons).[1] Eight times this quantity is believed to exist based on actual surveys and theoretical extrapolation (Table 1).[2] The demonstrated base will most likely increase as more exploratory drill holes are made in western states. Coal output for domestic use and export in 1975 was a record 640,000,000 tons (580,598,000 metric tons).[3] Coal production in the United States increased from its inception in the eighteenth century until the 1920s. After that the disadvantages of coal such as dirt, inconvenience, bulk, relatively low-energy content, production dangers, and high extraction and transport costs caused a decline in coal's contribution to national energy consumption relative to petroleum and natural gas. World War II energy demands resulted in a brief rise in output, but absolute and comparative attrition in coal production continued until the early 1960s. Coal output has risen since 1960 to current record levels, although coal's relative contribution to total national energy

From the *Annals of the Association of American Geographers*, vol. 67, no. 1, March 1977, pp. 109-125. Reprinted by permission.

Dr. Clements is Assistant Professor in the Department of Earth Sciences and Planning at Southern Illinois University in Edwardsville, Illinois 62026.

TABLE 1.—DEMONSTRATED COAL RESERVE BASE OF
THE UNITED STATES ON JANUARY 1, 1974, BY AREA,
RANK AND POTENTIAL METHOD OF MINING

	Anthra-cite	Bitumi-nous	Sub-bitumi-nous	Lig-nite	Total[a]
		(Billions of Short Tons)			
Underground					
East of the Mississippi River	7	162	0	0	169
West of the Mississippi River	<1/2	31	98	0	129
Total	7	192	98	0	297
Surface					
East of the Mississippi River	<1/2	33	0	1	34
West of the Mississippi River	0	8	67	27	103
Total	<1/2	41	67	28	137
Total[a]	7	233	165	28	434

Source: *Mineral Industry Surveys*, op. cit., footnote 1.
[a] Totals may not add because of rounding.

consumption continues to decrease.[4] In large measure this has been a response to direct and indirect governmental intervention. Wellhead price ceilings set by the Federal Power Commission produce artificially low prices on petroleum products and natural gas. Other elements include the removal of oil import quotas for East Coast users of residual oil, the 1970 mandate to the Environmental Protection Agency (EPA) to reduce coal-induced atmospheric pollutants, and, to a lesser degree, the Atomic Energy Commission's encouragement of nuclear power plant construction. Coal's contribution to energy production declined from nearly ninety percent in 1900 to approximately seventeen percent today (Tables 2 and 3).[5] Recent shortages, international difficulties, uncertain availability of preferred fuels, technical difficulties in expanding nuclear energy production, and increasing prices for alternative fuels have inspired a reassessment of the role coal should play in satisfying current and future energy demands. A rapid and massive return to coal is frequently promoted as a feasible solution to the national energy shortage. Federal regulations on air contaminants and miners' safety, state land reclamation requirements, strikes, and market uncertainties are claimed to be impediments to this transition.

TABLE 2.—UNITED STATES COAL CONSUMPTION
(Millions of Short Tons)

Sector	1965	1973	Percentage Change
Electrical power	244.9	387.6	+58
Coke plants	95.3	94.1	− 1
Industrial	104.1	68.1	−35
Retail	22.1	11.1	−50
Exports	52.3	52.9	+ 1
Total	518.7	613.8	+18

Source: Federal Energy Administration, *Project Independence Report* (November, 1974), p. 100.

TABLE 3.—FUEL CONSUMPTION FOR ELECTRICITY
GENERATION
(Million Short Tons of Coal Equivalent)

Fuel	1965	1973	Percentage Change
Coal	245	388	+ 58
Oil	28	121	+ 332
Gas	96	154	+ 60
Nuclear	2	36	+1800
Total	371	699	+ 88

Source: Federal Energy Administration, *Project Independence Report* (November, 1974), p. 100.

Air Pollution Abatement

Federal involvement with air pollution abatement commenced in 1955 with amendments in 1963, 1965, and 1967, but was limited largely to a technical, advisory, and persuasive role in assisting state and local governments with existing pollution control programs. The landmark Clean Air Act Amendments (P.L. 91-604) of December 31, 1970, placed the responsibility for setting standards solely in the hands of the federal government and specifically under the newly created EPA. This legislation provided a series of amendments which were couched in mandatory terms including a series of primary standards necessary to protect public health which were to be met by July 1975. The separate states were to develop their own implementation programs to comply with federal standards. The EPA monitors the abatement processes and may impose its own programs if it is not satisfied with the plans and progress made. Since 1955 the federal role has gradually proceeded from attempts to stimulate local and state pollution-control machinery to more rigorous federal intervention, complete with the coercive force of jail terms and fines. This is considerably more humane than the 1306 A.D. royal edict of Edward I of England which meted out death by hanging for persons producing pestilential odors from coal! By July, 1975, the 1970 Amendments had been somewhat successful in reaching their goals, but numerous variances and postponements retarded their achievement considerably. This law also set secondary air-quality standards which pertain to quality of life including effects on flora and fauna, structures, and atmospheric visibility. These are to be achieved by 1977, but the indeterminate nature of this phase of the law will make it more difficult to enforce.[6]

The major concern with the combustion of coal is the control of oxides of sulfur. Other pollutants such as oxides of nitrogen, carbon monoxide, and hydrocarbons can be controlled effectively by a combination of duration and temperature of combustion, available air, rate of emission cooling, and fuel-air contact and turbulence. Sulfur is an element which accompanies coals in varying degrees. American coals contain from 0.2 to 7 percent sulfur, but values from 1 to 2 percent are most frequently encountered (Figure 1).[7] High-sulfur coals are of marine origin and are overlain by fossiliferous shales and limestones, whereas the low-sulfur coals are roofed by nonmarine shales which bear plant fossils.[8]

Sulfur is an undesirable component of coal. It lowers the quality of coke as a reduction agent in the production of steel, and it enhances corrosion, air pollution, and boiler deposits. Unfortunately, some coal-burning operations must be redesigned or modified to accept the cleaner low-sulfur coals because of the higher ash-softening temperatures required. The fly ash from high-sulfur coals is more readily ionized, however, and thereby more amenable to removal in electrical precipitators.

The federal performance standard for sulfur dioxide (SO_2) is 1.2 pounds (0.54 kg.) of SO_2 per 1,000,000 Btu (251,996,000 g. cal.) of energy production for new emission sources. This is frequently termed the New Source Performance Standard for large boilers (NSPS). Pollution sources that existed prior to 1970 ("old" sources) are controlled by state statutes as required under federal law and these sources frequently have less stringent standards, such as intermittent use of scrubbers under adverse meteorological conditions. "Old" sources in metropolitan areas such as St. Louis, Chicago, and Peoria must conform to a maximum of 1.8 pounds (0.82 kg.) of SO_2 per 1,000,000 Btu, whereas those in areas beyond large cities are subjected to a less stringent maximum of 6 pounds (2.72 kg.) per 1,000,000 Btu. This is the equivalent of approximately three percent sulfur in average bituminous coal.[9] The location of "old" plants thus strongly influences their choice of fuel.

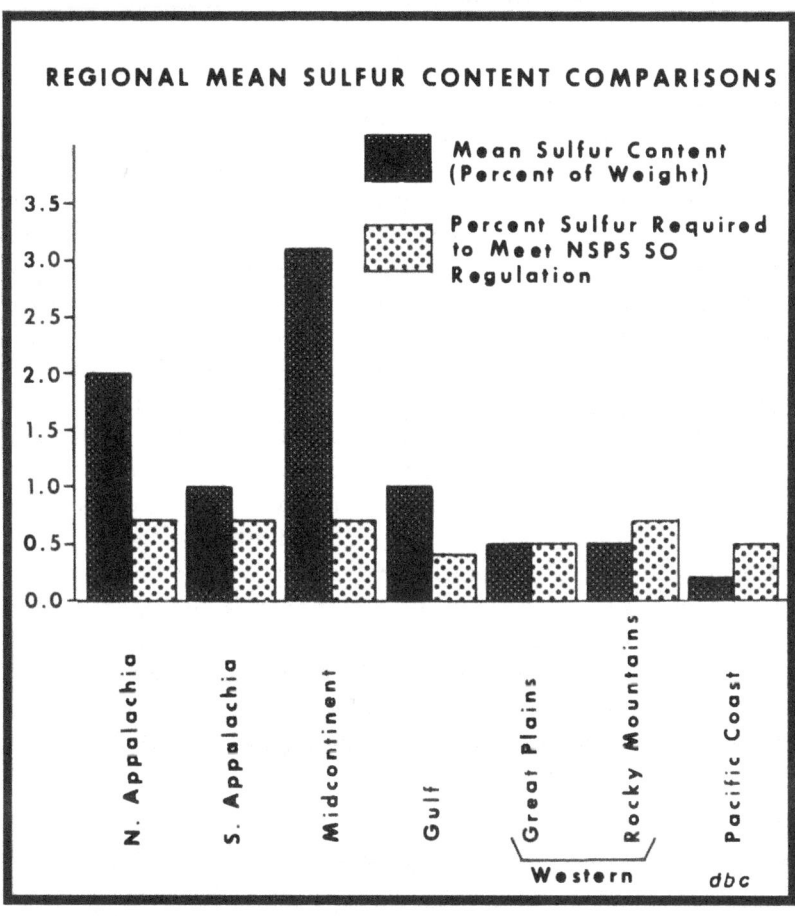

Figure 1. Regional differences in percentage of sulfur required to meet NSPS for SO₂ result from variations in energy content of coals. Source: Adapted from Federal Energy Administration, op. cit., footnote 7, p. 104.

Acceptable regional ambient air-quality conditions can be achieved only by regulating point-source emissions. The Act specifically directs that flue emission measurements be taken at the stack. As ninety-five percent of available sulfur normally is converted to SO_2 during combustion, this requirement must be met by reducing fuel-sulfur content or by cleansing gaseous emissions.[10] The U.S. Bureau of Mines reports that only thirty-five percent of coals east of the Mississippi can be cleansed to a total sulfur content of less than one percent. The 1970 EPA mandated NSPS is even more stringent.

The Appalachian Region

Three major coal-producing regions exist in the United States. These are Appalachia, Interior or Midcontinent, and Western (Figure 2). In recent years production emphasis has been in this regional order with Appalachia responsible for approximately two-thirds and the Midcontinent approximately one-quarter of national output. Thus far, yields from Gulf, Pacific, and Alaskan Regions have been relatively insignificant (Figure 3). Current regional output does not reflect actual reserves but is a response to past and current economic factors (Figure 4). Each region is unique in kinds of coals, mining problems, and use potentials. Appalachian bituminous coals are usually characterized by high energy content and are in demand for heating and power production (steam) in the industrialized, urban, northeastern states. Considerable metallurgical-quality, bituminous coals are also extracted, primarily for steel producers in the United States, Japan and the European Economic Community. In recent years approximately ten percent of United States coal production has been exported. Almost all of this was of Appalachian origin and consisted mainly of coking varieties. Most coals sold to Canada, the second ranking importer of United States coals, are of steam quality. If electrical companies in southern Ontario maintain a present trend of conversion to oil and natural gas, they will represent a diminishing market for these coals.[11]

Only approximately twenty percent of Appalachia's feasibly available coals have less than the 0.7 percent sulfur-by-weight required to satisfy the 1970 Clean Air Act's SO_2 NSPS primary goals.[12] The southern portion of the region contains only one-third of the total coal energy, but its coals average one percent sulfur content compared to two percent elsewhere in the region. The area surrounding the junction of Virginia, West Virginia, and eastern Kentucky has coals with little ash, low-sulfur content, and appropriate volatile materials which collectively rank them among the best coking coals on earth. These qualities have decidedly influenced regional production. Between 1965 and 1973 North Appalachian production declined by seven percent while South Appalachian output rose by five percent (Table 4). During this period Kentucky assumed West Virginia's position as the nation's premier coal-producing state. The decline in North Appalachian output has resulted from comparatively higher production costs in this older mining area and from conversions to alternate "clean" fuels by consumers in mid-Atlantic and northeastern states.[13] Almost one-half of the Appalachian low-sulfur coal reserves is employed or consigned by contract as coking coal for domestic or foreign customers. The remaining one-half could be applied to steam production, but high production costs in the deep seams generally prohibit this. Most new or

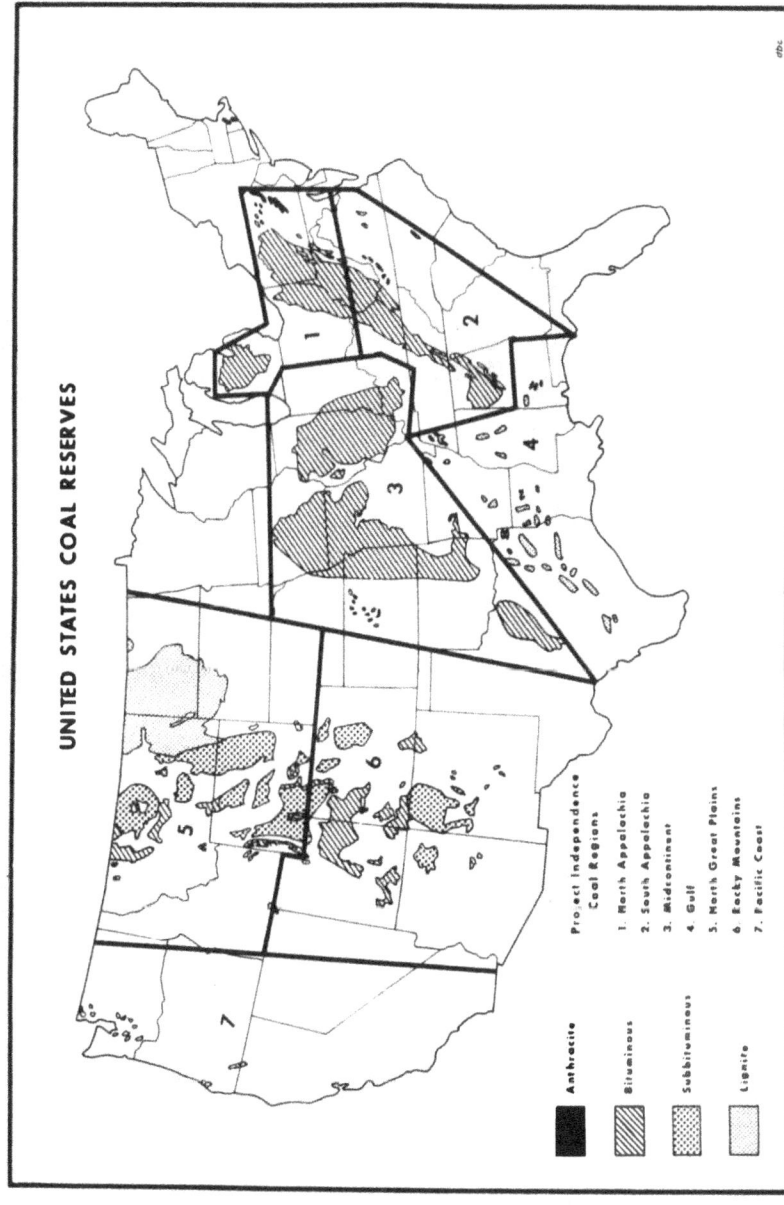

Figure 2. The Northern Great Plains and Rocky Mountains are collectively termed the Western Region. Sources: Adapted from Federal Energy Administration, op. cit., footnote 7, p. 10 ; and U.S. Geological Survey, *Bulletin 1275.* 1969.

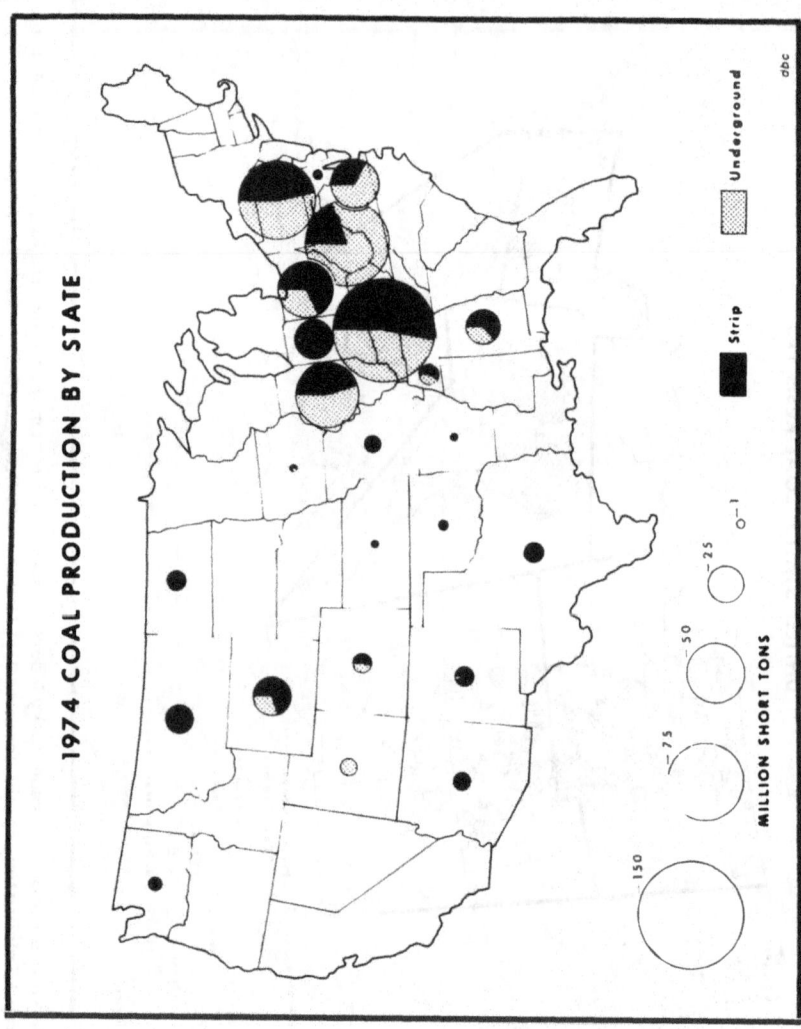

Figure 3. Source: Adapted from U.S. Department of the Interior, Bureau of Mines, *Mineral Industry Surveys: Weekly Coal Report No. 3036* (November 21, 1975), p. 6.

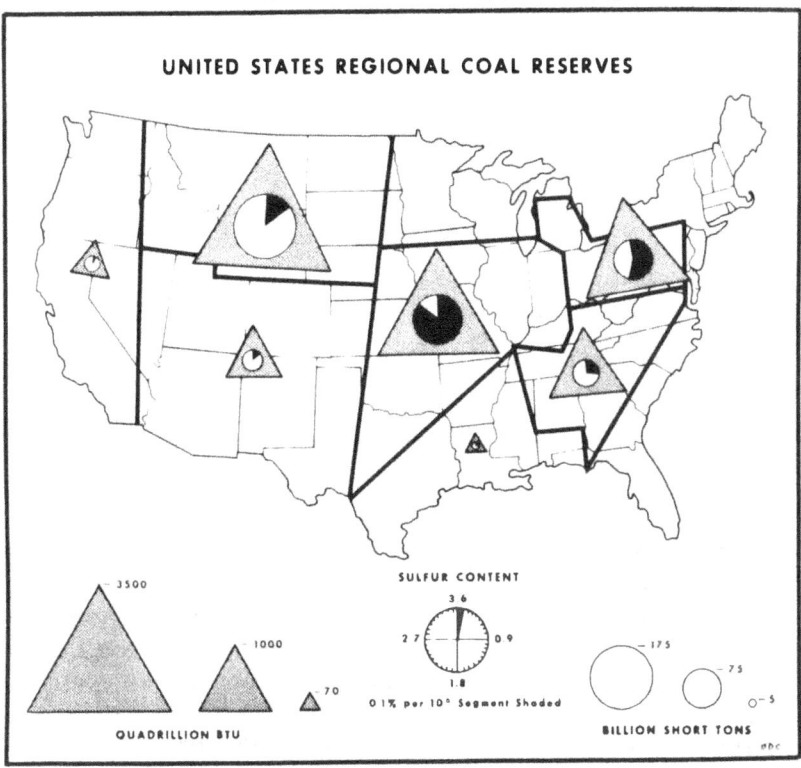

Figure 4. Source: Adapted from Federal Energy Administration, op. cit., footnote 7, p. 103.

planned additional regional capacity is for metallurgical use.[14]

Although not all Appalachian coals are compatible with the 1970 Clean Air Act, they have not been idle. Accelerated national demands for energy of all types, variances permitting high rates of power plant SO_2 emissions, use of scrubbers, and federal action forcing many electrical utilities to convert to coal have stimulated a recent boom in coal mining.[15] For decades Appalachia languished in economic doldrums; f.o.b. coal prices averaged below $5 a ton (0.91 metric ton) from 1948 to 1969. By 1975, when national unemployment was approximately nine percent, sharply increased demands and prices for coal caused a drop in the jobless rate in portions of Appalachia to less than five percent.[16] Old mines were reactivated and new sites were brought on stream. There were nearly 150 new millionaires among Appalachian mine operators in 1974 and 1975,

TABLE 4.—UNITED STATES COAL PRODUCTION
(Million of Tons)

Regions	1965	1973	Percentage Change
Appalachia	386	382	ᵃ
North Appalachia	191	177	− 7
South Appalachia	195	205	+ 5
Midcontinent	121	150	+ 24
Gulf	ᵃ	7	>1400
Western	19	56	+ 195
North Great Plains	6	32	+ 433
Rocky Mountain	13	24	+ 85
Pacific Coast	1	4	+ 300
Total	527	599	+ 14

Source: Modified from *Project Independence Report*, op. cit., footnote 7, p. 101.
ᵃ negligible

and even average pay rates for miners exceeded $50 per day by late 1975. These conditions influenced some regional expatriates to return in order to take advantage of new job opportunities. Although the Appalachian Region continues to lag behind national per capita income levels, the gap is diminishing. Certainly the new-found prosperity of coal enterprises is playing a major role.[17]

The excessive demand for coal slackened by May 1975, in part a response to earlier panic stockpiling by utilities and other consumers. This caused a reduction in coal prices and sporadic layoffs for some mining personnel. Notwithstanding these short-term fluctuations, it appears that future demands for Appalachian coals will be strong, especially for low-sulfur and metallurgical varieties. As other regions increase output Appalachia's relative contribution will decline, although projections indicate a regional yield of 1,100,000,000 tons (997,903,000 metric tons) by 1990, a 167 percent increase over 1973.[18]

The Midcontinent Region

Production in the Midcontinent Region is concentrated in Illinois and in the western portions of Indiana and Kentucky; some output is derived from portions of Missouri, Kansas, Nebraska, Oklahoma, Arkansas, and northern Texas. Although subsurface mining is practiced, these areas are frequently suited for stripping. In Indiana almost all output is from surface mines. Midcontinent coals generally contain less energy than Appalachian coals and almost all of the Midcontinent output is used for steam production. Midcontinent coals have a high sulfur content (averaging five percent) which limits or even curtails their usage under pollution abatement regulations. As is true of Appalachia, Midcontinent low-sulfur coals frequently are located deeper than other varieties and occasionally are not even properly mapped. Nevertheless, between 1965 and 1973 the regional coal yield increased from 121,000,000 to 150,000,000 tons (109,769,000 to 136,078,000 metric tons), a twenty-four percent rise (Table 4). Continued growth in Midcontinent coal usage appears to be contingent on developments in scrubber and synthetic coal-fuels techniques.

Illinois has the greatest coal production of this group of states. Coal-bearing Pennsylvanian strata are found beneath sixty-five percent of the state. Illinois has the greatest reserves of bituminous coal, although North Dakota and Montana individually have greater overall total coal reserves in energy terms. Surface mining in Illinois has been deemphasized relative to subsurface operations in recent years, while nationally the converse has been true. This results in part from the development of new, underground facilities which are required to reach the state's limited, deep, low-sulfur coals. Current production is from the "Quality Circle" low-sulfur area in Williamson, Franklin, and Jefferson Counties and is consigned largely for use as metallurgical coke. Much of this coal is owned by the United States Steel Co. and is thus not available to the commercial market. About fifteen percent of all coal mined in the United States is from such captive mines which frequently produce highly desirable, low-sulfur, metallurgical coals.

Low-sulfur fuel use is mandatory within the St. Louis and Chicago metropolitan airsheds. Unfortunately, Illinois, Western Kentucky, and other nearby low-sulfur coals are not available in sufficient quantity. Even the very lenient nonmetropolitan, "old" plant emission standard prohibits use of most Midcontinent coals, including eighty-one percent of Illinois coals.[19] This often necessitates high-cost purchases from western states. Although f.o.b. costs for Western Region coals are usually low, delivery costs are high. Consumers of Western coals in St. Louis and Chicago pay up to twice the delivered price required for local coals. It is argued, however, that on the basis of kilowatt production and with scrubbing included, costs for Western coals are currently at or near par with local costs.[20]

Low-sulfur coal frequently is purchased to provide a satisfactory mix with high-sulfur material to affect an overall sulfur reduction. Alternatively, dual-fuel systems employ natural gas or low-sulfur oil "resid" in combination with high-sulfur coals.[21] An interesting variation is the intended burning of a garbage-coal mix by the Union Electric Co. of St. Louis. The required facility will be operative in 1977 with refuse gathered from throughout the St. Louis SMSA. Not only will this novel input provide ten percent of the unit's total energy requirements but it will be a non-sulfur fuel which will permit coals of a slightly higher sulfur content to be consumed.[22]

The Western Region

Extractable coals are found scattered throughout the Rocky Mountains and High Plains. In the main, these are subbituminous and lignite coals which develop relatively low energy, often exhibit high moisture content, and have a propensity for spontaneous combustion. Higher-energy bituminous varieties including metallurgical types are also found in Utah, Colorado, and Montana where underground mining predominates. Western coals are remotely located relative to most American consumers, and for many years these deposits remained unused or satisfied only local needs. Contemporary events are producing rapid changes in their utility. The low-sulfur content (0.5 percent) of many Western fuels is compatible with the EPA mandated SO_2 regulations, even on an energy basis. Additionally, many of these coals exist in thick seams with relatively shallow overburden. Three-quarters of the strippable coal in the United States is located west of the Mississippi River. Carboniferous layers are frequently 100 feet (30.5 m.) thick with less than 50 feet (15.2 m.) of overburden, which allows strip mining with ninety-eight percent coal recovery (Figure 5). This permits far cheaper and more efficient extraction than is generally possible elsewhere.[23] Notwithstanding these shallow, thick seams, low-energy content often results in small maximum allowable ratios of overburden to seam thickness—as low as 1.5 to 1 in portions of Wyoming.[24] Although extremely thick seams are advantageous when close enough to the surface to permit stripping, they can be wasted if found at depths requiring underground mining.

Large quantities of Western coals are already supplying producers of steam in the Middle West and South-Central states. Although local Mid-continent coals are prodigious in quantity, easily obtained, and cheaper, their high-sulfur content renders them less attractive than coals often more than 1,000 miles (1,610 km.) distant.[25] Railroads have suddenly added low-sulfur coals as a transport commodity to their Great Plains systems.[26] These unit trains frequently exceed 100 cars in length and are efficient bulk transportation systems; unit train hopper cars often average five times the ton-miles hauled by equivalent cars in general service (Figure 6). The resulting economies of scale are enhanced by the low point-to-point rates permitted by this kind of operation.

Residents of the Western Region hold contrasting viewpoints concerning coal mining and usage, but a majority accept the premise that mining

Figure 5. Stripping low-sulfur coal at Black Mesa, Arizona. Photo courtesy of Peabody Coal Co.

Figure 6. A Denver and Rio Grande Western Railway unit train unloading low-sulfur Colorado coal at Illinois Power's Wood River electrical power plant.

is inevitable and that it must be accomplished prudently and be economically advantageous to the area. Advocates of Western coal development emphasize the economic importance of coal and stress that the amount of disturbed, unreclaimed soil at any time will be nominal. By example, the feasibly strippable area of Montana is less than one percent of the state, while five to ten percent of western North Dakota is potentially strippable.[27]

Even though Western reserves are impressive, only 71,000,000 tons (64,410,000 metric tons) were produced in 1974 or twelve percent of the national total; this was a significant increase of twenty-four percent over 1965. There is no existing surplus mine capacity in the Western Region, and a customer requiring these coals is forced to contract with a mine to expand capacity or actually develop a new facility. This necessitates major capitalization of long-term contracts and as much as three years in which to bring the operation on stream. Even so, projections indicate impressive future expansion in this region. An Exxon Co. nationwide projection predicts that nationally 870,000,000 tons (789,251,000 metric tons) of coal will be mined in 1980 with one-third coming from Western Region states.[28] A comprehensive nationwide survey of planned new coal mining capacity undertaken in 1973 indicated that by 1983 the Western Region will receive 50.2 percent of national new mine capacity with Wyoming contributing the major portion of this new productivity (Figure 7).

Western coals are not all shipped directly to neighboring regions for use but are frequently converted to electrical energy. This practice will likely become even more commonplace as technologies for long-distance transmission of electrical power become more efficient. Mine-mouth conversions to electricity or "clean" synthetic fuels appear to be practical means of employing Western Region lignites as their low energy-to-weight ratios usually prohibit long-distance transportation (some exceed thirty-five percent water and ash). New Mexico was the first state to introduce an electricity tax on exported power. The tax is ostensibly levied to all users, but it is essentially a tax on exported power as New Mexicans may credit the charge against the existing state electricity tax.[29] In addition to the transportation economies realized by mine-mouth power generation, there are the advantages of local rural environments over which airborne effluvia may be ejected without the most dire consequences of the EPA's 1970 mandate.

Water transportation modes are also beginning to play a role in the massive eastern movement of Western coals.[30] Until recently most coal shipments on the Great Lakes consisted of modest amounts moving on the lower lakes generally from American and Canadian producers and largely destined for Canadian consumers. Quantities are expected to in-

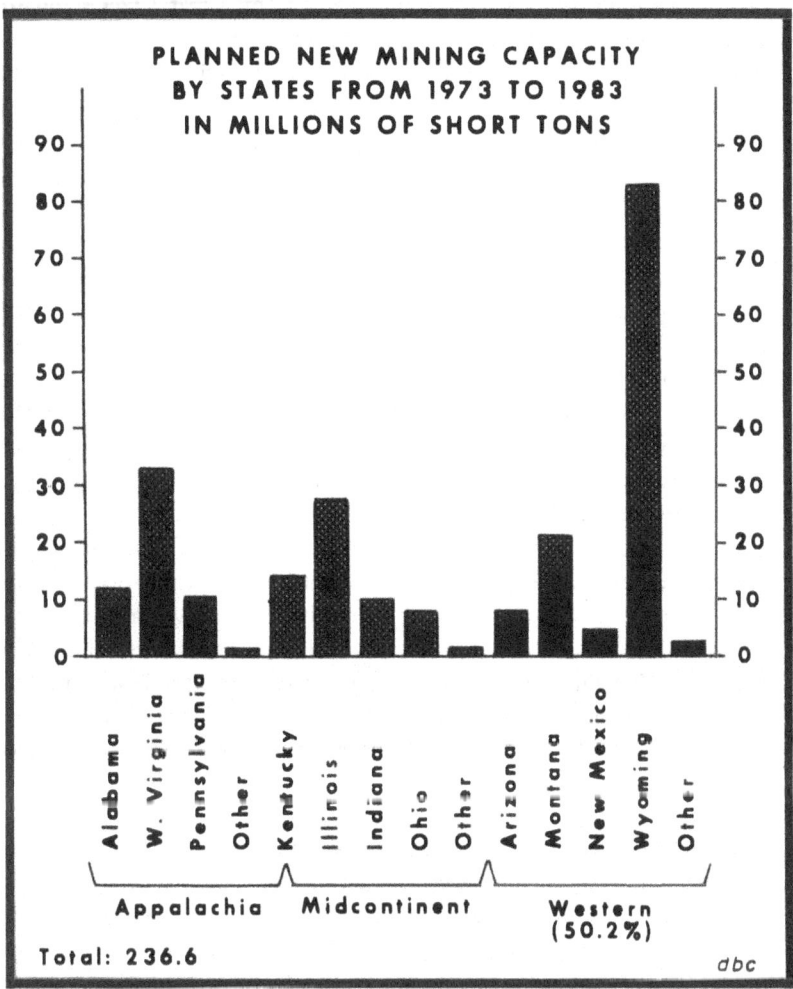

Figure 7. Source: Adapted from G. F. Nielsen, "Coal Mine Development Survey Shows 236.6 Million Tons of New Capacity," *Coal Age*, Vol. 80, No. 2 (February 1975), p. 132.

crease considerably as Western coals flow to United States power plants along the lower lakes, thus producing a net west-to-east transfer. Movement of some coals from Western mines to a special Lake Superior terminal near Duluth is accomplished by rail in 10,000 ton (9,072 metric ton) units. Final delivery by water involves new 62,000 ton (56,245 metric ton) American Steamship Company ships. These self-unloading, up to 1,000 feet

(304.8 m.) long, floating containers are large by Great Lakes standards. Terminal storage facilities must also be substantial in order to accommodate year-round rail deliveries with the short navigation season of the upper Great Lakes.[31]

Barge movement of Midcontinent coals on the Mississippi River is an established fact, but Western coal is a new commodity. A 10,000,000 ton-per-year (9,072,000 metric ton) handling facility on the Mississippi River at St. Louis is scheduled for operation in 1978. The port will be supplied with Montana coal by rail and cheaper barge transportation will then complete delivery to power plants in Indiana, Ohio, and the South. Demand for coal in the South is increasing as utility plants along the Mississippi-Gulf waterways system are converting from natural gas to coal.[32] St. Louis is the closest Mississippi River port to the Montana mining area, downstream from which the river remains ice-free, is unencumbered by locks, and commands a minimum channel depth of 9 feet (2.7 m.) to the Gulf.[33]

Future eastward movement of Western coals may be promoted by coal-slurry pipelines exceeding 1,000 miles in length. Advocated benefits from slurry pipeline use include relief for a heavily used rail system and a reduction of accidents and human stress resulting from noise and surface traffic congestion. Pipelines require more capital investment than labor, and they are more immune to inflationary forces. Pipeline advocates do have the thorny problem of satisfying their large water requirements in a water-scarce region. Approximately 200 gallons (757 l.) of water are required for each short ton of pulverized coal to achieve the desired fluid consistency. When compared with alternate coal uses such as on-site gasification, however, slurry technology does not appear too inefficient in water use as far as the arid, producing region is concerned.[34] Long slurry pipelines are feasible. Since 1970, the Black Mesa Pipeline Co. has operated the longest slurry pipeline in the United States. Its 18-inch (0.46 m.) system moves 5,000,000 tons (4,536,000 metric tons) of coal annually from Black Mesa near Kayenta in northeastern Arizona over 273 miles (439.4 km.) of rough topography to the Mohave generating plant in southern Nevada (Figure 8). Comparatively low rail rates and relatively high costs for water, slurry preparation, and dewatering render short slurry pipelines infeasible, but they prove to be efficient systems as distances and throughput increase, particularly in areas which lack existing rail transportation (Figure 9).[35] One of the greatest impediments to their development is the difficulty of receiving rights of eminent domain from state and federal legislative bodies.

The rapid development of Western coal deposits has produced a variety of local impacts in Montana and Wyoming. Energy-related jobs are trans-

Figure 8. Input terminal of coal slurry pipeline at Black Mesa, Arizona. The line passes beneath 273 miles of rough Arizona and Nevada landscape. Photo courtesy of Peabody Coal Co.

forming the existing ranching economy. Near Rock Springs, Wyoming, energy companies have inflated grazing land costs from $60 to as much as $1,000 an acre (0.4 hectare) within a year. A population influx and concomitant demands for amenities have overcrowded schools and hospitals, strained the housing supply, and have caused excessive inflation rates in Rock Springs and Gillette. The trailer has been adopted by many newcomers in response to the lack of conventional housing and high mortgage interest rates. Although many local businessmen are euphoric over the rapid influx of money, other lifelong residents are less than ebullient about the obvious change in their way of life. Not all new growth is focussed in existing communities. Colstrip, Montana, a Western Energy Co. new town, houses about 3,000 newcomers. There, a combination of housing types and social amenities satisfy most residents as the extremely low worker turnover rate attests. Some Western mines are sited on Indian treaty lands. These include the Crow Indian properties at Absolka, Montana, and the Navajo and Hopi Reservations in northeastern Arizona. In addition to royalties paid to the tribal organizations, coal mining is pro-

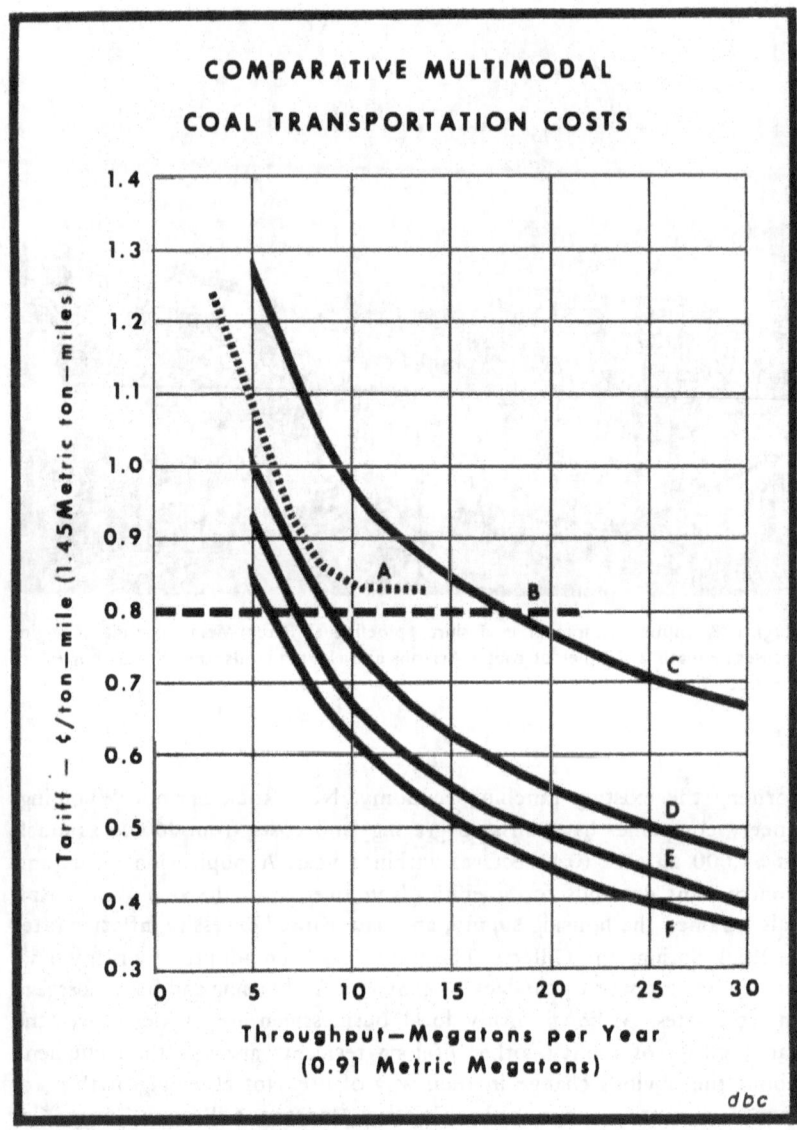

Figure 9. (A) Extra high voltage electrical transmission at 1,000 miles; (B) unit train at 1,000 miles; (C) slurry pipeline at 500 miles; (D) slurry pipeline at 1,000 miles; (E) slurry pipeline at 1,500 miles; and (F) slurry pipeline at 2,000 miles. Pipeline costs include slurry preparation, water supply, and dewatering. Source: Adapted from "Rail Transport Dominates . . . ," *Coal Age*, Vol. 80, No. 6 (Mid-May 1975), p. 112.

viding some training and jobs for Indians in areas of chronic unemployment.

Although demand for all low-sulfur fuels has been brisk, development of Western and other suitable coal mines has been slower than expected. This is the result of uncertainties in future demands relative to the twenty to twenty-five years required to amortize initial mine-investment capital. These uncertainties derive from several widely divergent sources including variation in the application of the Clean Air Act Amendments. Although the EPA stresses enforcement, numerous variances have been permitted. Incertitude is also created by the inability to predict future prices of oil and natural gas resulting from the caprices of the OPEC and vacillations of federal intent regarding wellhead price regulations, import quotas, and tariffs. Further confusion derives from the validity of electricity demand projections. Many electric utility spokesmen believe that demands will continue to double every ten to fifteen years.[36] The National Electric Reliability Council projects electric utility coal consumption at 780,000,000 tons (707,604,000 metric tons) by 1984. In recent years the United States manufacturing sector has been gradually shifting from the direct use of coal to greater reliance on electrical power. Although this sector's total energy demands have increased, the rate of increase has been declining in recent years in response to sharply increased energy costs. Greater efficiencies in energy use in manufacturing processes are expected to emphasize this trend.[37] Further restrictions to an even greater market for coal are: the encroachment of federally encouraged nuclear energy development and projections of its rapid ascendency; a continuing uncertainty over the substance of an ultimate, federally-administered, strip-mine reclamation bill; and the reluctance of the Interior Department to lease more coal-bearing lands.[38] Unlike coals elsewhere in the United States, eighty percent of Western coals are found on federal lands (Figure 10). Also, several environmental groups including the powerful Sierra Club have won court battles which restrict production in some portions of the Western Region. Even so, Western coal production increased from approximately 19,000,000 tons (17,000,000 metric tons) in 1965 to 71,000,000 tons (54,000,000 metric tons) in 1974.

The Coal Mine Health and Safety Act

December 30, 1969, marked the implementation of the Federal Coal Mine Health and Safety Act (P.L. 91-173). This set the most stringent terms for subsurface mining hitherto encountered by the industry. Dust was to be ultimately limited to 2 mg. per cubic meter of air, and carbon dioxide and methane to no more than 0.5 and 1 percent of respirable air

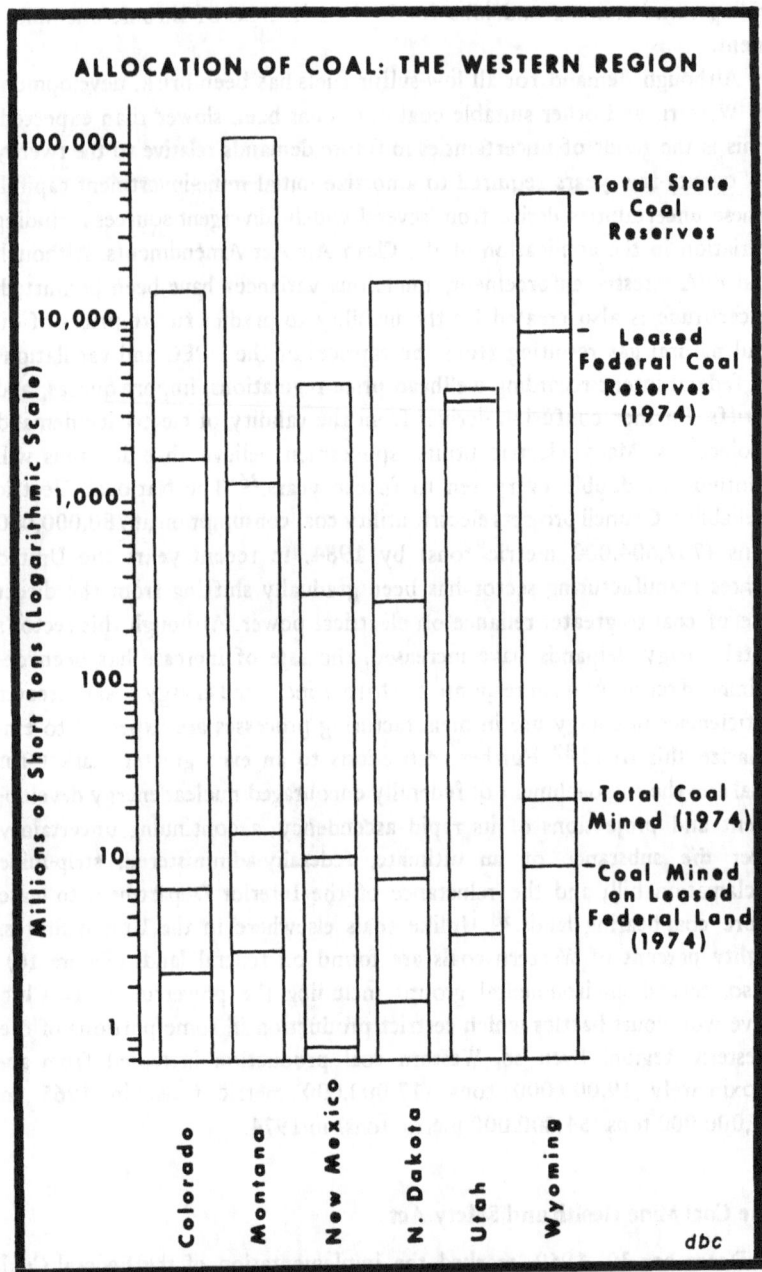

Figure 10. Sources: Adapted from *Mineral Industry Surveys,* op. cit., footnote 1, p. 4; U.S. Department of the Interior, Bureau of Mines, *Mineral Industry Surveys: Weekly Coal Report No. 3036* (November 21, 1975), p. 6; and U.S. Department of Land Management, *Public Land Statistics 1974,* pp. 104-05.

by volume, respectively. The Act specified appropriate approaches to crossmembering and roof bolting. Specific directives were also given regarding acceptable electrical equipment, blasting, hoisting, noise reduction, and fireproofing techniques. Thus, coal mining firms have been required to invest considerable capital in new equipment. Because of the highly technical nature of this Act's provisions, compliant mine operation has required a greater complement of inspection and supervisory personnel.

Implementation of P.L. 91-173 has been an important factor in a decrease in productivity in the United States coal mining industry. Miner productivity declined from 15.6 tons (14.2 metric tons) per day in 1969 to about 11 tons (10 metric tons) in 1973, while daily costs for miners increased about twenty-four percent. The effects have been pronounced on underground mines, especially small-scale operations. These small-volume units have been unable to comply with the law's requirements and remain profitable, and they have declined in number since the Act's issuance. For example, the number of independent mines in Pike County, Kentucky declined from 636 to 205 by July, 1971. After a four-year "grace period" on March 31, 1974, the use of explosion-proof machinery became mandatory in all mines. This aspect of the 1969 Act was the last to be fully implemented and immediately forced more than 500 small mines, mainly nonunionized "dogholes," to cease operations—a loss of about five percent of the nation's coal production capacity and up to 6,000 mining jobs. Most of these operations were located in central Appalachia, specifically in eastern Kentucky and southwestern Virginia.[39]

A 1972 nationwide study included data from thirty-four companies responsible for seventy-five percent of the output from all underground mines exceeding 500,000 tons (454,000 metric tons) per year. It indicated several effects of P.L. 91-173. Appalachian production costs rose $1.49 per ton (0.9 metric ton) and productivity declined twenty-four percent. Midcontinent producers faced increased costs of $1.00 per ton and a productivity decline of twenty-one percent, while firms in the Western Region faced cost increases of $1.00 but a productivity loss of only eleven percent.[40] Notwithstanding these productivity losses the United States still leads the world's major coal producing nations in average daily mine worker efficiency and the gap is expected to widen.[41]

Predictably, there has been much criticism of the Act's contribution to decline in mine productivity and attrition of small mines. This is made more poignant by the frequency of disasters which largely derive from lack of continual inspection and adherence to the safety regulations. A common element in criticisms of the Act is its universal applicability with no regard for spatial variation in physical conditions. Those affected vehemently argue that it was written for conditions in a portion of Appalachia.[42]

They maintain that the requirements are frequently excessive. For example, many mines (usually non-Appalachian) produce little methane and some produce coals that are not brittle enough to create excessive dust, but nevertheless, must conform to the uniform set of regulations. Operators insist that the less rigorous state rules are more sensitive to local conditions and should prevail.

This Act has induced sharply greater costs for subsurface mine operations relative to open cost operations. Gordon indicates that it "clearly provided a major impetus for strip mining since compliance was much more expensive underground."[43] It might be further concluded that because of relative stripping costs the Act has particularly stimulated stripping in Midcontinent and Western Regions.

Strip Mined Land Reclamation

Costs of reclamation generally are greater in the Midcontinent and Appalachian areas where state legislators have had more time in which to respond to the problem. Greater physical relief adds to reclamation difficulties and costs throughout most of Appalachia. West Virginia, eastern Virginia, and Kentucky have greater reclamation costs than Pennsylvania. Nationally, current land-reclamation costs vary widely, but there is little evidence to support the notion that these costs curtail mining operations. Under present conditions, the necessary price increase to electric utilities resulting from reclamation ranges between a nominal one cent to two cents per 1,000,000 Btu (251,996,000 g. cal.).[44]

Several states have surface mining laws which require varying degrees of land reclamation. These include: Illinois, Indiana, Kentucky, Maryland, Ohio, Pennsylvania, Virginia, West Virginia, Wyoming, and Montana. The language and degree of enforcement in each state code varies but there are elements in common which include: elimination of spoil-bank peaks, especially that of the highwall; impoundment of acidic runoff water; filling of depressions, especially the final cut, unless a lake is to be created there; and replanting with vegetation. License fees provide for inspection and for some redress of past excesses of the mining industry. Actual compliance with these statutes is achieved by a universal performance bond requirement. Bond values vary but usually include a fixed minimum plus additional amounts for each acre (0.4 hectare) stripped.[45]

The coal mining and consuming industries generally are opposed to stringent reclamation requirements and believe their implementation would seriously impede or even curtail expansion of strip mining, especially in portions of the Western Region. A concurring report from the National Academy of Sciences indicates that although areas with a minimum of ten inches (25 cm.) of precipitation may be revegetated easily, drier areas will require major inputs of water, fertilizer, and management. The report also indicates that in these extremely dry regions energy distribution systems may pose greater ecological danger than the actual mines.[46] Some critics of reclamation laws contend that the main effect will impinge on small, stripping operations in Appalachia whose operators are relatively less able to afford the costs. Additionally, when applied to reclamation of Appalachian contour-stripped slopes, currently proposed standards would necessitate excessive costs, thus invalidating the practice. Fortunately, relatively little of total national coal production is derived from this mining technique.

Although stringent reclamation measures are onerous to mine operators, many welcome passage of a national law in order to remove operational uncertainties and provide a means to improve their public image. Moreover, those that are currently involved with reclamation are discovering that it can have profitable consequences.

Rehabilitated lands are now utilized for recreation, forestry, range and pasture, cropland, and even urban uses. Mining companies are becoming acutely aware of potential profits and seek to minimize "downtime" in converting mined-out lands to producing farmland (Figure 11).[47] Tree farming is particularly profitable and suitable in the East. Species that can survive in highly acidic soils include Chinese Chestnuts and the rapidly growing, hybrid poplars. Evergreens such as European pines and Douglas fir hide furrowed land even in winter. Revegetation in western states is critically dependent on soil-water retention and soil elements such as nitrogen, phosphorus and sodium. Grasses employed to revegetate mined lands in the Western Region include: wheatgrasses, needlegrasses, foxtail barley, and bluestem. Species are selected for their survival ability in this environment and are mixed to provide year-round pasture. (There has been some local resistance to the introduction of exotic species.) Economical grass production can be realized by the second growing season following seeding.[48]

Summary

Federal laws enacted in 1969 and 1970 have strongly influenced sectoral and spatial aspects of the United States coal industry. The 1970 Clean Air Act Amendments have severely restricted use of coals with high-sulfur content. A consequence has been loss of some New England markets for northern Appalachian mines and expansion of output from low-sulfur mines in southern Appalachia. Unfortunately, the latter are limited in extent and are further limited by ownership and contract. Notwithstanding

Figure 11. Reclaimed stripped land supporting winter wheat at Lynville, Indiana. Stripping continues in the background. Photo courtesy of Peabody Coal Co.

some impediments to production, portions of Appalachia have experienced a surprising economic renaissance with a decline in unemployment. Use of Midcontinent coals is particularly limited because of excessively high sulfur content. Regional utilities must either provide for coal cleansing and flue-gas scrubbing or import low-sulfur coals from the Western Region. Emission sources that existed prior to 1970 are permitted to emit more SO_2, especially in rural areas. Although there is a national interest in coal gasification and liquefaction, Midcontinent advocates of coal use are especially aware of the need for conversion to clean-burning fuels. The 1970 Act has had a precipitious effect on the development of Western coals. These low-sulfur but low-energy fuels are transferred over long distances either as electrical power or via the world's longest unit trains. Massive transfers of Western coals to Middle Western and Southern users require new railroad and storage facilities. This transfer has also necessitated new ports on the Great Lakes and Mississippi River. Coal slurry pipelines exceeding 1,000 miles (1609 km) may also be constructed for this task. Extensive gasification of Western coals may become a reality if local opposition and scarcity of water can be overcome. Predictions indicate continued rapid growth of coal production in the Western region. Some Western communities are already experiencing consequent boom conditions.

The 1969 Coal Mining Health and Safety Act resulted in a nationwide decline in underground mine productivity. This was most acutely felt by small-scale operations, and the majority of these were located in Appalachia, particularly within the central portion of the region. To some degree this Act stimulated strip mining, particularly in Midcontinent and Western Regions.

Various states require degrees of reclamation for strip-mined land. Largely because of adverse topography, Appalachia generally experiences the highest current reclamation costs, although costs for revegetation of very dry Western areas also may be prohibitive. A proposed national reclamation law is not welcomed by mining companies, but these operators are becoming aware of potential profits from reclaimed lands.

The effects of legislation can be far-reaching. The purposes of laws such as those discussed here are altruistic in seeking to protect public life, health, and property, although it is unlikely that the wide range of consequences was anticipated.

Notes

1. U.S. Department of the Interior, Bureau of Mines, *Mineral Industry Surveys: Demonstrated Coal Reserve Base of the United States on January 1, 1974* (June 1974), p. 1.

2. *Mineral Industry Survey,* op. cit., footnote 1, pp. 1-3.

3. U.S. Department of the Interior, Bureau of Mines, *Mineral Industry Surveys: Weekly Coal Report No. 3044* (January 16, 1976), p. 3.

4. H. E. Risser, *The U.S. Energy Dilemma: The Gap Between Today's Requirements and Tomorrow's Potential,* Illinois State Geological Survey, Environmental Geology Notes Number 64 (Urbana, Ill., July, 1973), pp. 3-4.

5. H. E. Risser, *Energy Supply Problems for the 1970s and Beyond,* Illinois State Geological Survey, Environmental Geology Notes Number 62 (Urbana, Ill., May 1973), p. 6.

6. J. E. Krier, *Environmental Law and Policy: Readings, Material and Notes on Air Pollution and Related Problems* (New York: Bobbs-Merrill Co., 1971).

7. Federal Energy Administration, *Project Independence Report* (Washington, D.C.: U.S. Government Printing Office, November 1974), pp. 103-04.

8. M. E. Hopkins and J. A. Simon, *Coal Resources of Illinois,* Illinois

State Geological Survey, Illinois Minerals Note 53 (Urbana, Ill., January 1974), pp. 8-9.

9. R. L. Major, *Fuels and Energy Situation in the Midwest Industrial Market*, Illinois State Geological Survey, Illinois Minerals Note 52 (Urbana, Ill., December 1973), p. 11.

10. The experimental sulfurtain process promises to reduce the conversion of sulfur to oxides during coal combustion. This is a unique approach in which coal is pretreated with limestone. Tests indicate that a significant proportion of sulfur is incorporated into the solid wastes following combustion; W. H. Buttermore, "Trapping out Sulfur Dioxide while Burning Pulverized Coal," *Coal Age*, Vol. 80, No. 2 (February 1975), pp. 148-49.

11. "Appalachia—The King of Metallurgical Coal Exports," *Coal Age*, Vol. 80, No. 6 (Mid-May 1975), pp. 114-18.

12. The 1.2 pounds (0.544 kg.) SO_2 per 1,000,000 Btu (252,000,000 g. cal.) new plant guideline is also roughly expressed as 0.7 percent total sulfur by weight for coals with 12,000 Btu (3,024,000 g. cal.) per pound (0.45 kg.); Risser, op. cit., footnote 4, pp. 40-41.

13. Federal Energy Administration, op. cit., footnote 7, p. 101.

14. T. R. Scollon, "Coal in 1972," *Mining Congress Journal* (February 1973), p. 101.

15. In late 1973, Congress considered the Coal Conversion Act (S2652) which would have required existing industrial installations and base-load power plants (other than auxiliary units used for peak-load generation) and all future plants to use coal as the primary fuel. The administration was eventually given authority to force selected power plants to convert to coal. For example, in August of 1974, the Federal Energy Administration ordered seventy-four operating power plants to convert to coal, and forty-seven units under construction were required to have coal-burning capabilities. It was further required that all changes should satisfy the EPA in terms of pollution effects. It is important to note that conversions from coal to oil or gas are relatively easily accomplished but retrofitting to coal is difficult because of boiler and fuel handling requirements.

16. In 1975, coal prices ranged from $7 to as high as $70 per ton (0.9 metric ton) for some top spot prices, although 1974 and 1975 f.o.b. mine prices averaged approximately $16 per ton; U.S. Department of the Interior, Bureau of Mines, *Minerals and Materials* (December 1975), p. 13.

17. J. Pickard, "Per Capita Income Gap Between Appalachia and U.S. Diminishes," *Appalachia*, Vol. 8, No. 6 (June-July 1975), pp. 28-41.

18. W. H. Miernyk, "Coal and Future of the Appalachian Economy," *Appalachia*, Vol. 9, No. 2 (October-November 1975), p. 32.

19. Major, op. cit., footnote 9, p. 11.

20. Risser, op. cit., footnote 4, p. 43. The U.S. Department of the

Interior North Great Plains Resource Program (1975) noted only a five percent price differential for Western over local coals used in Illinois, but scrubber costs were included. Costs may actually be at par if electrical power output is used as a basis of comparison.

21. Although some Middle Western plants consume resid in toto, costs and the difficulties of moving imported, high viscosity resid inland largely confine use of this fuel to coastal states.

22. "Garbage, the Cinderella Fuel," *Coal Age*, Vol. 79, No. 4 (April 1974), p. 109.

23. It is only feasible to extract lignites by low-cost stripping of shallow, thick seams.

24. R. L. Gordon, *U.S. Coal and the Electric Power Industry* (Baltimore, Md.: Johns Hopkins University Press, 1975), p. 102.

25. Delivered costs of Western coals at Middle Western sites are generally greater than for Midcontinent coals, but actual Western f.o.b. prices are remarkably low. In a long-term contract concluded in January 1975, the Tennessee Valley Authority agreed to purchase Montana subbituminous coal to supply its Shawnee power plant in Western Kentucky. Although the delivered cost is approximately $19, $14 of this is for rail transportation. As some Western coals have water content levels as high as twenty-five percent of total weight, thermal drying is sometimes employed for freezeproofing and weight reduction; "This Week in Coal," *Coal Age*, Vol. 80, No. 1 (January 1975), p. 33.

26. Sixty-three percent of all United States coal was transported by rail in 1973, whereas twenty-two percent was moved over inland waterways; "Transportation for Coal," *Coal Age*, Vol. 79, No. 7 (July 1974), pp. 102-30.

27. A. A. Thornburg, *Surface Mine Reclamation in Montana*, Report presented at Mined-Land Reclamation Symposium, Louisville, Kentucky (October 22-24, 1974), p. 1.

28. *Energy Outlook 1975-1990*, Report from Exxon Co., Houston, Texas (1975).

29. "New Mexico Strings Power Lines—But There's a Rub," *The Salt Lake Tribune* (June 15, 1975), p. 6A.

30. Transportation charges for hauling bituminous coal on United States inland rivers are approximately one-third that of rail rates, while lake and ocean transport tariffs are only about one-fifth; "Using Waterways to Ship Coal," *Coal Age*, Vol. 79, No. 7 (July 1974), pp. 122-30.

31. "Large Coal Terminal Going Up in Superior," *Coal Age*, Vol. 80, No. 1 (January 1975), p. 30.

32. "Using Waterways to Ship Coal," op. cit., footnote 30, pp. 122-24.

33. The recently completed Kaskaskia River Canal flows through coal-

bearing southern Illinois to the Mississippi River south of St. Louis and was built by the U.S. Corps of Engineers with a public investment approximating $130,000,000. Initially conceived as a means of providing virtual mine-mouth water transportation access to the Mississippi-Ohio River system, it is now questionable as to whether the 1970 Clean Air Amendments will permit the mining of sufficient Kaskaskia Valley coals to justify this investment.

34. The Keller Corporation of Dallas, Texas has developed a new liquid for use in slurry pipelines. Their "methacoal" is a mixture of alcohols (derived from coal) and coal and requires only twenty percent of the water used in conventional slurrying. The company also claims that not only does this fluid perform better as a pipeline "vehicle" but ultimately requires no dewatering and may be burned in toto; "Methanol Sub for Water in Slurry Pipelines," *Coal Age*, Vol. 81, No. 1 (January 1976), pp. 52-57.

35. "Coal Slurry Pipelines . . . A Rapidly Growing Technique," *Coal Age*, Vol. 79, No. 7 (July 1974), pp. 96-98.

36. H. E. Risser, *Power and the Environment—A Potential Crisis in Energy Supply*, Illinois State Geological Survey, Environmental Geology Notes Number 40 (Urbana, Ill., December 1970), p. 2.

37. Major, op. cit., footnote 9, p. 2.

38. Recent Bureau of Land Management (BLM) considerations would tighten the requirements for leasing federal coal lands. At the present time the bulk of leased domain is not being operated because much is being held by coal brokers rather than producers. The BLM proposes that after a reasonable period all lease holders should demonstrate production from a "logical mining unit" (LMU); *Federal Register*, Vol. 39, No. 239 (December 11, 1974), p. 43229.

39. "This Month in Coal," *Coal Age*, Vol. 79, No. 4 (April 1974), p. 49.

40. "Effect of 1969 Coal Act Examined at the Greenbrier," *Coal Age*, Vol. 78, No. 1 (January 1973), pp. 69-70.

41. *Bank of London and South America Review*, Vol. 9, No. 10 (October 1975), pp. 562-63.

42. In large measure this legislation was promoted by the Farmington, West Virginia mine disaster.

43. Gordon, op. cit., footnote 24, p. 109.

44. *The Economic Impact of Public Policy on the Appalachian Coal Industry and the Regional Economy*, A report prepared by Charles River Associates of Cambridge, Massachusetts for the Appalachian Regional Commission (1972).

45. Illinois Department of Mines and Minerals, *Rules and Regulations Pertaining to the Surface-Mined Land Conservation and Reclaiming Act* (Springfield, Ill., April 4, 1972); and U.S. Department of the Interior,

Surface Mining and Our Environment (Washington, D.C.: U.S. Government Printing Office, 1967), pp. 96-102, 118-21.

46. National Academy of Sciences and National Academy of Engineering, *Rehabilitation Potential of Western Coal Lands* (Washington, D.C.: U.S. Government Printing Office, 1975).

47. "At Amax. Environmental Engineers Find Better Uses for Reclaimed Land," *Coal Age,* Vol 79, No. 10 (October 1974), pp. 132-38.

48. U.S. Department of Agriculture Research Service, *Progress Report of Research on Reclamation of Strip-Mined Land in the Northern Great Plains* (Washington, D.C.: U.S. Government Printing Office, 1975), p. 2.

2

Surface Mining and Reclamation

Coal Age

The major problems of surface mining in the Appalachian region stem from geologic conditions and the steepness of slope. This situation is magnified in states like West Virginia, where the average median slope is 14°. "And in the specific case of McDowell County," points out Ben E. Lusk, president of the West Virginia Surface Mining & Reclamation Association, "which is the largest coal-producing county in the U.S., 90 percent of the land surface lies on mountainsides that are in excess of 20°."

Many of the innovations in surface mining methods now being tried out in many areas of Appalachia stem from work done by various companies belonging to WVSMRA, and by the association itself. These innovations include controlled placement techniques, including haulback, flat-top or mountaintop removal methods, and longwall stripping (which should be called shortwall stripping). Information on the status of these techniques is available from Lusk, who also recently gave a talk on the subject at a Society of Mining Engineers meeting in Acapulco, Mexico.

Strict Surface Mining Law

The stimulus for improving the surface mining methods has come in the form of new state surface-mining laws. In West Virginia, for example, the law, passed in 1967 and amended in 1971, requires that complete engineering and pre-planning be performed before a permit is issued. Also required are bonding fees that can reach $1000 per acre; complete drainage and water treatment systems installed before mining begins; progressive reclamation or backfilling, grading and seeding within 60 or 90 days after removal of the coal; regrading to the approximate original contour on slopes less than 14°; regrading the highwall to 30 ft. on slopes from 14° to 25°; and on slopes from 25° to 33°, the operator must apply the controlled placement method of mining.

From *Coal Age*, vol. 80, no. 6, mid-May 1975, pp. 265-274. Reprinted by permission.

Regarding reclamation, the law requires herbaceous as well as woody revegetation, a mandatory two-growing-season waiting period before bond release can be considered, and quantitative revegetative success before final bond release. If state inspectors are not satisfied with the results after two growing seasons, the area must be kept under bond, and the operator is responsible for reseeding until the area passes state inspection.

During the first two years of operating under the stringent requirements of the comprehensive law, the industry suffered heavy losses. "Regardless of the progress we made through our reclamation research projects, it seemed that many companies simply did not have the ingredients needed to satisfy all the state's requirements and still remain in business," Lusk recalls.

Controlled Placement

With both houses of the U.S. Congress also passing legislation that calls for total highwall elimination with no spoil on the outslope, operators began developing methods that could comply with such standards. Until this time, conventional contour surface mining had consisted of cutting away the overburden, pushing at least some of the spoil material over the outslope, removing the coal, regrading the bench back to the vertical high-wall, and then revegetating. But in order to insure continued mining in the mountainous areas, with protection of aesthetic beauty, things really had to change. And so the new steep slope mining theory of "controlled placement" was born.

This is a broad general term which means the placement of spoil in steep areas where its condition can be completely stabilized and revege-tated. It covers methods such as haulback, mountaintop removal, over the shoulder, block cut, and any other method where overburden is delivered, hauled or compacted so that there are no adverse aftereffects when mining has been completed.

Under controlled placement, the whole approach to mining in Appala-chia has changed, Lusk points out. Before any displacement of soil or un-covering of the coal seam transpires, a complete drainage system must be engineered and installed. This includes silt ponds, silt retarding dams, acid treatment facilities, if necessary, and haulroad water control through use of drains and culverts. These drainage systems guarantee that siltation does not leave the disturbed area and find its way into the main streams and tributaries of the watershed below the mining operation. Thus reclamation actually begins before mining. Siltation structures can either be "dug out" or excavated below the natural level; or dams can be erected above the natural level.

These ponds are riprapped with stone material on both the input or upstream waterways and the downstream waterway. Riprapping the spillways insures that they will not undercut and wash away during periods of heavy rainfall.

Water Retention Techniques

In West Virginia, the silt basins are normally excavated by dozers, but smaller ponds and drainways can be installed with equipment such as the Gradeall. This type of machinery is also effective in cleaning and maintaining these drainage structures throughout the duration of the operation. A small dragline or clamshell can also be used for the purpose. Some of the methods of water retention and filtration are built aboveground, such as the gabion dam, which is constructed of rock-filled wire baskets. This type of structure is relatively new to surface mine drainage systems, and has been found to be very effective so far.

To comply with drainage requirements as the mining progresses and the disturbed area increases, a series of siltation structures is developed that make up the total drainage system. Surface water in this area must pass through not only one, but sometimes three and four of these ponds, fully insuring that no contaminated or muddy water enters major waterways below the mining site.

Only after the drainage system is completed by the operator and passes state inspection is the permit granted so that he can start the actual production cycle. However, in most cases the operator utilizes valley fill or head of hollow fill.

To construct a valley fill, the head of the particular hollow is cleared of all existing vegetation and timber before the disposition of material begins. A rock core is built to accommodate the flow of water which is anticipated from elevations above the operation, and also to insure that the structure has a good solid base. The material is then deposited over the side with gravity action, allowing the rougher materials to find their way to the bottom to create the base of the valley fill.

Haulback Techniques

Following completion of the drainage system and state approval, the initial cut is disposed of in the valley fill. Now the operator should have enough room to begin surface mining by the haulback or lateral movement method.

On initiating the second cut, reclamation is once again the first priority. All topsoils and subsoils that are suitable for supporting vegetation are segregated and stockpiled so that they can be brought back in for final regrading. After the sandstone and shale materials have been drilled and shot, they are loaded into off-road haulers and begin their lateral movement along the bench to the fill area. At this time, all black, shaley material that will not support vegetation, and all pyritic materials that could cause acid drainage are also segregated and backfilled into the pit first. These materials form what eventually will be the base of the highwall backfill.

Recovery of the coal is then accomplished by a front-end loader and highway trucks, which haul the mineral off the hill. This stripping sequence will repeat itself until the coal-to-overburden ratio gets to the point where it would not be economical to take another cut into the highwall. At this time, if coal seam conditions are acceptable, further coal recovery can be realized through augering.

There are different avenues of thought on augering a coal seam, particularly in terms of locking up coal reserves from future production, but the auger production from most jobs has helped significantly in offsetting the added costs of the haulback method. Augering must be completed in time for the backfill of spoil material from the pit. If this timing cannot be consistent, there will be a loss of man-hours—and the advantages of augering.

The haulback method is actually a modification of several methods developed earlier. These include the boxcut and modified blockcut, which simply are not adaptable to steep-slope mining because of equipment utilization requirements and geologic conditions. However, the basic concept of the different methods is the same.

Mountaintop Removal

Another relatively new method in Appalachia that fits under the general heading of controlled placement is flat-top mining or mountaintop removal. As the name implies, the basic idea behind this type of mining involves removing 100 percent of the overburden covering the coal seam, in order to recover 100 percent of the coal. The excess spoil material that cannot be backfilled in the mined-out pit is hauled or dumped into the head of a nearby hollow for construction of a valley fill. The end result is a large expanse of relatively level land, where only a rugged mountaintop existed before.

The most notable mountaintop projects, Lusk reveals, are located at Cannelton and Welch, W. Va., where future land use is already planned, and at Falcon Coal's operation in Eastern Kentucky. This type of mining is gaining popularity with many companies that see the advantage of long-term, single-permit operations that afford total recovery of available reserves.

Multiple-Seam Mining

Even multiple-seam contour mining has received a face-lift through experimentations by the Tennessee Valley Authority. TVA's triple-seam project presently features elimination of both the highwall and spoil piles, and is not causing the long-term environmental damage normally associated with this practice. Experiments to date have proven costly, but officials believe the work has been justified because of the high rate of production in seams that could not be mined otherwise.

Longwall Stripping

A unique approach to contour mining that eliminates some of the disadvantages of other contour mining methods will be tried on property owned by Southern Appalachian Coal Co. near Julian, W. Va. The technique is being funded by EPA and WVSMRA.

First a narrow bench is opened running parallel to the coal outcrop and about 10-15 ft. from the actual outcrop. Then, perpendicular to this bench, an open trench or entry is driven into the coal seam. This perpendicular trench or driven entry will be the longwall face that will continuously advance along the coal outcrop.

The bench will be advanced at a rate that would allow continuous progression of the longwall system and the continuously moving open outby end. The material removed from the advancing bench will be returned behind the area previously mined. It is thought that this initial face will be from 200 to 250 ft. long from inby to outby end.

Recovery of coal will hit nearly 100 percent along the longwall face, as opposed to 30-40 percent for auger mining. The face will run about 200 ft. straight into the seam, with the direction of travel parallel to the highwall, and from right to left. A 60-ft-wide bench will be used for the operation, but this will sometimes drop to 35 ft., the minimum needed for maneuvering the equipment. Seam thickness is expected to average 60 in.

Longwall stripping is a misnomer for the system, William Piper of WVSMRA says, since it is neither stripping nor longwall. Rather, it is a shortwall application utilizing such equipment as Hempscheidt 558-ton-capacity chocks, a LeeNorse 285 H continuous miner, and a 17-in. armored face conveyor that will carry the coal back to a surface conveyor.

Production is expected to hit 500-600 tpd. If the production rate gets up to around 1000 tpd, WVSMRA would then consider the project highly attractive from an economic standpoint. The best use for the new method is in areas where the cover is too shallow or poor for deep mining.

Reclaiming Coal from Refuse Ponds

When Peabody Coal Co.'s Bee-Veer mine began operations back in 1937, its processing equipment was incapable of separating all of the coal fines from the waste. The waste materials were pumped into a slurry settling pond which today is about 100 acres in size. In just this pond alone, it is estimated that there are 1 million tons of reclaimable coal, and there are innumerable such ponds scattered throughout Appalachia.

How to recover this coal quickly and easily? Peabody finds that each pond should be analyzed to select the most suitable method. The most intriguing method developed so far involves the use of a versatile two-man mini-dredge called a Mud Cat, developed by National Car Rental System, Inc.

The machine, launched in the water end of the slurry area, literally eats its way into the compacted waste lying at the bottom of the pond, much like a continuous miner working the face of a coal seam underground. This material is pumped back to the Bee-Veer preparation plant, now equipped with newer processing machinery, which recovers 40 tph of coal. When the operation is completed the unsightly slurry pond will be replaced by a freshwater lake, and thousands of tons of valuable coal will have been recovered.

Small as it is—the Mud Cat is only 8 ft. wide by 39 ft. long—it is in reality a powerful dredge equipped with a 175-hp Detroit Diesel engine which drives a suction pump capable of sucking up, in a single pass, coal or sediment deposits 18 in. deep and 8 ft. wide. The dredge can reach down to 15 ft. below the surface of the water, "and by lowering the water level, the Mud Cat will be able to go to the bottom on Peabody's 60-ft-deep slurry lake," reports Eric Seagren, Mud Cat's district sales manager.

The key mechanism in the dredge is a novel 8-ft-wide underwater auger attached to a hydraulic boom at the front end of the machine. As the auger rotates, it chews up the coal waste deposits, forcing the material into an intake tube. A suction pump takes the material through the tube and out a discharge pipe, which in this case is 3000 ft. long.

The Mud Cat is propelled by a winch with the cable anchored on shore, and moved from side to side by means of pullover cables. Only two men are needed for the operation—one to run the dredge itself, and the other to remain on shore and move the cables as necessary. At the Peabody Bee-Veer mine, the dredge is kept running 7 hr. per day.

Although the dredge has been involved in applications where it had to eat its way into solid material, it works best in 21 in. of water.

3

Bureau Seeks New Surface Mining and Reclamation Techniques

Donald L. Donner
Edward Kruse

If domestic coal production is to increase in response to the growing energy demands of the U.S., surface mining must assume a fundamental role. However, in order to increase surface mine productivity, new technology must be developed to: 1) improve existing overburden handling, coal extraction and reclamation practices, 2) accelerate the mine planning and development process necessary to bring new operations on line, 3) minimize the impact of mining on land and water resources, and 4) facilitate compliance with local and federal environmental quality and safety standards.

In response to these requirements, the Bureau initiated a research and development effort to improve surface mining, extraction and reclamation technology in order to increase coal production and productivity in an environmentally responsive manner.

The resultant Bureau of Mines' "Surface Mining, Extraction and Reclamation" program has as its objectives the development of improved systems and techniques for the removal and handling of overburden, the extraction of coal from single and multiple seams, and the reclamation of the land disturbed by mining. The Bureau is placing major research emphasis on: 1) developing improved equipment scheduling and utilization methodologies, 2) optimization of mine design, 3) overburden removal and deposition, 4) coal extraction, 5) materials transportation, 6) stabilization and reclamation of mined lands, and 7) minimization of the impact of mining on land and water resources.

The prevailing goal of this research effort will be the integration of mining and reclamation activities into a total, contemporaneous mining system, while increasing productivity and enhancing reclamation in a cost-effective manner. Research activities under this program have been channeled into engineering and environmental analysis, mine systems development, mining and reclamation planning, and spoil stabilization and reclamation.

From *Coal Age*, vol. 80, no. 8, July 1975, pp. 127-129. Reprinted by permission.

Engineering and Environmental Analysis

Two nationwide studies are underway to compile reclamation cost data and determine the current state-of-the-art of overburden handling and reclamation practices. Results obtained from these surveys will give an accurate picture of reclamation costs as well as the current overburden handling and reclamation practices employed throughout the major coal-producing regions of the country.

Surface-Mine Systems Development

Conceptual mine systems will be developed and analyzed in terms of their technical and economic feasibility. Those concepts showing the most promise will then be subjected to field demonstrations. In this vein, the Bureau has initiated an interagency agreement with the Tennessee Valley Authority and the U.S. Forest Service to demonstrate and evaluate three methods of contour surface mining and reclamation (whereby spoil is returned to the approximate original contour), determine the environmental impact of each method, and conduct an economic analysis of each mining method. This three-year demonstration project, which is totally funded by the Bureau, will be located in Breathitt County, Ky.

Other research efforts currently underway call for development of conceptual contour mining systems or techniques to:

- Reduce or totally eliminate highwalls under varying geologic and topographic conditions.
- Develop retreat-area surface mining concepts.
- Investigate the concept of integrating area surface mining entry techniques with longwall mining methods into an area-longwall mining system for the recovery of shallow, flat-lying coal deposits.
- Develop surface mining systems using belt conveyor haulage to promote continuous overburden and coal extraction activities for contour blockcut operation.
- Develop new methods and techniques for coal extraction and overburden handling to facilitate mining and reclamation operations.

A key obstacle to development of new surface mining operations is the lengthy process of developing and obtaining the approval of mining and reclamation plans. In this regard, the Bureau has initiated a research effort to develop and demonstrate a complete mining and reclamation planning rationale that surface mine operators may use to insure that all pertinent mine-site data are gathered and properly evaluated. Parallel to this effort, the Bureau is exploring the use of remote sensing as a tool for mining and reclamation planning, as well as developing improved blasting techniques to reduce the environmental effects resulting from air-blast and ground vibrations.

Future research efforts will be devoted to determining the effect of surface mining on surface and groundwater hydrology, development and evaluation of various soil amendments to facilitate revegetation of mine spoil, development of spoil stabilization techniques to minimize erosion and subsequently reduce siltation of adjacent waters, and field demonstrations of new mountaintop mining concepts and their impact on the environment, as well as research on other conceptual mining systems.

4

Land Utilization and Reclamation in the Mining Industry, 1930-71

James Paone
John L. Morning
Leo Giorgetti

The total land mass in the United States comprises about 2.27 billion acres. Of this area, the domestic minerals industry, in a 42-year period extending from 1930 through 1971, utilized 3.65 million acres, or 0.16 percent (Table 1). Also during this period 40 percent of the land utilized, or 1.46 million acres, was reclaimed (Table 2). In 1971, 206,000 acres were utilized and 163,000 acres reclaimed (Table 3). Thus the ratio between land used and land reclaimed doubled in 1971, compared with the ratio for the 42 years.

Land used by the minerals industry for the 1930-71 period is similar in areas to that used by railroads or by airports in operation at the end of 1971. Highways during this period accounted for 22.7 million acres, or over six times the area used by mining. Other comparisons of estimated land utilized in the United States are shown in Table 4. On the basis of national needs for surface area, mining requirements have been relatively minor. However, unlike many of the other uses of land surface, locations of mining operations are not amenable to optimum site selection; they are dependent on the natural location of the mineral deposits.

Of the total 3.65 million acres utilized by the minerals industry, about 59 percent was accounted for by the area of excavation, 20 percent by disposal of overburden and other mine wastes from surface mining, 13 percent by disposal of mill or processing wastes, and 5 percent by disposal of underground mine waste; the remaining 3 percent was subsided or disturbed as a result of underground workings (Tables 5 and 6).

Reclamation of 95 percent of the 1.46 million reclaimed acres was distributed as follows: area of excavation only, 68 percent, and area used for disposal of overburden and other mine wastes from surface mining, 27 percent. The remaining 5 percent was distributed among mill waste areas,

Reprinted from Bureau of Mines Information Circular 8642, U.S. Government Printing Office, Washington, D.C., June 1974, pp. 10-19.

Table 1.

Land utilized and reclaimed by the mining industry[1] in the United States

in 1930-71[2] by State

State	Total State land area thousand acres	Percent of total land area used for mining	Total area utilized,[3] acres	Total area reclaimed,[3] acres	% Reclaimed
Alabama	32,678	0.20	65,100	28,500	43.8
Alaska	365,482	.01	29,600	10,600	35.8
Arizona	72,688	.14	102,000	6,850	6.7
Arkansas	33,599	.09	29,500	9,040	30.6
California	100,207	.23	227,000	43,900	19.3
Colorado	66,486	.07	48,800	14,000	28.7
Connecticut	3,135	.39	12,300	3,410	27.7
Delaware	1,266	.10	1,330	370	27.8
Florida	34,721	.26	88,800	17,100	19.3
Georgia	37,295	.09	34,300	9,650	28.1
Hawaii	4,106	.12	4,810	1,160	24.1
Idaho	52,933	.08	41,300	8,660	21.0
Illinois	35,795	.83	297,000	188,000	63.3
Indiana	23,158	.76	175,000	113,000	64.6
Iowa	35,860	.15	55,300	18,300	33.1
Kansas	52,511	.08	44,000	21,500	48.9
Kentucky	25,512	.92	234,000	150,000	64.1
Louisiana	28,868	.06	18,200	5,210	28.6
Maine	19,848	.05	10,500	3,170	30.2
Maryland	6,319	.41	25,600	9,170	35.8
Massachusetts	5,035	.40	20,300	5,610	27.6
Michigan	36,492	.27	99,500	24,100	24.2
Minnesota	51,206	.27	136,000	13,000	9.6
Mississippi	30,223	.04	10,700	3,310	30.9
Missouri	44,248	.23	102,000	41,400	40.6
Montana	93,271	.05	42,800	10,600	24.8

Table 1 continued

Nebraska	49,032	.03	12,800	3,720	29.1
Nevada	70,264	.06	41,100	4,020	9.8
New Hampshire	5,769	.09	5,300	1,590	30.0
New Jersey	4,813	.59	28,400	7,470	26.3
New Mexico	77,766	.06	47,800	9,800	20.5
New York	30,681	.31	96,300	24,600	25.5
North Carolina	31,403	.12	36,600	9,640	26.3
North Dakota	44,452	.08	35,100	23,900	68.1
Ohio	26,222	1.11	292,000	181,000	62.0
Oklahoma	44,088	.08	35,500	16,500	46.5
Oregon	61,599	.06	34,000	8,940	26.3
Pennsylvania	28,805	1.32	381,000	186,000	48.8
Rhode Island	677	.34	2,330	540	23.2
South Carolina	19,374	.07	14,500	4,110	28.3
South Dakota	48,882	.03	16,500	4,650	28.2
Tennessee	26,728	.25	67,800	23,400	34.5
Texas	168,218	.05	78,000	20,500	26.3
Utah	52,697	.13	66,700	6,390	9.6
Vermont	5,937	.12	7,380	1,200	16.3
Virginia	25,496	.31	78,800	28,900	36.7
Washington	42,694	.08	35,900	9,740	27.1
West Virginia	15,411	1.36	210,000	105,000	50.0
Wisconsin	35,011	.13	46,900	12,400	26.4
Wyoming	62,343	.05	28,300	8,800	31.4
Total[4]	2,271,304	.16	3,650,000	1,460,000	40.0

1. Excludes oil and gas operations

2. U.S. Department of Commerce. Statistical Abstract of the United States. 1972, p. 196

3. Includes area of surface mine excavation, area used for disposal of surface mine waste, surface area subsided or disturbed as a result of underground workings, surface area used for disposal of underground waste, and surface area used for disposal of mill or processing waste.

4. Data may not add to totals shown because of independent rounding.

Table 2.

Land reclaimed by the mining industry[1] in the

United States in 1930-71, by State

(Acres)

State	Surface mining Mined area	Waste area	Underground mining Subsided or disturbed area	Surface waste area	Milling, surface waste area	Total land reclaimed[2]
Alabama	19,100	8,210	80	580	600	28,500
Alaska	7,650	1,940	-	10	850	10,600
Arizona	3,450	1,780	70	-	1,550	6,850
Arkansas	5,290	3,230	30	40	450	9,040
California	23,800	13,000	370	640	6,060	43,900
Colorado	9,470	3,160	130	310	950	14,000
Connecticut	2,080	1,130	-	-	200	3,410
Delaware	250	100	-	-	20	370
Florida	12,500	4,030	-	-	530	17,100
Georgia	6,230	2,980	-	10	430	9,650
Hawaii	620	480	-	-	60	1,160
Idaho	4,710	1,560	100	280	2,020	8,660
Illinois	133,000	51,600	280	1,730	1,820	188,000
Indiana	80,300	31,500	50	350	960	113,000
Iowa	11,500	5,960	10	70	740	18,300
Kansas	14,900	5,710	10	30	800	21,500
Kentucky	104,000	41,600	350	2,630	1,300	150,000
Louisana	3,320	1,570	-	-	320	5,210
Maine	2,170	810	-	-	190	3,170
Maryland	5,980	2,770	10	50	360	9,170
Massachusetts	3,530	1,750	-	-	340	5,610
Michigan	14,300	8,050	-	90	1,610	24,100
Minnesota	7,250	4,050	-	-	1,680	13,000

Table 2 Continued

Mississippi	2,280	850	-	-	180	3,310
Missouri	26,500	11,300	10	200	3,340	41,400
Montana	6,500	3,220	80	150	650	10,600
Nebraska	2,430	1,060	-	-	230	3,720
Nevada	1,720	890	360	50	1,000	4,020
New Hampshire	1,070	420	-	-	100	1,590
New Jersey	4,470	2,300	150	-	550	7,470
New Mexico	6,110	3,240	10	50	390	9,800
New York	16,900	6,190	130	20	1,370	24,600
North Carolina	5,810	3,280	-	-	560	9,640
North Dakota	17,100	6,610	-	20	160	23,900
Ohio	128,000	50,400	120	840	1,880	181,000
Oklahoma	11,400	4,700	10	40	390	16,500
Oregon	5,100	2,660	-	-	1,180	8,940
Pennsylvania	131,000	45,600	2,140	5,470	2,060	186,000
Rhode Island	340	170	-	-	30	540
South Carolina	2,650	1,260	-	-	200	4,110
South Dakota	3,050	1,280	30	10	280	4,650
Tennessee	14,800	7,230	190	250	920	23,400
Texas	12,800	6,420	80	20	1,260	20,500
Utah	2,970	1,430	40	290	1,660	6,390
Vermont	580	330	-	10	280	1,200
Virginia	18,400	8,460	140	1,010	900	28,900
Washington	6,020	3,000	50	50	630	9,740
West Virginia	70,000	25,900	800	6,170	1,960	105,000
Wisconsin	7,550	3,970	40	-	840	12,400
Wyoming	5,440	3,070	20	120	240	8,890
Total	987,000	402,000	5,870	21,600	47,100	1,460,000

1. Excludes oil and gas.

2. Data may not add to totals shown because of independent rounding.

Table 3.

Land[1] utilized and reclaimed by the mining industry in the United States

in 1971, by State and commodity group

(Acres)

State	Metals		Nonmetals		Fossil Fuels[2]		Total[3]	
	Utilized	Reclaimed	Utilized	Reclaimed	Utilized	Reclaimed	Utilized	Reclaimed
Alabama	180	170	1,000	590	3,170	2,160	4,360	2,930
Alaska	130	380	680	490	210	750	1,020	1,630
Arizona	10,900	670	2,260	650	W	W	13,200	1,320
Arkansas	140	80	1,120	900	150	170	1,420	1,150
California	1,760	2,060	10,300	6,610	W	W	12,000	8,670
Colorado	300	340	4,460	970	380	470	5,140	1,780
Connecticut	-	-	540	400	-	-	540	400
Delaware	-	-	200	60	-	-	200	60
Florida	850	120	11,000	2,240	20	4	11,900	2,360
Georgia	410	180	1,930	1,030	1	1	2,340	1,210
Hawaii	-	-	330	80	-	-	330	80
Idaho	360	1,000	1,070	560	W	W	1,430	1,560
Illinois	W	W	3,170	2,410	7,600	12,900	11,300	15,300
Indiana	-	-	2,090	1,340	4,350	8,030	6,440	9,380

Table 3 continued

	Utilized	Reclaimed	Utilized	Reclaimed	Utilized	Reclaimed	Utilized	Reclaimed
Iowa	-	-	1,620	1,320	120	420	1,740	1,740
Kansas	-	240	1,170	760	W	W	1.170	1,010
Kentucky	W	W	950	540	13,500	14,100	14,500	14,600
Louisiana	-	-	1,230	670	-	-	1,240	670
Maine	W	W	650	580	50	50	700	630
Maryland	-	-	1,210	740	250	370	1,460	1,110
Massachusetts	-	-	1,450	850	W	-	1,450	850
Michigan	960	180	3,840	2,660	100	150	4,900	3,000
Minnesota	5,400	1,640	2,130	1,550	W	W	7,540	3,190
Mississippi	-	-	700	410	-	-	700	410
Missouri	W	W	1,780	1,030	W	W	1,780	1,030
Montana	1,510	270	2,070	1,430	740	460	4,320	2,170
Nebraska	-	-	940	620	-	-	940	620
Nevada	2,190	790	950	440	-	-	3,140	1,230
New Hampshire	-	-	330	240	-	-	330	240
New Jersey	W	W	1,180	970	60	4	1,240	970
New Mexico	2,790	330	1,000	430	W	W	3,790	750
New York	360	150	2,330	2,030	20	-	2,710	2,180

Table 3 continued

	Utilized	Reclaimed	Utilized	Reclaimed	Utilized	Reclaimed	Utilized	Reclaimed
North Carolina	W	-	2,240	960	-	-	2,240	960
North Dakota	-	1	550	420	1,000	1,590	1,550	2,010
Ohio	-	-	3,450	2,740	9,560	13,000	13,000	15,800
Oklahoma	W	W	690	490	W	W	690	490
Oregon	20	320	1,620	980	-	1	1,640	1,300
Pennsylvania	110	40	2,510	1,260	11,800	17,900	14,400	19,200
Rhode Island	-	-	180	80	-	-	180	80
South Carolina	-	-	790	510	10	-	800	510
South Dakota	80	20	910	700	-	10	1,000	740
Tennessee	W	W	1,560	1,120	2,070	2,050	3,630	3,170
Texas	360	250	4,830	2,500	W	W	5,190	2,750
Utah	3,140	850	990	640	W	W	4,130	1,490
Vermont	-	110	250	120	-	-	250	240
Virginia	W	W	2,260	1,100	2,900	2,830	5,160	3,930
Washington	60	60	2,000	1,380	W	W	2,060	1,440
West Virginia	-	-	500	360	11,100	12,500	11,600	12,900
Wisconsin	W	W	2,130	1,540	W	W	2,130	1,540
Wyoming	1,690	190	1,510	740	750	540	3,940	1,460

Table 3 continued

	Utilized	Reclaimed	Utilized	Reclaimed	Utilized	Reclaimed	Utilized	Reclaimed
Undistributed	2,640	2,140	-	-	5,230	6,440	7,870	8,590
Total[3]	36,400	12,600	95,100	53,200	74,900	96,900	206,000	163,000

W - Withheld to avoid disclosing individual company confidential data; included with "Undistributed"

1. Includes area of surface mine excavation, area used for disposal of surface mine waste, surface area subsided or disturbed as a result of underground workings, surface area used for disposal of underground waste, and surface area used for mill or processing waste.

2. Excludes oil and gas operations.

3. Data may not add to totals shown because of independent rounding.

Table 4.

Comparison of land utilized by the United States

in 1971, by various types of use[1]

Activity	Million Acres[2]
Total United States	2,271.3
Agriculture	1,283.0
Cropland	472.1
Grassland pasture and range	609.6
Forest land grazed	198.0
Farmsteads grazed	8.4
Forest land not grazed	525.5
Urban areas	34.6
National Park system	29.6
Highways	22.7
State Park system	8.6
Mining[3]	3.7
Airports	3.3
Railroads	3.2
Municipal and county park and recreational areas	1.0

1. Estimates based primarily on reports and records of the Bureau of Census and Federal and State agencies.

2. 1969 data

3. Land utilized 1930-71.

Table 5.

Land utilized by the mining industry[1] in the United States

in 1930-71, by State and function

(Acres)

State	Surface mining		Underground mining		Milling, surface waste area	Total land utilized[2]
	Mined area	Waste area	Subsided or disturbed area	Surface waste area		
Alabama	39,700	12,900	2,080	4,830	5,510	65,100
Alaska	22,300	3,550	100	60	3,610	29,600
Arizona	26,600	34,900	2,910	360	37,600	102,000
Arkansas	18,700	6,180	260	290	4,130	29,500
California	105,000	57,100	2,230	12,200	50,400	227,000
Colorado	30,200	5,750	1,320	3,310	8,200	48,800
Connecticut	8,730	2,180	-	-	1,290	12,300
Delaware	980	230	-	-	130	1,330
Florida	71,500	10,300	-	-	7,040	88,800
Georgia	23,600	5,420	10	30	5,240	34,300
Hawaii	3,460	810	-	-	540	4,810
Idaho	16,700	3,980	520	3,010	17,100	41,300
Illinois	201,000	63,800	6,320	14,600	10,500	297,000
Indiana	125,000	39,300	1,270	2,930	6,060	175,000

Table 5 Continued

Iowa	38,300	10,100	260	570	6,030	55,300
Kansas	27,500	7,970	70	150	8,290	44,000
Kentucky	146,000	49,000	9,470	22,100	7,030	234,000
Louisana	12,900	3,260	-	10	2,070	18,200
Maine	7,620	1,900	-	-	1,020	10,500
Maryland	17,800	4,760	180	400	2,540	25,600
Massachusetts	14,600	3,650	-	-	2,100	20,300
Michigan	64,600	17,100	1,690	840	15,300	99,500
Minnesota	72,300	37,900	-	-	26,100	136,000
mississippi	7,680	1,890	-	-	1,140	10,700
Missouri	55,300	15,500	180	610	30,700	102,000
Montana	22,200	4,100	350	600	5,450	42,800
Nebraska	9,170	2,310	-	10	1,360	12,800
Nevada	12,000	11,100	2,160	2,170	13,700	41,100
New Hampshire	3,750	1,020	-	-	530	5,300
New Jersey	18,500	4,590	210	100	5,000	28,400
New Mexico	19,600	13,900	4,310	2,140	7,840	47,800
New York	55,800	12,600	190	190	27,400	96,300
North Carolina	24,100	6,550	-	-	6,020	36,600
North Dakota	25,700	8,350	70	150	770	35,100

Table 5 Continued

Ohio	206,000	63,400	2,980	6,860	12,600	292,000
Oklahoma	24,800	6,900	130	300	3,410	35,500
Oregon	22,100	5,290	-	-	6,600	34,000
Pennsylvania	221,000	64,500	35,600	43,600	16,500	381,000
Rhode Island	1,730	370	-	-	220	2,330
South Carolina	9,840	2,300	-	-	2,390	14,500
South Dakota	10,900	2,980	100	120	2,420	16,500
Tennessee	40,500	12,300	1,430	2,200	11,300	67,800
Texas	54,000	13,300	520	570	9,650	78,000
Utah	18,900	19,200	920	2,200	25,500	66,700
Vermont	2,870	860	-	30	3,630	7,380
Virginia	42,600	13,100	3,620	8,410	11,100	78,800
Washington	24,500	6,470	260	400	4,350	35,900
West Virginia	96,500	30,600	22,200	51,800	8,770	210,000
Wisconsin	32,200	8,730	60	30	5,780	46,900
Wyoming	14,800	9,040	620	1,450	2,410	28,300
Total[2]	2,170,000	733,000	105,000	190,000	454,000	3,650,000

1. Excludes oil and gas.

2. Data may not add to totals shown because of independent rounding.

Table 6.

Land utilized and reclaimed by the mining industry[1] in the United States in 1930-71, by area of mining activity

(Acres)

Type of Use	Metals Utilized	Metals Reclaimed	Nonmetals Utilized	Nonmetals Reclaimed	Fossil Fuels Utilized	Fossil Fuels Reclaimed	Total Utilized	Total Reclaimed
Surface area mined (area of excavation only)............	145,000	17,400	1,060,000	253,000	966,000	716,000	2,170,000	987,000
Area used for disposal of overburden and other mine waste from surface mining.....	123,000	5,270	291,000	129,000	320,000	286,000	733,000	402,000
Surface area subsided or disturbed as a result of underground workings.........	12,200	1,780	4,570	100	87,900	4,000	105,000	5,870
Surface area used for disposal of underground mine waste..............	21,900	1,500	2,080	180	166,000	20,000	190,000	21,600
Surface area used for disposal of mill or processing waste............	221,000	17,300	201,000	23,300	31,900	6,480	454,000	47,100
Total[2]	524,000	43,300	1,560,000	406,000	1,570,000	1,010,000	3,650,000	1,460,000

1. Excludes oil and gas operations.

2. Data may not add to totals shown because of independent rounding.

waste from underground mines, and subsided areas.

Fossil fuels and nonmetals each accounted for 43 percent of surface land used during 1930-71; the metal mining industry utilized the remaining 14 percent. On a commodity basis, production of bituminous coal accounted for 40 percent; sand and gravel, 18 percent; crushed and broken stone, 14 percent; clays and copper, 5 percent each; iron ore, 3 percent; phosphate rock, 2 percent; and all other commodities combined, 13 percent.

The fossil fuels industry reclaimed over 1 million acres, or 69 percent of the total land reclaimed; nonmetals, 28 percent; and metals, 3 percent. The largest single item of reclamation, about one-half of total reclaimed area, was reclamation of surface area mined for bituminous coal.

Part 2
Environmental Impacts

5
Coal and the Environment

H. Beecher Charmbury

Extraction of coal from the earth today should be accomplished with minimum environmental consequences. If it is not, tomorrow it will be required by law.

For the past ten years the coal mining industry has made great efforts to preserve and improve our environment during deep and strip mining operations. This has helped the industry to improve its image, but even greater efforts must be made by all operators. A few bad operations can spoil the entire image and far overshadow the good ones. It is the bad operations that receive the publicity. Some of the efforts to protect the environment have been voluntary on the part of industry and others have been required by strong state-level legislation.

For 100 years, prior to the middle 1960s, the coal industry was permitted to operate without any laws to protect the environment. The industry was in business to mine coal at a profit and at a time when this country needed the coal to help its economic growth and industrial development and to help the country fight two major world wars. Because of the need for the coal, the damage done to the environment was considered a necessary evil. There was land devastation from strip mining and deep mine caving. There was stream pollution from deep and strip mine drainage and there was air pollution from burning refuse banks and underground mine fires.

All of these helped to create a depressed environment wherever coal was mined so that as mining operations were completed, the area was practically useless. Other industries could not be attracted to the areas because of the devastated environment.

From the middle 60s until today, times have been changing. We are living in an environment-conscious society. People are demanding beautiful

From *Coal Mining & Processing*, vol. 2, no. 1, Jan. 1975, pp. 69-70. Copyright 1974 by Maclean-Hunter Publishing Corporation. Reprinted by permission.

land, clean air, and pure water, not just in the coal mining areas but in all areas and from all industries. Consequently, even though there is probably just as great a demand for coal now as there was during certain periods in the past, it is most doubtful that the citizenry, even in the coal producing areas, will tolerate the devastation to the environment that has occurred in the past. Furthermore, the legislators today are listening more to the people's demands.

Most segments of the coal mining industry have shown good faith in attempting not only to prevent the destruction of the environment, but also to clean up the previous destruction. The progressive coal company tries to become a good industrial community member, as well as a corporate citizen. This is being done by better housekeeping practices and applying modern engineering and technology developments and good judgment and common sense to the environmental aspects of its mining operations.

There are many things that the coal industry can do today to prevent legislation tomorrow. The industry must make every effort to police itself so that it does not attract attention to bad practices. Great care must be taken to store refuse properly, to prevent stream pollution, and to stop land devastation.

Stop Air and Water Pollution from Refuse

Coal refuse can be divided into two types—coarse and fine—or refuse banks and slime ponds.

The banks and ponds must be formed in such a manner to minimize the tendency for firing to create an air pollution problem and to prevent water pollution and to insure stability. There are definite steps that all deep and strip coal mining companies could take to prevent this. Some of these are listed below:

1. Prevention and control of air pollution.
   a. The disposal area should be cleared of vegetation and combustible material.
   b. Precautions to minimize ignition include layering, compacting, covering, limiting of slopes and such other measures as are necessary.
   c. Coal refuse should not be deposited on or near any coal refuse disposal area known to be burning.
   d. Other waste materials with low ignition points, including, but not limited to, wood, cloth, waste paper, oil, grease and garbage, should not be deposited on or near any coal refuse disposal area.

 e. Highly reactive coal refuse should be mixed with inert material in appropriate portions to minimize the ignition of coal refuse.

 f. If mine rock cannot be adequately compacted when mixed with other coal refuse or if it may cause size segregation when mixed with other coal refuse, it should be disposed of separately.

 g. A plan for detecting and extinguishing hot spots or fires should be established.

 h. A plan for controlling airborne dust from dried out silt basins should be established.

2. Prevention and control of water pollution.

 a. No discharge of waters from coal refuse disposal areas should have a total acidity exceeding total alkalinity.

 b. The pH of waters discharged from the refuse disposal area should be maintained between 6.0 and 9.0.

 c. No discharge of waters from a refuse disposal area should contain suspended solids in excess of 200 ppm.

 d. No discharge of waters from a refuse disposal area should contain more than 7.0 ppm total iron.

 e. Runoff from the top of the coal refuse disposal pile should not be allowed to discharge freely onto the slopes. Open ditches should be constructed to allow the controlled flow of this runoff. The top of the pile should be graded with shallow slopes falling away from the edge to concentrate the runoff in these ditches. Water runoff from the adjacent area should be directed away from the coal refuse pile.

3. Design requirements to insure stability.

 a. Impoundments should not be located on coal refuse disposal areas.

 b. A minimum clear space of 50 ft. should be provided from the outer perimeter of any coal refuse disposal area.

 c. Coal refuse should not be deposited in layers exceeding two feet in depth.

 d. Slopes of the sides of coal refuse disposal areas should not exceed 15 percent.

 e. The height of coal refuse disposal areas should be limited and other measures should be taken as necessary to insure the stability of slopes.

Stop Water Pollution from Mine Drainage

Mine drainage can range from alkaline to as much as 35,000 ppm of acid. The acid in mine drainage is usually sulfuric, but in some cases may be hydrogen sulfide. Mine drainage may also contain heavy metals, mostly iron, which pollute and discolor the streams. Other heavy metals are manganese and magnesium. Other metallic ions are aluminum and calcium. All of these metallic ions are generally associated with sulfate to form soluble salts. Mine drainage may also contain suspended solids such as coal silt, soil or oxides of iron. It may also contain certain types of bacteria.

There are certain steps that can be taken to stop water pollution from mine drainage. Some of these are:

1. Prevent it. It takes three substances to make acid mine drainage—air, water, and pyrite. The only practical one to control is water. Therefore keep as much water as possible from getting into the mines. Prevent subsidence which permits surface water or streams from getting into deep mines and conduct concurrent backfilling operations in strip mining with a minimum open cut.

2. If water cannot be prevented from getting into the mine and coming in contact with air and pyrite then minimize it or get it out of the mine as quickly as possible. Prevent or minimize the time of contact the water has to react with the air and pyrite. If water is transferred from sump to sump in the deep mine before it is pumped to the surface, use pumps and pipe rather than gravity flow on the floor of the mine. This will reduce acid formation and the formation of soluble salts.

3. If pollution of mine drainage cannot be prevented then treat it so as not to pollute the receiving stream. The water to be discharged into the receiving stream, regardless of the condition of that stream, should be free of excess acid, the total iron content should be less than 7 ppm, the pH should be in the range of 6.0 to 9.0, and the suspended solids should be less than 200 ppm. The type of treatment will of course depend upon the nature of the drainage:

   a. For acid control the drainage must be neutralized with such substances as hydrated or calcined lime, limestone, caustic soda, or soda ash. The agent selected will depend primarily upon its availability and price.

   b. For iron control the drainage must be oxidized by aeration, microbiological oxidation, or ozonolysis. Aeration is the most practical at present and consists of bubbling air through the drainage or spraying it into the atmosphere. The oxidation process converts the ferrous iron to the ferric state which precipitates it into a yellow or reddish-brown solid.

   c. For solids control the drainage must have sufficient time for the solids to settle so that only the clear fluid is released into the stream. This can be done in large settling lagoons or in thickeners. One of the big problems is the disposal of the settled solids or sludge. It is generally recognized that the sludge be concentrated first by filtering or centrifuging and then the concentrated sludge can be disposed of on the refuse bank, or in the abandoned strip or deep mine.

Stop Land Devastation from Strip Mining

It is a little late to discuss this subject since the Congress is determined to pass strong strip mining legislation. This has been brought about by the bad stripping practices in the past. The industry was warned about this on many occasions but failed to police its own operations. It can only be hoped that Congress will now use good judgment and common sense in passing legislation to permit the sincere strip mine operator to remain in business and provide the coal so vitally needed today. Basically, the thrust of strip mining reclamation is to put the land back to some useful purpose rather than to leave it in a useless devastated condition.

Coal is vital to our nation, thus coal mining, whether it be deep or strip, is a must. However, the sooner the industry recognizes that the mining must take place with minimum environmental consequences, the better off it will be to operate freely with the minimum number of ridiculous controls as a result of poor legislation. The industry would be much better off to appoint its own environmental police force to control its own operations rather than to spend its time and money fighting legislation. Otherwise, the industry is not going to win its fight against environmental controls.

6
A Crop for Mine Spoils?

Edward R. Buckner
J. S. Kring

The recent report by the Secretary of the Interior to the Appalachia Regional Commission (6) on strip and surface mining in Appalachia states that there are 740,000 acres in the Appalachia region that have been disturbed by strip mining and that 31,000 acres are added to this total each year. Approximately three-fourths (548,000 acres) of this is contour stripping which has left over 20,000 miles of the "highwall-spoil bank" contour as a prominent feature of the regional landscape. While this is a relatively small percentage of the total land area (0.6 percent) these disturbances are generally visible over great distances and directly or indirectly cause damage to water and recreational resources many miles from the actual mining sites. Damage to surrounding areas is in the form of increased flood peaks, stream acidification and siltation with resulting damage to fish and the loss of recreational potential due to water pollution and the loss of scenic values.

To minimize the damage resulting from strip mining, six of the twelve Appalachia states have passed regulatory laws. Although generally referred to as measures to reclaim disturbed land, the provisions of most of these laws are such that soil stabilization through grading and filling, the establishment of a vegetative cover, and control of acid pollution are the primary accomplishments. The productive use of these reclaimed lands is left to the landowner. In those states where corrective measures are not required and raw spoils are not reclaimed, there is little opportunity for productive use. According to the Appalachia report, reclamation varies from state to state (Table 1).

Rehabilitating wasteland implies more than simply soil stabilization to prevent damage to adjoining areas as is generally the case with strip mine

From *Keep Tennessee Green Journal*, vol. 7, no. 1, Spring 1967, pp. 14-18. Reprinted by permission.

Edward R. Buckner is Assistant Professor of Forestry and J. S. Kring is Associate Professor of Forestry, Department of Forestry, College of Agriculture, The University of Tennessee.

Table 1

Status of strip and surface mined coal lands in Appalachia

as of January 1, 1965 as reported by State authorities

State	Unreclaimed		Partially Reclaimed	Completely Reclaimed	Total Disturbed	% Partially or Completely Reclaimed
Alabama	2,200		11,700	5,000	18,900	88
Georgia	---		75	225	300	100
Kentucky	31,487		4,439	12,363	48,289	35
Maryland	494		753	995	2,242	78
New York	---	---	---	---	---	---
North Carolina	---	---	---	---	---	---
Ohio	33,540		21,900	123,816	179,256	81
Pennsylvania	---	208,500	---	92,600	301,100	31*
South Carolina	---	---	---	---	---	---
Tennessee	25,387		275	1,098	26,760	5
Virginia	15,014		13,549	503	29,066	48
West Virginia	---	145,718	---	46,320	192,038	24*
Total	108,122	354,218	52,691	282,920	797,951	42

*Percentage of total that is completely reclaimed

reclamation work. At lower elevations where open pit and area stripping are practiced the spoils can often be reclaimed for farming, pasture, recreational sites, and wildlife management areas. On the steeper slopes where contour stripping is practiced the narrow belt of relatively level land left following reclamation does not lend itself to a variety of uses. Following grading the most common practice is to plant trees or, where reclamation is not required, the raw spoil is left to reseed naturally. In either case it is not likely that a reasonable return will be realized by the landowner. Natural seeding following abandonment generally results in poorly stocked stands of undesirable species. Planted stands are not likely to give a return on the investment over the long rotations required for most timber crops on these poor sites.

Interest in stabilizing and managing spoil areas might be stimulated if a profitable, short-rotation crop could be found suitable to these extreme site conditions. Christmas trees generally meet these criteria provided suitable species can be found that would develop properly on these sites. Possible advantages of the strip mine planting site for Christmas trees include: 1) vandalism is less likely as most of these areas are not along major highways and access roads are easily controlled, 2) mine roads were constructed for year-round use providing easy access for the intensive management required in the production of high-quality Christmas trees and for the winter harvest, and 3) different exposures, slopes and altitudes provide sufficient site variation so that several species can be used. Poor chemical and physical properties plus the absence of profile development are characteristic of these mine spoils and generally result in slow tree growth. Very rapid growth, however, is not necessarily an advantage in Christmas tree production as it often results in poorly formed trees.

The extreme variation in spoil properties generally found on these sites is a major problem. Changes in chemical and physical properties are usually abrupt and may cause marked survival and growth differences among trees even at the close spacings used with Christmas trees. A second problem is the tendency of many species to become chlorotic when growing on poor sites. Proper fertilizer applications may largely overcome both of these problems.

A Pilot Study

To test the feasibility of growing Christmas trees on unreclaimed strip mine spoils, white pine (*pinus strobus*), Scotch pine (*Pinus sylvestris*), and Norway spruce (*Picea abies*) seedlings were planted on an abandoned strip mine on Little Brushy Mountain in Morgan County, Tennessee, in the spring of 1964. These species are considered desirable for the commercial production of Christmas trees (7,10).

The planting site is at an elevation of 1,900 feet on a dry southeast facing ridge. Raw spoil was left in a series of mounds and depressions with a maximum relief of approximately 15 feet. The upper portions of the mounds were free of vegetation. Black locust (*Robinia pseudoacacia*) and polkweed (*Phytolacca americana*) were present in the depressions. Rocks in the area are early Pennsylvania in age (13), flat bedded and dip slightly to the east. Spoil at the planting site is silty shale material removed as overburden from the coal. This "shale interval" forms the cap of Little Brushy Mountain and contains thin interbedded sandstone seams (Figure 1). The shale disintegrates rapidly upon exposure.

Two separate mining cuts were made in this area. The spoil from the second cut is partially supported by the thick sandstone bench that was exposed by the first cut. The remainder was pushed off the sandstone bench creating a steep downslope spoil slide. The spoil bank covering the first cut ranges from 15 to 30 feet higher than the depression created by the second cut. Spoil depth over the sandstone base in the second cut ranges from a few inches to three to five feet. Occasional sandstone boulders from a thin interbedded sandstone layer several feet above the coal

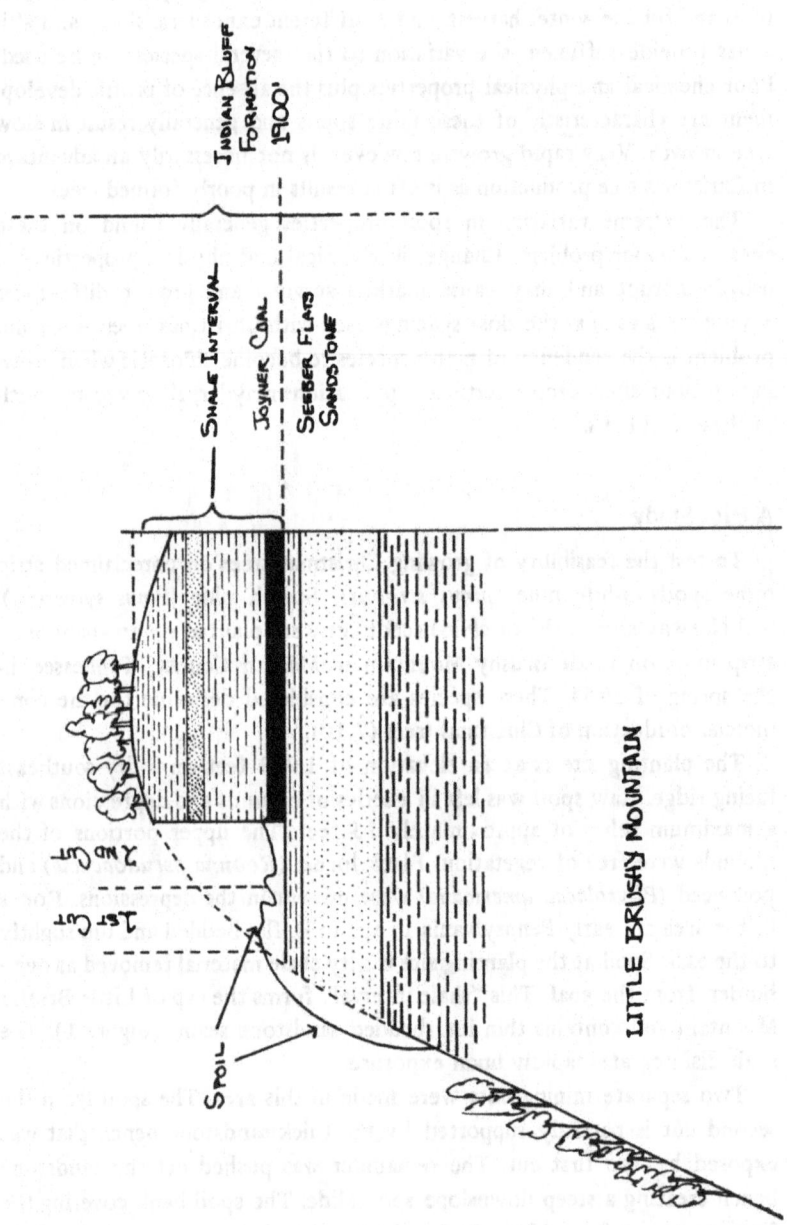

Figure 1. Stratigraphy of the Pennsylvanian rocks at the mining site on Little Brushy Mountain.

seam were scattered over the surface. Though highly variable, the resulting spoil is largely of silty clay texture.

The planting site is poorer than the average unreclaimed spoil found on this stripping operation, which extends for several miles and completely encircles the mountain. The absence of natural vegetation at the time of planting when the more productive spoils had seeded naturally to black locust and shortleaf pine (*Pinus echinata*) attests to the severity of the site for native species.

Analysis of 35 soil samples randomly selected from the planting area gave the following values:

	Average	Range
pH	4.7	4.2-5.2
P	14 lbs/acre	2- 90 lbs/acre
K	161 lbs/acre	100-280 lbs/acre

When pH values drop below 4.0 the site is generally considered too toxic for the growth of most tree species (1, 3, 4, 5). Only four of the 35 soil samples were 5.0 or above (pH 4.0-5.0 is considered strongly acid). Phosphorus was low to very low according to agronomic standards. Only eight of the samples tested above 15 lbs/acre. Potassium tested higher and there was less variation among samples than was the case for phosphorus. Although tests were not made for nitrogen, levels were obviously low as there was no visible organic content in the spoils. These soil data are in close agreement with those obtained by Thor and Kring (11) on adjacent spoil banks.

Due to space limitations only three replications were established. A factorial arrangement with a randomized block design was used in which each of the three species was tested for its response to a fertilizer application. Fifty grams of a slowly available fertilizer, 6-40-5 (6-18-4, by elemental analysis) containing Mg (14) were placed in the closing hole at the time of planting. Trees were planted at a 5 x 5 foot spacing. The few black locusts present on the planting sites were girdled and poisoned.

Survival

Poor planting stock resulted in Scotch pine failure the first growing season (survival in most treatments was under 20 percent). This species was replanted in the spring of 1965.

Survival was higher than might be expected on this extreme site (Table 2). White pine survival (89 percent) was exceptional when compared with other strip mine planting studies (1, 2, 3, 4, 5). Burton (2)

Table 2

Survival (percent) and height (feet)

at the end of the 1966 growing season

	White Pine Survival	Height	Norway Spruce Survival	Height	Scotch Pine Survival	Height
	3 Growing Seasons				2 Growing Seasons	
Fertilized	89%	1.7 ft.	69%	1.2 ft.	67%	0.9 ft.
Non-fertilized	83%	1.4 ft.	65%	0.9 ft.	69%	0.9 ft.
Average	89%	1.5 ft.	67%	1.1 ft.	68%	0.9 ft.

found white pine survival to be only 26 percent when planted on other strip mined areas in this region.

Norway spruce survival (67 percent) as well as survival of Scotch pine (68 percent), both exotic species, compared favorably with that found for loblolly pine (64 percent) on a nearby spoil (11). Two plots of Norway spruce located adjacent to the highwall (one fertilized and one not fertilized) were considered complete failures due to the following factors which reduced survival and growth: (1) shallow soil to sandstone bedrock (6 to 8 inches), (2) poor drainage with acid water standing during wet periods, and (3) partial covering or burial of seedlings due to sloughing from the highwall. Average survival in the other Norway spruce plots was 78 percent.

Scotch pine is extremely sensitive to shading from competing vegetation. Aggressive competition from black locust and polkweed reduced survival in all Scotch pine plots. Although not aggressive on spoil mounds these two species grew rapidly in the depressions. Even though competing black locust was girdled and poisoned in the spring, new growth would overtop planted seedlings by mid-season. Average Scotch pine survival was 68 percent.

Thor (8) obtained the following survivals for these same three species planted on a nearby old field site: white pine—99 percent; Norway spruce—97 percent; and Scotch pine—65 percent.

Development

The objective of Christmas tree fertilization is to promote the production of vigorous, dense foliage without excessive height growth (7). Although mean heights (Table 2) for fertilized white pine and Norway spruce were, respectively, 22 percent and 33 percent greater than mean heights for unfertilized plots, the differences were not significant. The wide site variation among treatments and the small number of replications apparently account for the failure to show significant differences.

Of the three species only white pine is native to the planting region. Early growth of this species is notoriously slow (12). After three growing seasons the mean height was 1.5 feet. This compares with an average three-year height of approximately 1.9 feet on a nearby old-field site (8). Average three-year height for the Norway spruce plots was 1.1 feet. This compares with approximately 1.5 feet for this species in the above mentioned old-field planting. White pine and Norway spruce generally do not begin rapid height growth until four to six years after planting thus the performance of these species may improve.

There was wide color variation in white pine and Norway spruce in both treatments. While genetic differences could account for some of this variation, the more plausible explanation is variation in the spoil in which they were planted. Approximately half of these trees would not make acceptable Christmas trees due to poor color. Fertilizer applied at planting time did not improve color after three growing seasons.

Height growth of Scotch pine was highly variable. When situated away from competing vegetation and on topographic positions other than the sun-baked southern exposure of spoil mounds, growth was satisfactory. After two growing seasons the height of Scotch pine averaged 0.9 feet. This compares with an average height of 1.1 feet after two growing seasons in the nearby old-field planting. The performance of seven-year-old Scotch pine on an adjacent mine spoil indicates that this species is suitable for planting on strip mines.

In contrast to white pine and Norway spruce, the color of Scotch pine was uniformly excellent. The good color characteristic of this seed source (French D'Auvergne) does not hold for trees from other sources (9). Using 19 different sources, Wright et al. (14) found that fertilizer applications did not affect color intensity of Scotch pine.

Although average height growth and survival of these species on mine spoils appears acceptable for Christmas tree production, the wide variation among trees in both growth rate and color is discouraging. Where trees chanced to be located in favorable situations, either in regard to soil properties or topographic position their growth was rapid and their color generally good. Maximum heights for these species were: white pine—3.6 feet and Norway spruce—1.6 feet after three growing seasons and Scotch pine—2.4 feet after two growing seasons. Trees located close by or immediately adjacent to these may show little growth or even dieback.

Conclusions

The average performance of these species on unreclaimed strip mine spoil does not make the growing of Christmas trees on such sites an attractive venture. While the average growth is acceptable, the variation in growth and color due to variations in topography and soil properties produces an irregular and ragged stand. It is obvious that there will be additional mortality as well as a large percentage of the survivors that will not make acceptable Christmas trees. Several observations are possible from the performance of these species on this variable planting site:

1. Norway spruce made acceptable growth only when protected from direct insolation either by topographic position or by overtopping vegetation. Those trees making satisfactory growth all had acceptable color; chlorosis was characteristic of this species only when stunted or very slow growing.

2. Scotch pine must be free of competing vegetation in order to survive and grow satisfactorily. Color will not be a problem so long as the proper seed source is used.

3. White pine made satisfactory growth on all topographic positions except the exposed mound tops and in the shallow, highly acid spoil next to the highwall. Wide color variation is characteristic of this species and does not appear to be closely correlated with growth rate.

4. Fertilizer applications made at planting time did not reduce survival and probably stimulated early growth. No improvement in tree color was traceable to this treatment.

Even though many of these trees will apparently develop into good Christmas trees, the wide variation in growth, form and color is discouraging. Grading the spoil banks to level the mounds and fill the depression adjacent to the highwall, such as is done on reclaimed spoils, would remove much of the site variation. Further research is needed to determine the suitability of reclaimed spoils for Christmas trees.

Literature Cited

1. Brown, James P. 1962. Success of Tree Planting on Strip Mined Areas in West Virginia, Bulletin 473, Agricultural Experimental Station, West Virginia University, Morgantown, W. Va.

2. Burton, James D. 1965. Greening up Barren Grounds. KTG Journal 5(1).

3. Hart, George and William R. Byrnes. 1960. Trees for Strip Mined Lands. Station Paper No. 136. Northeastern Forest Experiment Station, Upper Darby, Pa.

4. Limstrom, G. A. 1960. Forestation of Strip Mined Land in the Central States. Ag. Handbook No. 166.

5. Limstrom, G. A., and C. H. Deitschman. 1951. Reclaiming Illi-Strip Coal Lands by Forest Planting. Bulletin 547 Agricultural Experiment Station University of Illihois, Urbana, Illinois.

6. Secretary of the Interior. 1966. Study of strip and surface mining in Appalachia. An Interior report by the Secretary of the Interior to the Appalachia Regional Commission, Supt. of Documents, Washington, D.C. 20402.

7. Stangel, Harvey J. 1966. Technical Manual for Christmas Trees. Nitrogen Division, Allied Chemical Corporation.

8. Thor, Eyvind. 1967. Personal communication.

9. Thor, Eyvind. 1966. Christmas Tree Research in Tennessee. American Christmas Tree Journal Volume X (3).

10. Thor, Eyvind, R. T. Britt, J. Sharp, J. A. Catlett. 1962. Christmas Tree Production and Marketing in East Tennessee. Ag. Extension Circular 598, The University of Tennessee, Knoxville, Tennessee.

11. Thor, Eyvind, and James S. Kring. 1964. Planting and Seeding of Loblolly Pine on Steep Spoil Banks Journal of Forestry, Vol. 62(8): 275-276.

12. Wilson, C. W., J. W. Jewell, and E. T. Luther. 1956. Pennsylvania Geology of the Cumberland Plateau. Division of Geology, Tenn. Dept. of Conservation, Nashville, Tennessee.

14. Wright, W., S. Pauley, R. Polk, J. Jokela and R. A. Real. 1966. Performance of Scotch Pine Varieties in the North Central Region. Silvae Genetica 15 (4):101-110, July-August.

7
Highwalls—
An Environmental Nightmare

William T. Plass

An Environmental Dilemma

The documentation of tangible damages is not enough in this modern age when environmental concern is a part of everyday life. Intangibles that so far have defied quantitative expression must be considered. This is particularly important in emotional issues. In these cases, the intangibles may cause more public concern than hard facts documented by careful research. Furthermore, the fact that we cannot quantify a condition does not give us the liberty to ignore it.

How then, should environmental problems be treated to reduce both tangible and intangible damages? How can we assign research priorities? If there are limited funds for revegetation, how can we determine which treatments deserve the highest priorities? Conventional methods, such as cost benefit ratios, often are inadequate for these decisions.

This is the dilemma we face in considering treatments for surface mine highwalls. If we recognize only tangible damages, the outslope and bench would certainly demand highest priority. However, if we consider the single intangible value—aesthetics—many people would assign the highest priority to treatments that would soften the starkness of the highwall. This would be particularly true if the vegetation on the bench and on the outslope were lush and green or if the disturbed areas were partially screened by the original forest. A compromise between the two extremes seems to offer the best solution.

For years, surface mine revegetation research priorities have emphasized projects concerned with tangible problems. This is understandable because variables having measurable properties lend themselves to conven-

Reprinted from *Proc.*, Revegetation & Economic Use of Surface-Mined Land and Mine Refuse Symp., Pipestem State Park, West Virginia, Dec. 2-4, 1971, pp. 9-13.

William T. Plass is Principal Plant Ecologist, Northeastern Forest Experiment Station, Forest Products Marketing Laboratory, Princeton, West Virginia.

tional experimental designs. The treatments proposed by this research are designed to correct pollution problems on the bench and on the outslope.

This places me in a difficult position because I know of no research concerned specifically with highwall screening. I have decided therefore to explore the problem as I do when I prepare a problem analysis. The hypothetical solutions I will propose must stand the test of formal research before they can be accepted.

The problem is that people see highwalls and react according to their individual concepts of natural beauty. The key word in this statement is the word "see." Therefore, we must analyse factors that determine who sees the highwall and how much of the highwall is visible.

Highwall Screening

The point from which the general public views the highwall is very important. From experience, I believe that most people in West Virginia see it from moving vehicles on highways. If this is the case, the viewing point is usually some distance away and probably several hundred feet below the level of the bench. In some localities, mining operations can be viewed from the same elevation or from above. Air travel permits a few to see the disturbance from a very high vantage point, and there are some rugged individuals who travel along the surface mine benches.

Each of these vantage points presents different highwall screening problems. First, I will consider the situation as it appears to a person viewing a highwall from below the mined area. Two conditions affect the percentage of the total highwall height that can be seen. One is sighting distance, and the other is vertical distance below the level of the bench.

In a hypothetical situation, a surface mine is situated on a 30-degree slope, 750 feet above a level base line. Lines of sight—1/2 mile, 3/4 mile, and 1 mile in length—are extended from the crest of the outslope to the point of intersection with the base line (Figure 1). The angles of interception are 17 degrees at 1/2 mile, 11 degrees at 3/4 of a mile, and 9 degrees at 1 mile.

When these lines of sight are extended beyond the crest of the outslope to their interception with the highwall, the percentage of highwall exposed to view can be computed (Figure 2). In this case, the percentages are as follows: 1/2 mile, 26 percent; 3/4 of a mile, 54 percent; and 1 mile, 65 percent.

Next, I extended lines of sight a mile in length from the crest of the outslope to horizontal planes located 250, 500, 750, and 1,000 feet below the level of the bench (Figure 3). The angle of interception with

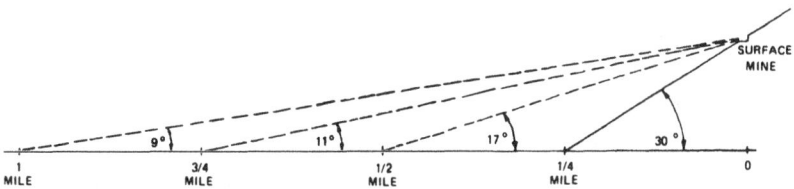

Figure 1. Angles of interception for lines of sight originating at various distances and 750 feet below the level of the surface mine bench.

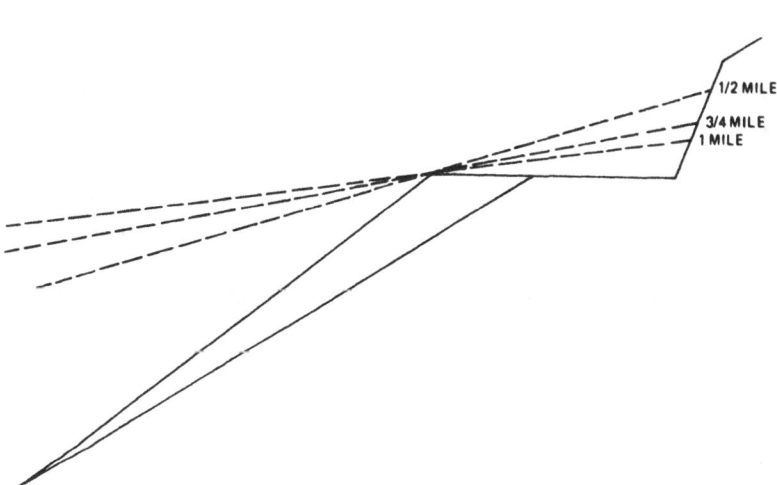

Figure 2. Points of interception on the highwall for lines of sight originating at various distances below the level of the surface mine bench.

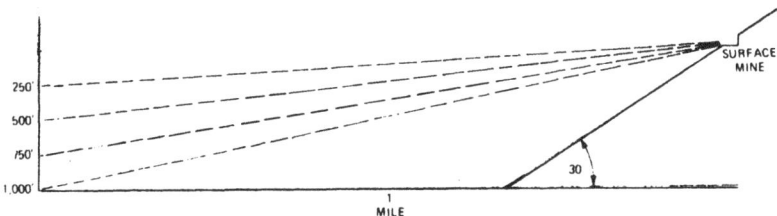

Figure 3. Angles of interception for lines of sight originating one mile from the surface mine and at various distances below the level of the bench.

these planes was 2 degrees, 250 feet below; 6 degrees, 500 feet below; 9 degrees, 750 feet below; and 11 degrees, 1,000 feet below. If these lines of sight are extended to the highwall, the percentage exposed is 86 percent, 250 feet below; 72 percent, 500 feet below; 65 percent, 750 feet below; and 52 percent, 1,000 feet below (Figure 4).

Proposals for screening the highwall must account for the need to reduce the percentage exposed as quickly as possible. Planted trees and shrubs will become more effective as they grow, but even the fastest growing species will require several years' growth before they significantly reduce the percentage exposed. Therefore, when the line of sight comes from below, regrading practices offer the best opportunities for immediate screening.

Although there may be many ways that the configuration of the bench cross section can be changed, a modification of the conventional reverse terrace regrading method is one alternative. Where it would not cause outslope stability problems, I propose to increase the slope angle to 10 degrees or more so that I can increase the elevation at the crest of the outslope (Figure 5). When the line of sight is a mile away and 750 feet below the coal outcrop, this grading method reduces the percentage of highwall exposed by almost one-third.

If a mound of spoil is stacked near the highwall, there is little or no decrease in the percentage of the highwall exposed. Grading the bench to the conventional 3- to 5-degree slope results in little reduction in highwall exposure.

The critical zones for revegetation are the slopes facing the viewing

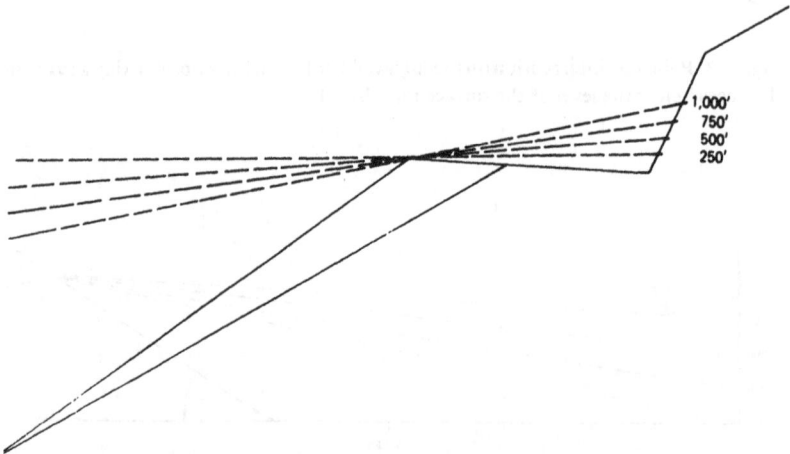

Figure 4. Points of interception on the highwall for lines of sight originating one mile from the surface mine and at various distances below the level of the bench.

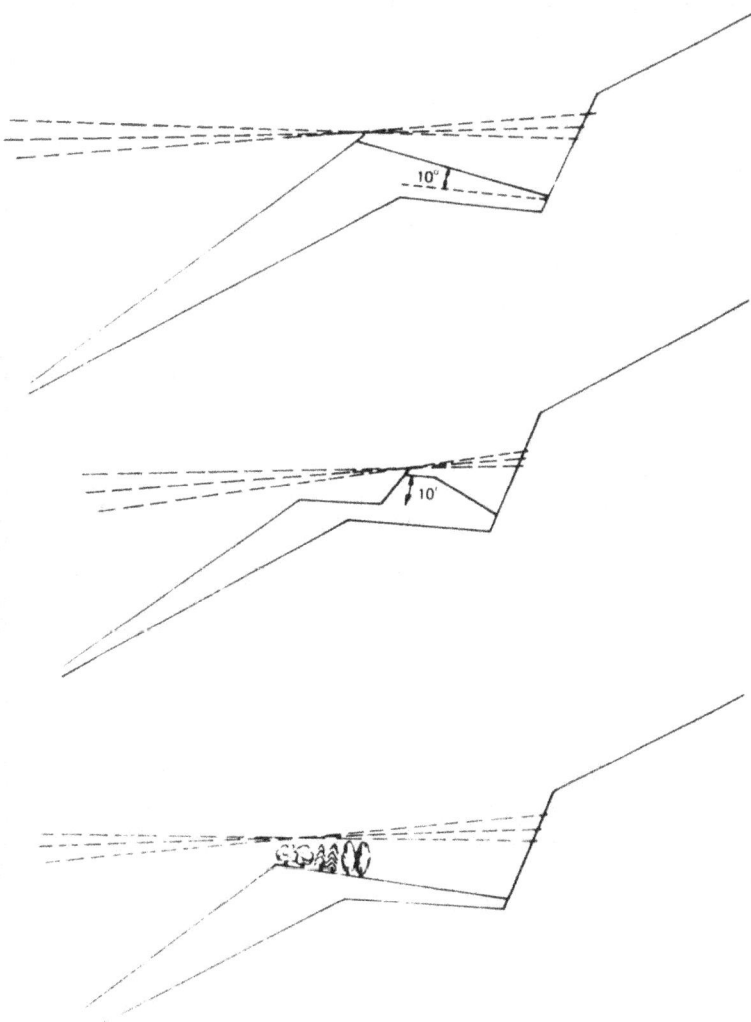

Figure 5. The influence of three regrading methods on highwall exposure for lines of sight originating below the level of the bench.

point and the highest points on the bench. Perennial grasses and legumes should be seeded on these exposed slopes. A green cover here may reduce the aesthetic impact of the disturbance even though the highwall is not completely hidden from view.

Several rows of trees should be planted parallel to the highwall along the highest points on the bench. At the crest of the outslope, two or more rows of shrubs or slow growing hardwoods should be planted. Shrubs may be preferred because many species develop bushy crowns that grow close to the ground. Then, two or more rows of conifers could be planted on the side facing the highwall. Conifers generally grow more slowly than most hardwoods, but they eventually provide a winter screen. Finally, two rows of hardwood species that will grow rapidly under favorable spoil and climatic conditions should be planted adjacent to the conifers.

When the line of sight comes from below the bench level, there is also the opportunity to modify mining methods to take full advantage of the screening provided by the natural forest. In the hypothetical example I have been using, a well-stocked stand of trees, 80 feet or more in height, would provide a screen for viewing points a half-mile away. More efficient use of the natural forest may be achieved by reducing the bench width and the height of the highwall.

Screening the highwall from vantage points at the same elevation or at a higher one will be the most difficult to accomplish. All of the highwall can be seen at any distance and at any elevation above the bench (Figures 6 and 7). The only practical approach appears to be grading and revegetation treatments at the base of the highwall.

Regrading the bench in the conventional manner with a slight slope to the highwall, and planting tree species known to make rapid growth at the base of the highwall will provide no immediate highwall screening (Figure 8). When the trees are 10 feet tall, 20 percent of the highwall may be screened.

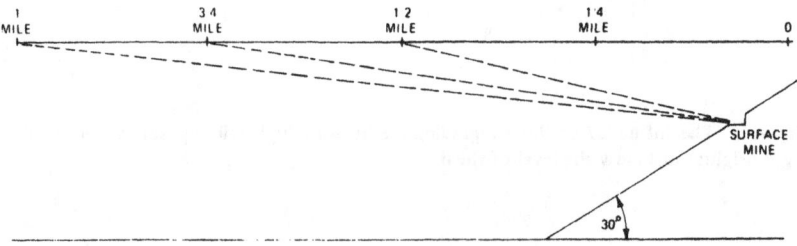

Figure 6. Angles of interception for lines of sight originating at various distances and 500 feet above the level of the bench.

However, if spoil material is stacked against the highwall to a depth of 10 feet before planting to the same species, about 20 percent of the highwall will be covered immediately. If the trees survive and grow as expected, they will cover another 20 percent of the highwall when they reach 10 feet in height. Stacking more spoil against the highwall, of course, will increase the percentage of the highwall screened from view.

A Second Method

There is another option that may have application in special situations. The occurrence of soft, easily fragmented rock strata in the upper third of the highwall is a prerequisite to considering this method. Because the costs may be high, the impact of the mining operation on regional aesthetics must also be severe enough to justify its consideration.

My proposal is to construct a 15- to 20-foot wide bench in the upper third of the highwall. The material removed would be pushed over the highwall onto the lower bench. After it was graded to a stable slope, this would cover 20 to 30 percent of the highwall. Material from above the highwall would then be pushed onto the upper bench.

This method will increase the height of the highwall and the width of the disturbed area. However, if the spoil is stacked against the highwall as I have described, 40 to 50 percent of the highwall will be covered immediately. The exposed areas will be two narrow bands separated vertically by strips of spoil covered by trees, shrubs, grasses, and legumes. The objective is to break the monotony of an unbroken, stark wall of rock.

The percentage of highwall exposed will decrease as the planted trees and shrubs increase in height. When these are 10 feet tall on both benches, 60 to 70 percent of the highwall will be screened.

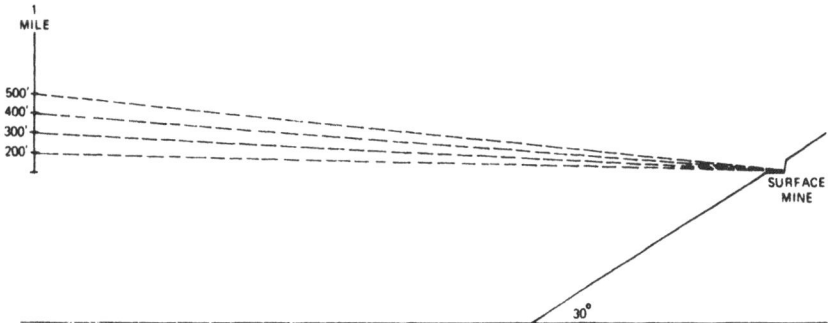

Figure 7. Angles of interception for lines of sight originating one mile from the surface mine and at various distances above the level of the bench.

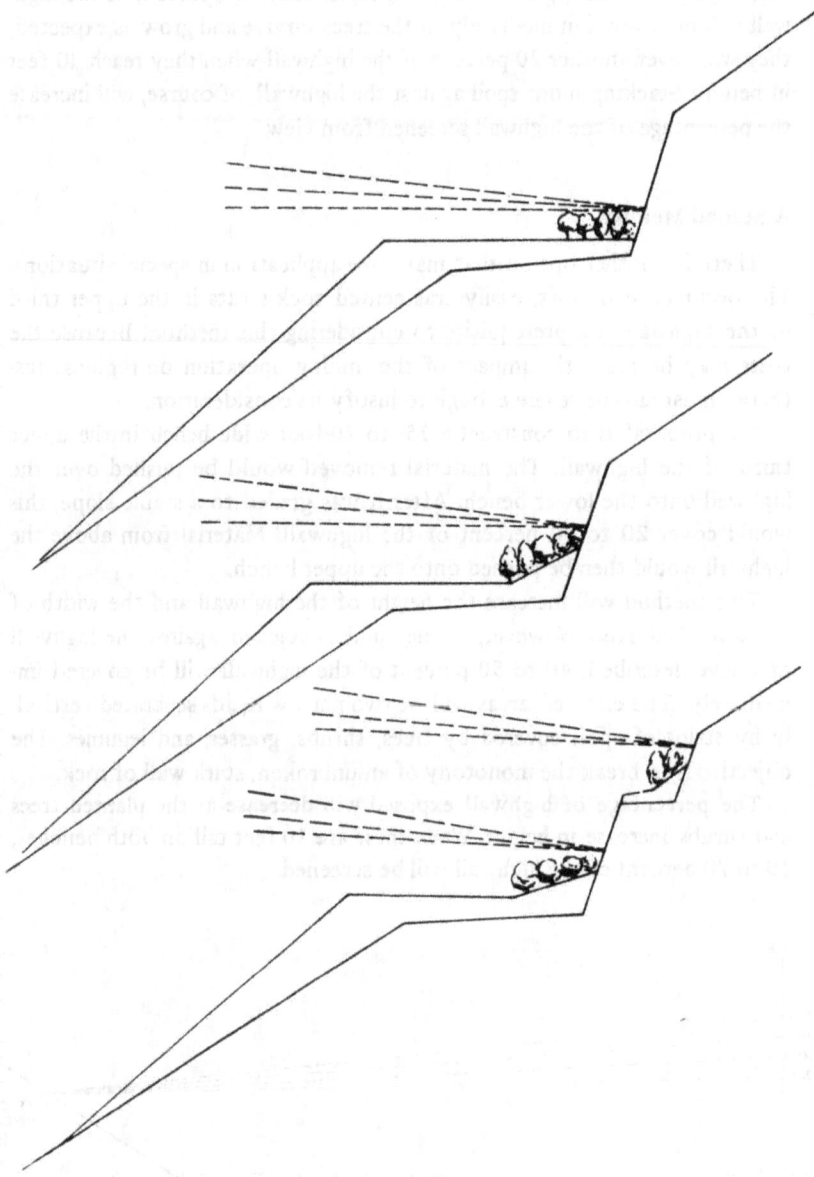

Figure 8. The influence of three regrading methods on highwall exposure for lines of sight originating above the level of the bench.

Other Methods

Another option that should be considered is the use of climbing vines for highwall screening. This has often been suggested, but I know of few examples where it has been used successfully. Native vines would have the greatest appeal, but planting stock is either not available or is difficult to obtain. Until research evaluates native vines for highwall screening, we can only guess at their usefulness.

Some exotic vines are attractive from the standpoint of their rapid growth. However, two that will grow in West Virginia, kudzu and Japanese honeysuckle, frequently damage nearby forest and agricultural crops. Control measures are expensive and unreliable. Therefore, I question and discourage their use.

Roadside screens have been suggested for use where highways overlook areas disturbed by strip mining.[1] A roadside screen planting refers to narrow strips of trees so placed that when they have attained 15 feet or more in height, they will shield or screen disturbed areas from view from a highway. This approach has been applied along main highways in the Anthracite Region of Pennsylvania.

Conclusions

This review of the problems and opportunities for highwall screening is based on hypothetical situations. Using this method, some obvious problems in highwall screening have been identified. Research priorities can be assigned to these problems, and we have the framework for a program to attack them.

We need facts to document the public's attitudes toward mining disturbance, placing special emphasis on highwalls. When viewed from a distance, what do people find objectionable about surface mining disturbance? What features of the highwall are offensive to their concept of natural beauty? Answers to questions such as these will provide a basis for developing treatments.

There is also a need for data on the percentage of surface mining disturbance exposed to public view. This would include descriptions of the viewing point and estimates of the number of people that may be involved. Because we are concerned with human values, the results should be used to develop treatments rather than for assigning research priorities.

This preliminary review of hypothetical situations clearly demonstrates that highwalls exposed to views from the same elevation or above will be the most difficult to screen. A variety of treatments needs to be developed and evaluated. These will involve both regrading methods and revegetation treatments. It is conceivable that in some localities the entire mining operation may have to be changed to reduce the aesthetic impact of mining.

There is cause for some optimism in this review of the problem. From personal experience, I would estimate a small percentage of the total area disturbed is exposed to public view. Much of the disturbance is hidden by topographic features, and present day revegetation methods provide a satisfactory cover. Also, the location of the state's highway system suggests that much of the disturbance that can be seen is viewed from below the level of the bench. These would be the easiest sites to screen.

It is apparent from the public outcry that highwalls are a major environmental problem. The reasons may be obscure, and we probably cannot express the objectionable factors quantitatively. Nevertheless, the problem must be recognized.

There also appears to be justification for considering research devoted specifically to highwall screening or beautification. Perhaps some research group will recognize its importance, and initiate appropriate projects.

Note

1. Frank, Robert M. 1963. A guide for screen and cover planting of trees on anthracite mine-spoil areas. NE. Forest Exp. Sta., Res. Paper NE-22.

8
Revegetating Surface-Mined Land

William T. Plass

Three years ago, the environmental revolution exploded, destroying many traditional concepts of land use. This revolution has affected the surface-mine industry, too: and the industry has had to adjust its mining and reclamation methods to achieve greater environmental protection.

Evidence that the industry can make this adjustment is reflected in the successful development of improved revegetation practices in the Midwest and Appalachian coal fields. This experience and knowledge can help guide efforts in the development of the western coal fields.

The purpose here is to briefly review some of the important advances in revegetation techniques that may have national or regional application, with emphasis on site evaluation, site preparation and revegetation techniques.

Site Evaluation

We must accept the fact that the prompt restoration of vegetation on areas disturbed by surface mining begins when the operator considers opening a new mine. The depth and character of the overburden will determine the cost of mining the coal. The chemical and physical characteristics of the rock strata above the coal will determine the cost and method of reclamation. These costs, plus the costs of additional treatments to achieve the land-use objectives, will decide the feasibility of opening the mine. The operator must know the physical and chemical characteristics of the overburden and modify his mining methods to make the most efficient use of these materials. This is a relatively new concept, so its full impact on reclamation costs has not been realized. It will become more widely understood and used in the future.

From the *Mining Congress Journal*, April 1974, pp. 53-59. Reprinted by permission.
  William T. Plass is Principal Plant Ecologist, Northeastern Forest Experiment Station, Forest Products Marketing Laboratory, Princeton, West Virginia.

Physical Characteristics

The physical characteristics of mine spoils alone are rarely limiting, but they now appear to be much more important than we originally thought. For example, I have shown in greenhouse tests that the emergence of pine was significantly higher when the seed was covered by hard, coarse fragments of rock. Fine soil-size material crusted over and restricted seedling emergence.

Grass and legume seed may be similarly affected. Van Lear (1) has shown in the greenhouse that the best fescue growth occurred on spoils having equal portions of fine and coarse material when no chemical properties limited plant growth. On spoils where chemical properties limited plant growth, the effects of the chemical factors were intensified on the finer textured spoils and fescue growth was reduced. Improved fescue growth occurred on the coarser textured material from the same spoil. Thus, toxic spoils containing coarser textured materials may be easier to revegetate than those made up of predominantly fine particles.

There is evidence that the color of the surface material influences soil temperature. In extreme cases, solar heating of black and dark gray surface material may kill vegetation in the seedling stage.

Chemical Variation

The chemical variation among rock strata in the stratigraphic section above the coal seam is a well-established fact. An old reliable axiom is to bury the black material.

This is founded on evidence in the Midwest and Appalachian coal fields, showing that bone and rider coals often contain high concentrations of pyrites, which can become extremely acid when weathered (2). Sandstones and shales may be acid or alkaline, and no specific rule can be applied to all situations. Regional studies of overburden characteristics may provide guidelines for judging chemical characteristics based on the composition, texture, or color of the rock strata.

Researchers at West Virginia University have made extensive studies of the Mahoning sandstone occurring above the Upper Freeport coal seam in northern West Virginia (3). They have identified a weathered zone with greatly reduced sulfur. This zone occurs in the upper horizons of this sandstone layer. Below this weathered zone, the sulfur content increases. These zones can be identified by using a Munsell color chart and freshly broken samples of weathered rock. The high chroma hues are low in sulfur; the low chroma hues have a high sulfur content.

Berg and May have shown that variations in plant-available phosphate occur between rock strata in the same stratigraphic column (4). It is pos-

sible that similar variations occur for other essential plant nutrients. Therefore, an operator may occasionally have the opportunity to use rock strata
high in phosphate to cover the spoil surface. A reduction in costs for fertilizer may result. Also, the vegetative cover may benefit for many years
by this careful planning of the mining operation.

The advantages of knowing the chemical characteristics of a highwall
were illustrated recently in northern West Virginia. This region of the state
has a history of acidic spoils and difficult revegetation problems. For
example, Allegheny Mining Co. was top-soiling a spoil from the Elk Lick
coal seam to meet state revegetation requirements. A rock strata having
low potential acidity and high phosphate was identified by sampling the
highwall. It was possible to modify the mining procedure and spread this
material on the surface. Top-soiling was no longer necessary, and a saving
in mining and reclamation costs resulted. This treatment also permitted
more flexibility in selecting appropriate quick-growing plants and obtaining a permanent cover.

The highwall analysis concept will apply to all regions. In the Midwest
and West, operators often mine seams with overburden that is uniform
from place to place: their operations extend through a period of several
years. In the Appalachian coal field, the opportunities to apply this concept may be limited. This region has many minable seams: there may be
large variations in the overburden above the same seam, and the life of
each operation is usually short.

Spoil Classification Systems

Once an area is mined and reshaped, there may be a need to evaluate
the plantability of the spoil surface. Years ago, pH was adopted as a reliable indicator. It has withstood the test of time and is still considered
the best single indicator in revegetation planning. Field tests are simple
and satisfactory for most situations. Forest Service research indicates that
there is a wide variation in the accuracy of the field kits now available and
that only one kit is satisfactory for use in the field (5). Glass electrode pH
meters, which are battery operated, are more accurate, but may not be
so convenient. There are several models available.

Sampling procedures and intensities have not been standardized. Limstrom's and Grandt's spoil-classification systems used pH classes (6,7). When
working with grasses and legumes, two additional classes should be recognized: pH 4.0-4.5 and pH 4.6-5.0. Knowledge of the pH scale and the site
requirements of grasses and legumes indicates that this is a logical sub-
division. Because these systems were developed in the Midwest where
acidity is a common problem, further modification may be necessary for
regions with alkaline soils.

Subdivisions may be used in conjunction with the pH classes to describe conditions relevant to a region. Grandt recognized three classes of spoil based on texture: chiefly sandy materials, chiefly loamy materials, and chiefly clayey materials. In West Virginia, we recognize classes of stoniness and steepness of slope. Hodder, working with alkaline spoils in Montana, suggested the use of salt hazard classes (8). He also suggested recognition of sodic or high sodium spoils. All of these serve a purpose by describing spoils in more specific terms.

Research scientists are identifying specific soil factors affecting vegetation establishment and growth. Complex laboratory analyses can be used to delineate specific spoil characteristics. This is necessary in research, but the techniques are not practical for most field applications. Selected use may be made of some more intensive laboratory tests for difficult revegetation problems or where land-management objectives justify the cost.

Preparation of the Site

Site preparation involves all operations needed to achieve the revegetation objectives. These may be very intensive or minimal, depending on the situation. Protection of the environment must be considered as well as appropriate land-use objectives. The options are as varied as the sites to be revegetated.

Reshaping

Reshaping offers the first opportunity to prepare the site for the vegetation treatments. The need to dispose of overburden materials that inhibit vegetation establishment is common to all regions. It is equally important to distribute over the surface some material that has physical and chemical properties suitable for vegetation growth. Consideration must also be given to reshaping systems that improve spoil stability, control surface runoff and recognize future land use. In the far West, surface shaping is used to reduce wind velocity, modify surface temperatures and trap precipitation. Agricultural machinery, range reseeding equipment and special mechanical devices are used to achieve a variety of surface configurations.

Top-soiling is an alternative on some difficult sites. It can be very effective if adequate consideration is given to the physical and chemical characteristics of the top-soiling material. However, it must be recognized that in some situations, native soils may be no more productive than the spoil itself. If this is the case, it is preferable to intensify treatments to the spoil to correct toxicities or nutrient deficiencies. These treatments probably would cost less than top-soiling and may be as effective.

Fertilization

Deficiencies in plant nutrients are as important as acidity-related problems on surface mine spoils. It is logical to assume that nutrient deficiencies occur to some degree on all spoils. For certain land uses, these deficiencies may be tolerated while other operations require their correction.

In general, nitrogen is very deficient on most fresh spoils. Phosphorus deficiencies have been noted in every coal region from east to west. Although the plant-available phosphorus in spoils is quite variable, most sites are classified in the low to very low range. Potassium is generally adequate for most revegetation objectives, but deficiencies can occur. Information on occurrence of specific deficiencies of other plant nutrients is limited. If deficiencies occur, they apparently are not critical for the revegetation practices now in use. Other nutrient deficiencies may be recognized under intensive management programs involving commercial crops.

Fertilizer technology is well advanced. Many different fertilizer formulations can be used to correct nutrient deficiencies. Rates of application depend on spoil characteristics, the crop to be grown and the land-use objectives. Ammonium nitrate, triple superphosphate and diammonium phosphate are high-analysis inorganic fertilizers often used in the Appalachian region. A single application at the time of seeding may be satisfactory for many vegetation plans. However, on sites that are difficult to vegetate or where intensive treatments are planned, two or more fertilizer applications should be considered.

Other Soil Amendments

When the establishment and growth of vegetation is restricted by conditions relating to the acidity of the spoil, consideration should be given to treatments that will neutralize the acidity. Lime, in a variety of forms, is the material commonly used. On many sites, physical properties prevent the use of scarifying equipment to work the lime into the spoil. Precipitation and weathering will, in some cases, distribute the lime below the surface without scarification on coarse-textured spoils.

Czapowskyj and Sowa (9) treated extremely acidic anthracite breaker-refuse with 2.5 tons of lime per acre. This surface application neutralized the surface 3 in., and its effects could be detected to a depth of 9 in.; it was effective for 7 years. However, there have been many cases where the effects of lime have been of short duration. It is advisable to apply surface treatments several weeks or months before seeding or planting and to work the lime into the spoil if at all possible.

Selected power plant fly ashes offer an alternative neutralizing material for several years (10). Bureau researchers have found that alkaline fly ashes are effective neutralizing materials when applied at rates of 150 tons or more per acre. At these rates, this waste product supplies significant amounts of several essential macro- and micro-nutrients. Tests have also shown that spoils treated with fly ash have higher infiltration rates and higher moisture levels at depths of 2 to 3 ft. The prospects of expanding the use of dry fly ash for surface mine revegetation treatments are good. Analyses of fly ashes from specific sources and the development of treatment guidelines will encourage greater use of the material.

Revegetation Techniques

Seeding and planting on the site, once it has been prepared, is becoming a complex and exacting science. Operators need to select and train competent personnel to plan and supervise their reclamation and revegetation programs. Mining, regrading and other site preparation treatments create the conditions that will determine the plant material options. In addition to site characteristics, the revegetation specialist must consider state regulations, land-use objectives, site protection and aesthetics. The operator who has planned his reclamation before mining and has created a site capable of supporting a wide variety of plant materials is in a favorable position.

Species Selection

Species selection may be based on personal preference or convention, but the most reliable method is selection on the basis of site characteristics. It is time that we expanded the use of various plant species beyond the traditional fescue-ryegrass-lespedeza mixtures. We must realize that many spoils are as good as or better than native soils. With proper treatment, they will support a wide variety of plant materials. There are opportunities for agricultural, range, forest, wildlife and horticultural crops. We must know the capabilities of a site and select plant materials that will provide the most economically desirable use of the land. There is a continuing need for plant material testing. Initially, we are interested in any grass, legume, forb, shrub, or tree that will survive and grow on sites with specific site characteristics. This testing is being done with many native species in the western coal fields today. Later, as we become more confident of success, species evaluations will involve plant materials with restricted site requirements that provide quick cover and improve the spoil, or yield products having a tangible economic value. The Midwest coal field has entered this phase, and the Appalachian region is giving more consideration to the opportunities for more intensive management. Species evaluation in all regions has been supported by the Soil Conservation Service plant materials program.

Genetics

The opportunity to improve growth or yield by genetics has not been fully realized. Some agricultural species have been developed through genetic selection and breeding to produce acceptable yields on acidic soils. The author has shown significant differences in the growth of Virginia pine seedlings collected from several seed sources and planted on an extremely acid soil (11). In each region, there are species that are accepted for surface-mine revegetation. Some of these could be improved through genetic manipulation to provide better site protection, higher crop yields, or a higher frequency of successful seeding or planting.

Species Compatibility

Species compatibility is a term that will be used more frequently in revegetation planning. The agronomist has recognized this for years in recommending grass and legume mixtures on agricultural soils. It is my belief that we have not fully utilized his experience in selecting species and rates for surface-mined lands. More recognition also needs to be given to the effect of grasses and legumes on the initial growth of seeded or planted trees and shrubs.

Vogel (12) has shown that a dense ground cover reduces the initial growth of planted trees and that later growth is affected by the species composition of the ground cover. The degree of response varies among tree species. Earlier Forest Service research has shown that trees interplanted with nitrogen-fixing tree species grow at a more rapid rate than trees without a nurse crop (13).

Time of Seeding

Seed germination and site protection often depend on the time of seeding. In western regions, where there is limited and seasonally-distributed rainfall, seedlings and plantings must be made when weather conditions are considered optimal for germination and growth. Often the options are very limited. In the Midwest and East, where the average annual precipitation is higher and well distributed during the growing season, more options are available to the revegetation specialist. Operators in the Appalachian region are using results from Forest Service research to seed grasses and legumes 8 or 9 months of the year. Summer seedings include warm-season annuals as a nurse or cover crop for perennial grasses and legumes. In the fall, cool-season annual cover crops and appropriate grasses and legumes have been successful.

Attempts have been made to develop methods allowing successful tree or shrub planting beyond the traditional spring and fall planting seasons. Several methods using seedlings growing in small containers have proved successful in experimental plantings. Technical problems and expense of these treatments have delayed acceptance of this technique on a large scale.

Other Methods

Materials and treatments that aid germination and early growth are being used in many regions. In the East, mulches and soil stabilizers are being used to reduce erosion losses while the vegetation becomes established. The writer has tested many different products and found several to be effective (14). Success depends, to some extent, on selecting a material that will provide the desired site protection for specific spoil and weather conditions. On acid spoils, where the opportunities for vegetation establishment are marginal, a heavy mulch may benefit plant establishment. Toxic salts come to the surface by capillary action during wetting and drying cycles. The mulch, by keeping the spoil surface cool and moist, causes the toxic salts to leach out. In the West, where annual precipitation is low, mulches and soil stabilizers applied after seeding become more useful to conserve moisture and reduce surface temperatures.

Waste products have been mentioned as mulching materials, but as yet they have only limited use. Bark and chipped wood waste products have been effective on experimental areas. Expanded use is restricted by a lack of suitable and economical application machinery. Composted municipal wastes have been tested successfully; but there is a stigma attached to their use, and public health problems must be resolved. Similar problems exist for solid wastes from feedlots, broiler houses, and municipal sanitation systems. It is difficult to predict whether or not the use of these materials will expand. They may be used in localized areas as the various constraints are removed.

The development of machinery and equipment for surface scarification, seed distribution, and planting of surface mine spoils has not kept pace with the advances in revegetation methods. There is no easy solution to this problem. I believe markets exist for this specialized machinery. Conventional farm and highway-construction equipment is not designed to operate under the conditions commonly found on surface-mined land.

This brief review of the accomplishments in revegetation is reassuring, but we cannot relax our efforts to improve current practices. The surface-mining industry, regulatory agencies and research organizations must continue and expand their cooperative efforts in all regions where surface mining is performed. All these groups must make inputs into the program so that results will concern relevant problems. Task groups should be formed to develop and test intensive or unusual treatments that may be applied in the future.

Research groups are found in all regions. Two developments, however, deserve special note. In the Appalachian region, a committee is being formed to stimulate interest in surface mining and reclamation research. This group will be composed of representatives from industry, state regulatory agencies, universities, and federal research agencies. They are expected to identify problems having regional significance and encourage appropriate research groups to concentrate on these problems. The regional committee will also provide a means of disseminating research results. The best research program is meaningless if the man in the field does not know relevant facts exist or if he cannot understand how they apply to his problems.

The Forest Service has proposed the SEAM (Surface Environment And Mining) project in the West. This is a research development and application program designed to provide land managers, the mining industry and the states with an innovative way of economical planning and reclamation alternatives, which satisfy both environmental and mineral needs. SEAM will be a cooperative effort and will pull together knowledge and skills from wherever they can be found. Projects relating to surface mining have been initiated in New Mexico, North Dakota, Wyoming, and Montana.

The next few years will see many changes in revegetation treatments—the direction of change will be toward more intensive management. A higher percentage of the area disturbed by surface mining will be returned to some land use that produces a tangible economic return. This is happening in the Midwest. The Appalachian region is giving more consideration to intensive revegetation treatments, this may also come about in the West after reliable revegetation treatments have been developed there.

References

1. Van Lear, David H. Effects of spoil texture on growth of K-31 tall fescue. USDA Forest Service. NE. Forest Exp. Sta. Res. Note NE-141. 1971.

2. May, Robert F. and William A. Berg. Overburden and bank acidity Eastern Kentucky strip mines. *Coal Age.* 1966.

3. Grube, W. E.; R. M. Smith; Singh; and A. A. Sobek. Characterization of coal overburden materials and minesoils in advance of surface mining. Symposium on Mined Land Reclamation, Bituminous Coal Research, 1973.

4. Berg, William A. and R. F. May. Acidity and plant-available phosphorus in strata overlying coal seams. *Min. Cong. J.* March 1969.

5. Berg, William A. Determining pH on strip-mine spoils. USDA Forest

6. Limstrom, G. A. Forestation of strip-mined land in the Central States. USDA Agr. Handbook 160., 1960.

7. Grandt, A. F. and A. L. Lang. Reclaiming Illinois strip coal land with legumes and grasses. Univ. Ill., Agr. Exp. Sta. Bull. 628. 1958.

8. Hodder, R. L. and B. W. Sindelar. Coal mine land reclamation research progress report 1971. Mont. Agr. Exp. Sta. Res. Rep. 21. 1972.

9. Czapowskyj, M. M. and E. A. Sowa. Lime retention in anthracite coal-breaker refuse. USDA Forest Serv. NE. Forest Exp. Sta. Res. Note NE-154. 1973.

10. Capp, J. P. and D. W. Gilmore. Soil-making potential of powerplant fly ash in mined-land reclamation. Symposium on Mined Land Re-claim. Bitum. Coal Research, Inc., 1973.

11. Plass, William T. Genetic variability in Virginia pine increase survival and growth. *In* R. J. Hutnik and G. Davis (ed.) Ecology and Reclamation of Devastated Land. Gordon and Breach, London, 1973.

12. Vogel, W. G. The effect of herbaceous vegetation on survival and growth of trees planted on coal mine spoil. Symposium on Mined Land Reclam.: 197-207. Bitum. Coal Research Inc., 1973.

13. Dale, Me. E. Interplant alder to increase growth on strip-mine planta-tions. USDA Forest Serv. Cent. State Forest Exp. Sta. Res. Note CS-14, 1963.

14. Plass, William T. Chemical soil stabilizers for surface mine reclamation. *In* Soil Erosion: Causes and mechanisms, prevention and control. Proc. Conf.-Workshop. High. Res. Board. Spec. Rep. 135. Washing-ton, D.C. 1973.

9

Coal Mining vs. Environment:
A Reconciliation in Pennsylvania

David R. Maneval

When the country was young and resources were considered boundless, Pennsylvania's coal powered the industrial growth of a nation.

But over a century of uncontrolled coal mining left an ugly legacy: thousands of acres covered by devastated and derelict mining land, hundreds of miles of streams contaminated by the acid water flowing out of abandoned mines, air polluted by mine fires that have smoldered for years, residential areas caving in because the mines beneath them have collapsed.

Appalled by this devastation, many Pennsylvanians—and conservationists throughout the United States—began a systematic campaign to pass state and federal legislation that would prevent further damage to the environment. Committees of the U.S. Congress have heard hundreds of hours of testimony and are at work studying literally dozens of bills that have been proposed to deal with this problem. State legislatures in many mining states are also wrestling with the same issues.

The questions involved are not simple. They involve our skyrocketing needs for energy, the changing economics of coal mining (today it's cheaper to strip mine than to deep mine many beds of coal) and a mining technology that is making the industry far more efficient—and potentially more destructive.

Take the economic issues alone. They're complex, and there's even controversy as to precisely what the issues are. Some say that the major question is whether coal can compete with other fuels if its price is increased to cover the cost of restoring the environment. Others say that the major issue is not price but scarcity, and that the consumers of coal (and other fuels such as oil, gas and nuclear power) should and can pay for repairing the environmental damage caused when they are produced. According to this argument, our national energy requirements will soon be so enormous that, even with such price increases, there will still be a

From *Appalachia*, vol. 5, no. 4, Feb.-Mar. 1972, pp. 10-40. Reprinted by permission.

large unmet demand for fuels of all kinds—so much excess demand that we may need to curtail fuel exports and ration domestic consumption to high-priority uses.

These economic issues are beyond the scope of this article. But they must be dealt with and resolved. There is no time to waste.

They are, however, not the only issues involved. There's also the question of how to go about reconciling coal mining and environment. It's not a simple job, and there is no unanimity as to how we can best do it—or indeed, whether it can be done at all.

Some believe that the wounds made by mining need be only temporary—that by using our engineering know-how to design programs of restoration, we can have our coal and save our environment too.

Others believe that there must be strong restrictive measures on deep mining and the outright banning of strip mining. They believe that no matter what we do to restore strip-mining lands, some of the damage is irreparable—that the earth can never fully recover from this kind of wound and that the ecological balance will be thrown more and more out of kilter if strip mining continues.

This is a subject on which honest men disagree, and to my knowledge there has been no definitive scientific study which settles the argument once and for all. Perhaps it can never be settled in yes-or-no terms. But certainly we should keep on trying to find the best possible answers, and studies to do just that should be high on our nation's list of priorities.

In the meantime, all would agree that as long as coal mining continues, every possible step should be taken to protect our legacy of land and water.

This means that some coal deposits—those where effective restoration is not possible—should not be mined at all.

It means that every operator should be required by law to restore land and preserve water purity in the areas he mines and that these requirements should be strictly enforced.

It means we must acknowledge that we are dealing with a national problem. Environment doesn't recognize man-made boundaries; acid water discharged from a mine in one state, for example, endangers the water in all the states downstream.

And it also means that restoration standards should be uniform for the entire nation, so that no coal producer in any state is put at a competitive disadvantage because he does what is required to save our environment.

THE PENNSYLVANIA PROGRAM

The Pennsylvania program, now nearly a decade old, was designed to translate these areas of agreement into everyday reality—and it has evolved as the needs have become clearer.

Many of the techniques used in Pennsylvania are directly applicable only where the topography is similar; different terrain means different problems. This is particularly true of the Commonwealth's work in restoring bituminous strip-mine sites. In Pennsylvania, most bituminous deposits are found in flat or gently rolling terrain; strip mining in this kind of landscape creates less damage than it does in mountainous areas where the slopes are steeper—and whatever damage is done is much easier to repair. So the Commonwealth's know-how can't automatically be exported for use in other states, and the costs of reclamation in Pennsylvania cannot be used precisely to predict what the costs would be elsewhere. And there is more to learn, even when tackling just Pennsylvania's problems.

But there is no question that Pennsylvania has been a real pioneer in restoration and reclamation, and that its experience should be called upon in designing any federal and state legislation and programs. The article which follows describes what Pennsylvania has done.

What's Needed?

If you're going to do anything about the environmental damage caused by coal mining, you must have three things: strong legislation, an effective system of enforcement and trained personnel. But you don't usually get any of these—and certainly not the combination of all three—unless you have something else: aroused and enlightened citizens who are determined to do something about the problem.

This is what Pennsylvania had—and has. Supported by the press, conservation groups, sportmen's clubs and schools, the state legislature has passed a series of increasingly stringent and detailed laws designed to halt the environmental rape and help repair the damage already done. On the basis of this legislation, the Commonwealth of Pennsylvania has developed what the *Denver Post* has called "probably the largest, most comprehensive effort of its kind in the nation."

All of the problems are not solved yet, as this article will emphasize. But Pennsylvania has found some answers already, and is still looking for more—witness the continuing research on revegetation and the new state law, passed last November, which sets higher reclamation standards and strengthens enforcement procedures.

The story which follows describes what has already been done to reconcile coal mining and environment in the Commonwealth of Pennsylvania—a past which is prologue.

Two Kinds of Problems

Until comparatively recently, Pennsylvania's—and the nation's—mining laws were keyed to preserving life and limb of the men who toil in the mines, a subject that still needs close and dedicated attention.

It was not until the 1940s, however, that there was real recognition that uncontrolled coal mining was also damaging the state's environment—and that the legislature could do something about it.

Actually the Commonwealth of Pennsylvania faced two separate though related problems in trying to reconcile coal mining and environment: how to prevent future damage from mining operations and how to rectify the damage already done.

Although some experts refer to the former as preservation and the latter as reclamation or restoration, in this article we make no such distinction in our use of words. Under the Pennsylvania program, preserving mine land does *not* mean that the land is not mined at all. It means rather that there are strict regulations as to what must be done, before and after the coal is extracted, to be sure that the land is restored to its previous appearance and usefulness. So restoration and reclamation are not just methods used to mend past damage; they are also techniques used to be sure that today's coal mining leaves no ugly scars.

But although many of the techniques are similar, there are still two distinct kinds of problems, and the differences have been recognized in the legislation, administration and funding of the Pennsylvania program. In this article we will begin by describing what is now done in Pennsylvania to *control current mining operations,* first what is done in strip mining and then the measures taken to control deep or underground mining. In the second part of this article, we will discuss what has been and is being done to *repair past ravages*: correction of mine drainage, extinguishing coal refuse bank fires and underground mine fires, backfilling strip-mining pits and treatment of mine subsidence.

All of the state-level regulatory, enforcement and restoration activities described in this article are performed by Pennsylvania's new (since 1971) Department of Environmental Resources (DER).

CURRENT STRIP-MINING OPERATIONS

What do we mean by "strip mining?" To many, the term conjures up a mental picture in which a few inches, or at most a few feet, of topsoil and rock (called "overburden") are taken off of a coal seam, and the coal then removed. Although this is sometimes the case, very frequently it is not. In some current strip-mining operations, the overburden removed is nearly 200 feet deep, and new mining technology may soon make it possible to scoop off as much as 2,000 feet. This kind of "progress" makes the environmental damage more serious—and harder to solve.

Another confusion in the public mind is between "strip mining" and "surface mining." Many people believe that the two terms are synonymous, and that which of the two you choose depends on whether you approve of this method of extracting coal (in which case you call it "surface mining") or whether you don't (in which case you call it "strip mining").

But this is not the case; the two terms are not synonymous. Surface mining is the broader term, and strip mining is only one type of surface mining, just as roasting is only one method of cooking meat. Technically speaking, surface mining is any kind of mining in which topsoil, rock and other strata are removed in order to get at underlying mineral or fuel deposits; the distinguishing characteristic of strip mining is that this removal is done a narrow band at a time. Since in Pennsylvania it is primarily strip mining which has damaged the environment, it is this type of surface mining which we will discuss in this section of the article.

Legislation on Strip Mining

Pennsylvania's program of preserving lands that are currently being strip mined is based on a network of laws which began with the Bituminous Coal Open Pit Mining Conservation Act of 1945 (Act 418). This act was followed by two laws passed in 1947, one requiring the planting of trees, seedlings, shrubs and vines (Act 333) and the other dealing with major revisions of the anthracite strip-mining law (Act 472). Numerous technical additions were passed by the legislature during the next 15 years, but the most significant new legislation was a set of amendments (to the 1945 Bituminous Coal Open Pit Mining Conservation Act) passed in August of 1963. Although the administrative organization set up to enforce this legislation has changed somewhat during the years since, it is these 1963 amendments which have been the basis of Pennsylvania's present program. The most recent law dealing with these issues is the Surface Mining Conservation and Reclamation Act; it was passed by the legislature in November 1971 and went into effect in January 1972.

Program for Control

We will discuss Pennsylvania's program for control of bituminous strip mining first—because the state's program is considered a model and because bituminous is where the action is—12,000 acres stripped and 17,200 acres restored in 1971, as compared with 350 acres stripped for anthracite and 411 acres restored.

Bituminous Mining

The Commonwealth's program for bituminous mining operations has four major components, each of which will be discussed separately:

1. licenses and permits which must be obtained by mine operators;
2. a requirement that each operator must post a bond to cover the cost of reclamation of each site he intends to strip mine;
3. a set of requirements which must be met by the reclamation project proposed by the mine operator; and
4. a system of strict enforcement of all the above requirements.

Licenses and Permits

In order to operate a bituminous strip mine within the Commonwealth, a mine operator must obtain an annual license which costs $300 per year and which may be revoked if he does not comply with all state regulations. During the latter part of the 1960s, the number of licensed operators declined (from 369 in 1965 to 300 in 1967) and then increased to 375 in 1971.

In addition, for each specific site to be mined, the operator must obtain two types of permits, each dealing with an important aspect of the environmental impact of strip mining.

A *mine-drainage permit* is necessary because coal + air + water = acid water; the oxygen in the air combines with the sulfur minerals in the coal to form ferrous sulfate, a chemical compound which is water soluble and which turns a clean stream into an acid one. (When iron compounds are also present, the streams turn red as well as acid—an infamous and unbeautiful by-product of the coal-mining economy.)

In strip mining, the union of coal, air and water is almost inevitable. Coal seams are exposed to the air during stripping operations, and rainfall furnishes the water, which frequently runs down the hillside into the area where the stripping is going on. Because of this inevitable union, an operator must obtain a mine-drainage permit from DER before he begins a stripping operation, and in order to obtain the permit he must file an application describing in detail the methods of water control and treatment he will use during the operation. DER sets the standards which must be met.

At present, water drainage from a site (1) may not have an iron content exceeding 7 milligrams per liter and (2) must have a pH reading of between 6 and 9. The pH scale, which runs from 0 to 14, indicates the degree of acidity or alkalinity of water; a pH of 7 is neutral, the lowest readings represent highly acid conditions and the highest readings represent high alkalinity. The range of 6 to 9 is that in which most aquatic life can survive.

If DER determines that the operator's plans will meet these water-quality standards, he is issued a mine-drainage permit for the mining site. The fee for the permit is $25.00.

A strip-mining permit must also be obtained by the operator from DER. In order to obtain the permit, the operator must file an application which describes the method of mining he will use at that particular site and the program of reclamation and reforestation he will carry out when the stripping is completed. (Details of these requirements are described later in this article.) If DER determines that his proposals will meet state standards, he is issued the permit, for which there is no fee.

In practice, a mine-drainage permit is usually issued for a rather large mining site; this large site is then subdivided into smaller portions, and a strip-mining permit is issued for only one of these subdivisions at a time. After—and only after—mining and subsequent reclamation have been completed and mine-drainage requirements have been met for one sub-division the operator's required bonds are released by DER and application is made for a permit for the second subdivision, and so on until the entire site has been mined. This consecutive method minimizes the amount of money that an operator must commit in bonding fees.

Under both the 1945 and 1971 laws, DER is empowered to deny mine-drainage or strip-mining permits if it feels that pollution of air, water or land would result during or after the mining operation. For example, certain areas of Clarion County are under a moratorium preventing fur-ther stripping because the acidic nature of the overburden in that area per-mits the production of acid runoff even after contour reclamation is completed. During 1971, 351 mine-drainage permits and 863 strip-mine permits were issued by DER.

Posting of Bond

Before beginning a strip-mining operation, the operator must also post a bond to cover the cost of reclamation of the site—sort of "pay now, fly later" approach to conservation. Before 1964, the bond required for bitu-minous strip mines ranged from $300 to $400 per acre, depending on the operator's past record of compliance with state requirements. From 1964 until 1971, the range of required bonding was from $500 to $1,000 an acre. Under the new 1971 law, the basis for determining the amount of bond has been changed, and there are no limits on what may be required.

Although some operators put up marketable securities to meet this requirement, most obtain their bonding from commercial surety companies. The fee charged by these companies is based on the size of the bond and is usually 1 percent per year. Since most bonding is for a five-year period, the cost for a $50,000 bond would be $500 per year, or a total of $2,500 over the five-year period. A portion of the premium may be returned, however, if the operator's bond is released quickly.

If an operator fails to carry out reclamation which meets state requirements, that is, if he forfeits his bond, either his marketable securities will be taken over by DER or his surety company will be informed of his failure to comply. In the case of forfeiture, the bonding company is given the option of either paying off the full value of the bond or of carrying out the required reclamation to meet state standards. In the former case (which happens frequently), the full value of the bond is paid into a special state fund, and DER contracts to have the reclamation completed. In most cases, however, the surety company prefers to make its own arrangements to have the reclamation work done (since it can usually be done for less than the amount of the required bond). When the work has been completed and approved by DER, the bond is released.

During 1971, a total of $8.3 million in surety and collateral bonds (representing 15,000 acres) was deposited with DER by bituminous strip-mine operators; during the same year, operators restored approximately 17,200 acres, and as a result $4.4 million in bonding was released. This performance record represents a marked improvement over the 1963-1966 period, when operators forfeited bonds in the amount of $1.2 million (out of a total of $17.0 million), a forfeiture which represented 1,122 acres which had to be restored by DER or surety companies.

Requirements for Reclamation Projects

In order to understand the requirements that a Pennsylvania coal mine operator must meet in his reclamation project for a given mine site, it's necessary first to understand what the mine site looks like, and this in turn depends on what kind of terrain is involved and what kind of mining has been done.

Since contour strip mining of bituminous coal accounts for the largest number of acres requiring reclamation in Pennsylvania, we'll discuss that type first and then follow with a description of how operators reclaim sites that have been disturbed by the area strip mining used to extract anthracite.

Contour Strip Mining of Bituminous. Figure 1 shows in diagrammatic fashion what happens when contour strip mining takes place, usually in a rolling or mountainous area. Note that the coal seam (shown in black) is relatively flat and horizontal, and that there is an outcropping of coal at ground level on the hillside (Point X). In contour strip mining the process

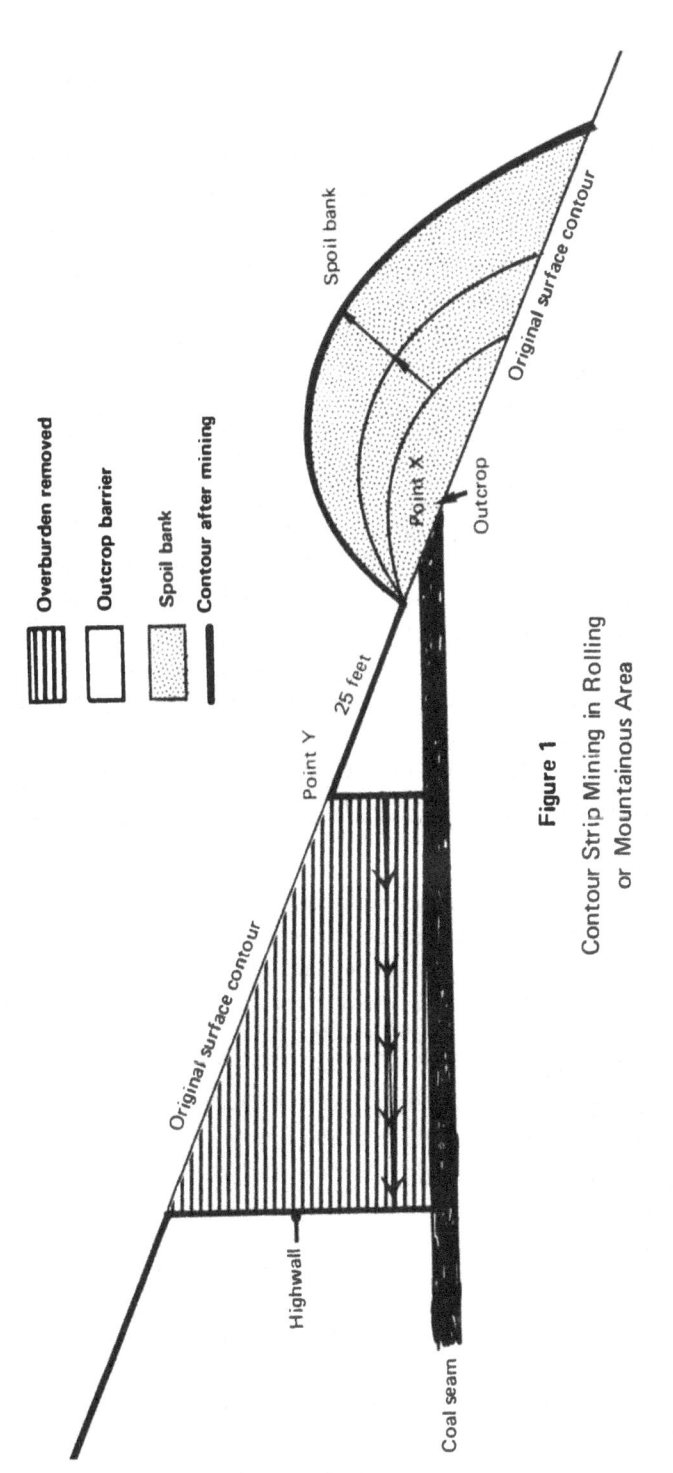

Figure 1

Contour Strip Mining in Rolling
or Mountainous Area

Overburden removed

Outcrop barrier

Spoil bank

Contour after mining

Spoil bank

Original surface contour

Point X

Outcrop

Point Y

25 feet

Original surface contour

Highwall

Coal seam

is one of cutting away the overburden and piling it in an increasingly larger spoil bank that overflows down the hillside. In Pennsylvania the operator is not permitted to begin mining at the point of outcrop (Point X). In order to contain any water which might accumulate in the pit floor—and this frequently continues to be a problem even after the pit has been filled during the restoration process—in Pennsylvania the strip-mine operator must leave the outcrop barrier in place (shown in gray in Figure 1), beginning his mining 25 feet higher up the slope at Point Y.

The mining operation continues, moving back into the hillside until the cost of removing overburden becomes so great that it's uneconomical to go farther. At that point we have the contour represented by the black lines in the top diagram in Figure 2 (Diagram A): a nearly vertical exposed highwall (which may rise 100 feet or more above the horizontal bench) and a spoil bank that has been cast down the hillside.

There are basically two ways to reclaim this kind of mine site: contour backfilling (shown in Diagram B) and terrace backfilling (shown in Diagram C). In general, Pennsylvania law requires that reclamation of bituminous strip mines must be to original contour (as shown in B), but that under special circumstances terrace backfilling (C) is permitted. Let's discuss them both in turn.

Contour Backfilling. In simplest terms, contour backfilling is a method in which the material in the spoil bank is used to fill in the big hole made during the mining. The 1963 Pennsylvania law requires that the material removed or disturbed during the extraction of the coal must be put back or replaced after the extraction is completed, and that this backfilling must be done to approximate original contour. This does not necessarily mean that the material must be put back in the same location from which it was removed, although this is desirable and may, as in the case of a one-cut operation, be the only way that an approximate original contour may be obtained.

Furthermore, the requirement for approximate original contour does not mean that the final surface of the restored area will necessarily have the exact elevation of the original ground surface. This is practically impossible since the size of the spoil bank seldom is the same as the size of the hole from which it came. Usually the spoil bank is larger, because the extracted material swells when it is exposed to air and because the material is not as tightly compacted after removal. However, the spoil bank may be smaller than the excavation if a particularly large amount of coal has been removed. Recognizing this difficulty, the Pennsylvania law requires that the restored area should have approximately the same contour as it had originally.

In some cases the new contour may even be an improvement on the

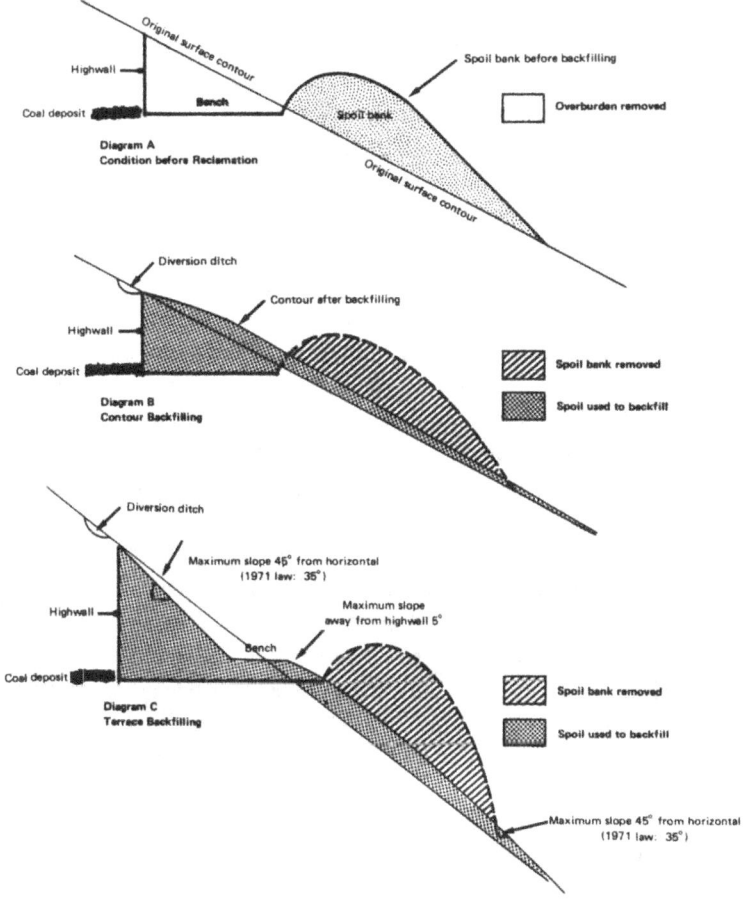

Figure 2
Contour Strip Mining

original condition of the land. For example, in some cases it is possible to strip mine—that is, to clip the tops off of—two adjacent hills and then to deposit the overburden in the valley in between; the result is that you have generated flat land where you never had any before. In a case like this, which is really a combination of terrace and contour backfilling, you finish up with new and more valuable real estate than when you started.

Another aspect of restoration of strip-mining sites—and this applies to both contour- and area-type operations—is the necessity for controlling the water that may accumulate at the site, water which can cause erosion and acid drainage. In order to prevent the accumulation of water, the operator is required to put a diversion ditch at the top of the highwall in order to divert upslope water away from the mining operation, that is, to keep the water from running downhill into the pit (see Diagrams B and C in Figure 2). In addition, if any water accumulates in the pit as the result of direct rainfall, the operator must routinely have the water pumped out and discharged through two sedimentation ponds in order to treat and settle any solids. If the water is acid, it must also be neutralized.

Terrace backfilling, which is illustrated in Diagram C of Figure 2, is generally permitted in Pennsylvania only in those instances where the land has a very steep contour. As a matter of fact, the state not only permits but actually prefers terrace backfilling in these situations, since you're apt to have more erosion problems with restored steep single slopes than you are with the gently sloped benches that you get when you use the terrace backfill method.

Terrace backfilling is essentially a method of restoration in which part of the spoil bank is used to fill in the highwall and part is used to reduce the slope of the spoil bank (see Diagram C). The resulting gentle slopes have a long, relatively flat table area (called the bench) in between. In Pennsylvania the guidelines for terrace backfilling are as follows:

• The steepest slope of the highwall and of the outer slope of the spoil bank shall be no greater than 45 degrees from the horizontal. (Under the new law the permitted slope is reduced to 35 degrees.) The outer ridge of the spoil shall be rounded off with machinery.

• The table portion of the restored area shall be a terrace which is either flat or sloping downward away from the highwall at an angle of not more than 5 degrees.

• There shall be no depressions to hold water and no lateral depressions or drainage ditches of sufficient length to cause erosion of the restored area.

• Lateral drainage ditches shall have a minimum width of 12 feet.

Area Strip Mining of Bituminous Coal. Area strip mining is practiced on relatively flat terrain (that with a slope of less than 12 degrees) where there is no outcrop of the coal seam. Because of the Commonwealth's topography there is relatively little of this kind of strip mining in Pennsylvania.

As shown in Figure 3, a trench or "box cut" is made by removing the overburden, which is deposited in a spoil bank on the opposite side of the trench. The coal is then removed, and the operation moves on to make the next parallel cut. As shown in Figure 4, as each cut is made, the overburden

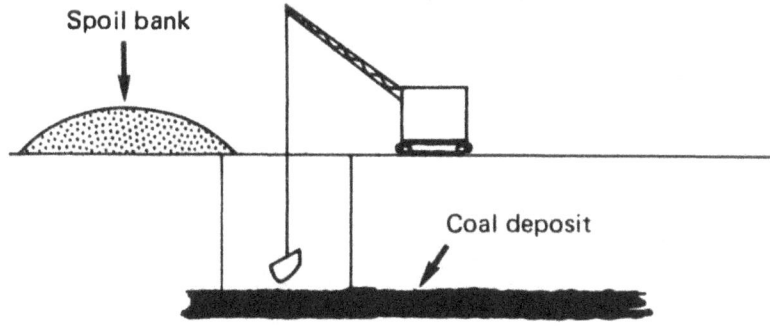

Figure 3

Area Strip Mining in Flat Terrain
(Slope less than 12°)

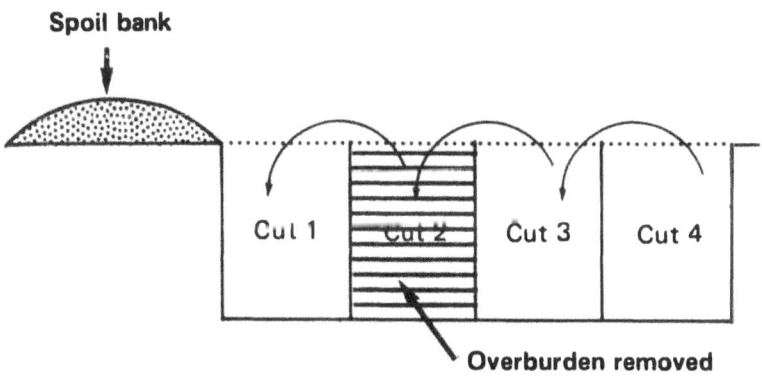

Figure 4

Successive Cuts in Area Strip Mining

is deposited in the cut previously excavated. However, since there is usually some leftover spoil (because of the swelling and lack of compaction of
the overburden), at the end of an area strip-mining operation the mine site
usually looks like a gigantic washboard (see Figure 5) with a large spoil
bank at the end of the first cut, smaller "leftover" banks between the succeeding cuts and an open trench at the last cut. There is frequently a large

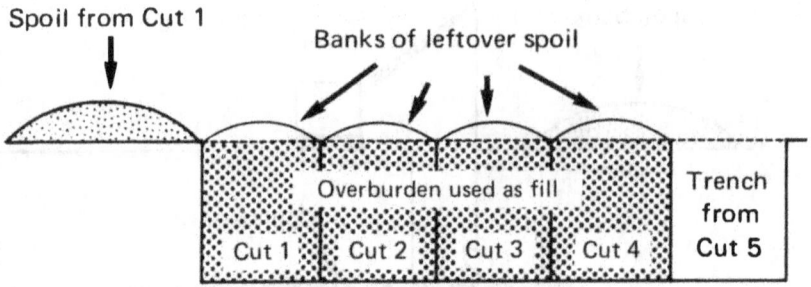

Figure 5

Results of Area Strip Mining

distance—sometimes a mile or more—between the first cut spoil bank and the last trench.

Restoration of this type of site is basically a process of leveling and grading. The first-cut spoil bank must be leveled so that the outside slope of the spoil will blend into the contour of the adjoining land. The smaller peaks from successive excavations are also leveled, and all this material is pushed toward the last cut, where it is deposited in the final trench. With this procedure, an approximate original contour will be obtained. Under Pennsylvania law, this work must be carried on while mining progresses; generally no more than 1,500 feet of cut may be open at any one time. The completed restoration may not have any highwall, any spoil peaks or any depressions that could hold water which might percolate through the spoil and produce acid drainage.

One of the major problems in strip mining is that it is difficult to be sure that the topsoil stays on top—and this difficulty becomes particularly evident in area strip mining. Unless preventive steps are taken, the soil layers deposited in Cut 1, for example, will be exactly the reverse of the way they come out in Cut 2: the topsoil will finish up on the bottom and the heavy rocky lower layers will be the last to be put in. The environmental problems resulting from this topsy-turvy deposition are enormous and must be avoided. This is done by carefully removing the topsoil first and storing it separately while the underlying layers of subsoil and rock are removed and stored and the coal taken out. When the mining is completed, the overburden can then be put back in its natural sequence and the topsoil can be redeposited where its name tells us it belongs—at the top.

Block-Cut Method of Strip Mining. When strip mining is done in mountainous areas, particularly in areas where the soil is very unstable, one of the most damaging parts of the operation occurs when the overburden is cast down the hillside, where it frequently slips and slides, uprooting trees, blocking streams and roads and occasionally avalanching into homes and farms.

In the mountainous areas of Pennsylvania, another method of strip mining—a modified block-cut method—has been used to help minimize this kind of damage. It is a method which may have applicability in the mountainous areas of the other mining states.

The block-cut method is basically the area method adapted for steep terrain. At a first-cut site which has been carefully chosen to minimize slide damage (shown diagramatically in Figure 6), an original box cut (Cut 1) is made into the hillside; the cut, which is customarily two to three times as wide as subsequent cuts, goes into the hill to the maximum depth that the operator intends to mine at any point around the hillside. A highwall is left, the overburden is dumped down the hillside in the traditional way and the coal is removed.

In all successive cuts, however, the overburden is *not* cascaded down the hillside; it is deposited in the void left by the previous cut. Once the

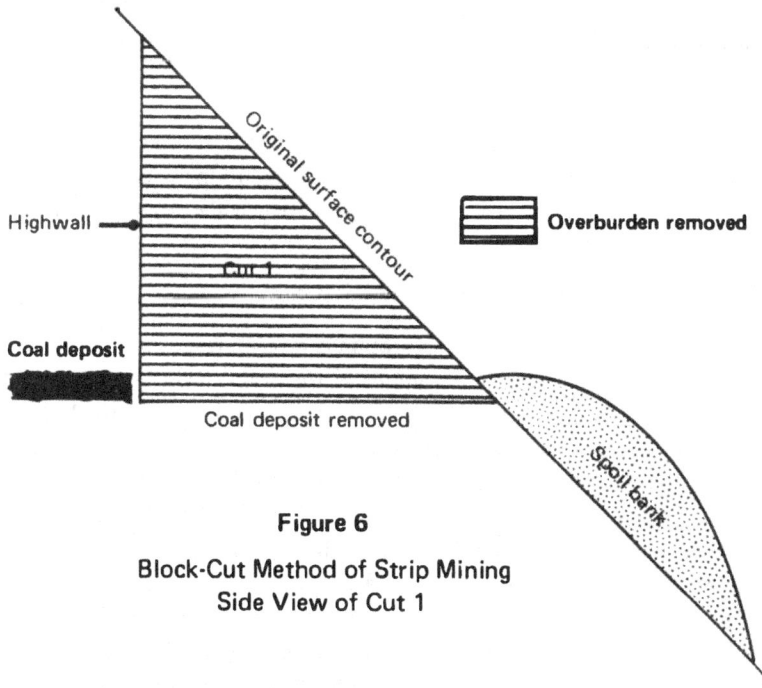

Figure 6

Block-Cut Method of Strip Mining
Side View of Cut 1

original box cut has been made, mining operations can be continuous working around the mountain in both directions from Cut 1. Figure 7, which shows a top view of a box-cut operation, illustrates how the mining operations move around the mountain; overburden from Cuts 2 and 3 is put into Cut 1 (Diagram C), from Cut 5 into Cut 3 (Diagram E) and so on. In all cuts after the first, an undisturbed mound or outcrop barrier is left to trap runaway spoil and mine water and prevent its movement down the hillside or into nearby streams. Figure 8, which shows a side view of Cut 2 after it has been filled with overburden from Cut 4, illustrates how these later cuts will look after the mining operation is completed.

The block-cut method is not without its problems. If only a small amount of coal is extracted and/or if the overburden is not easily compacted because the nature of the soil causes it to swell more than usual, there may be leftover spoil that can't be fitted into the cuts. In the case of area strip mining, where the same problem exists, the leftover hillocks can be "struck off" or smoothed over when the operation is finished. In mountainous areas, this won't work because the leftover spoil would be sliding down the hillside. In this situation the surplus spoil is contoured into terraces or carried away to other and flatter locations where it can be deposited and revegetated.

In block-cut mining the problem of saving topsoil is both more and less difficult than it is in the area strip mining done in flatter areas—more difficult because it's harder to find a place to store the topsoil, and less difficult because the layers of topsoil are usually much thinner in mountainous areas. In both area and block-cut mining, however, the topsoil must be stored for only a short time because of the continuous cut-and-fill type of mining procedure used in both. In mountainous areas the topsoil can be temporarily stored on the ledge between the top of the highwall and the diversion ditch.

Revegetation

The next step in mine site restoration is revegetation.

After a mine operator has completed the backfill work on a given bituminous mine site, and has received DER approval of that work, he can apply for and receive partial release of his bond. DER, however, retains a portion of the bond in order to assure that the operator will then revegetate the site properly. For this part of the restoration job the operator has two options: either paying the state to do the revegetation (in which case the remainder of his bond is immediately released) or doing it himself. Most operators opt for the latter. (The former option is no longer available to operators, since it was eliminated in the new 1971 law.)

In Pennsylvania the vegetation projects must meet the following requirements:

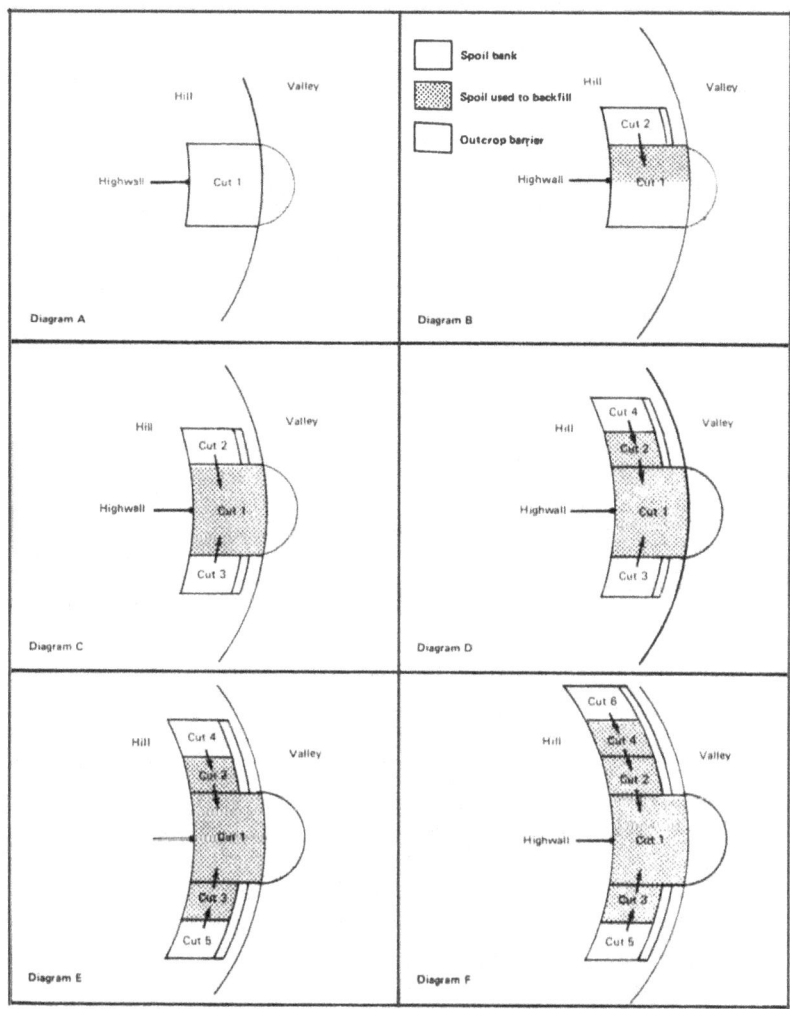

Figure 7
Block-Cut Method of Strip Mining
Top View

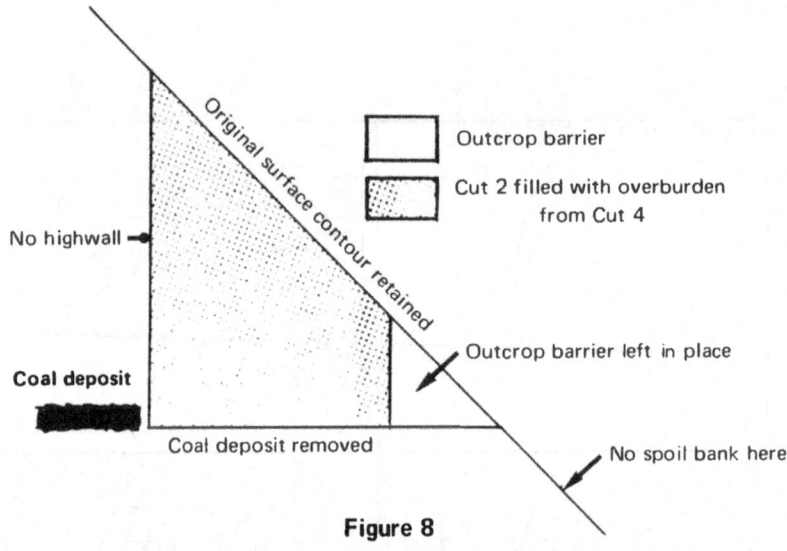

Figure 8
Block-Cut Method of Strip Mining
Side View of Cut 2 after
Filling with Overburden from Cut 4

1. Under the supervision of a DER forester, *the operator must supply analyses of soil samples from the mine site.* Analysis of a "grab sample" isn't enough to meet this requirement. The operator must furnish analyses of soil taken from a scientifically designed array of sampling points (usually at least five) to give DER a real feel for how acid the soil is and what sorts of nutrients are needed to restore it. The samples are tested, usually by the Pennsylvania State University, and on the basis of their evaluation DER stipulates which soil nutrients (commercial fertilizer containing potassium, phosphates and nitrates) and lime must be added, and in what quantities. The operator may choose any brand of commercial fertilizer he wishes, so long as it contains the proper quantities of nutrients and is applied so many pounds or tons to an acre, as specified by DER.

2. *The operator must then plant vegetation on the strip-mined site.* Prior to November of 1970, the operator had the option of planting either trees, grass, or a grazing crop such as clover. Since planting trees costs less than planting grass—approximately $45 per acre for trees (on the basis of spacing them six feet apart) as compared with $90 per acre for grass—most operators opted to plant trees, which have many advantages but which

do not prevent soil erosion during their early years of life. In November of 1970, however, the Pennsylvania legislature passed an act which had been drafted at the suggestion of the Soil Conservation Commission. Under this law, DER is empowered to require that the operator plant both grass and trees. This requirement might be applied, for example, in very hilly terrain, or in other situations where DER determines that it is necessary to do both types of planting in order to prevent erosion.

If DER requires the planting of grass, the whole mine site area must be seeded as soon as the backfilling operation has been completed. On the basis of recommendations from the forester, DER specifies the species and quantity of grass seed that must be planted. Since passage of the 1970 law, a mixture of annual and perennial grasses has been used. This type of mixture has a number of advantages. It furnishes "instant green," it keeps the soil from running down the hillsides and it does not have to be re-planted the following season.

If DER approves revegetation with trees (either alone or in combination with grass), the operator usually waits until spring to do his tree planting. The reason for the delay is that research has shown that trees, shrubs and bushes survive best if they are planted in spring (80 percent survival as compared with 65-70 percent if planted in the fall and only 50 percent in summer or winter). The operator has his choice of what he plants, as long as it's from the DER-approved list. (The list has been published on page 15 of *A Guide for Revegetating Bituminous Strip-mine Spoils in Pennsylvania*; copies of this booklet may be obtained free of charge by writing to the Mine Drainage Section, Division of Mine Reclamation, Department of Environmental Resources, P.O. Box 2063, Harrisburg, Pennsylvania 17105.)

The bushes and trees on this list have been selected for their acid resistance (since most of the soil in reclaimed mine sites may be acid), for their ability to survive on certain types of terrain, for their resistance to aluminum and iron compounds (which are often found in strip-mine spoils and which are toxic to plant life) and for their speed of growth. The choices are usually based on the results of DER-sponsored research, most of which has been done by the U.S. Forest Service and Penn State; DER has research contracts with both these organizations.

Pennsylvania is generally covered with deciduous hardwood trees—trees that drop their leaves every fall. So for many years the only trees on the DER list were softwoods such as pine and spruce; mine site reclamation was used as a method of getting more variety in the type of trees growing in the Commonwealth, and hence more variety in the type of pulpwood that could be produced commercially. More recently, however, research results have shown that some of the hardwoods grow faster, that they are in demand for pulp, that they make good cover and good food for game and that they help put nitrogen back in the soil. A number of hardwoods have therefore been added to the DER list during the last few years.

If the mine site has very porous soil and is in a particularly wet area, planting water-hungry vegetation can help prevent acid mine drainage. For this reason, DER may require that the operator plant something which has a great ability to draw moisture out of the soil and get it out into the atmosphere by transpiration (evaporation from the leaves). One new plant is the bristly locust, which was developed by the U.S. Soil Conservation Service during the last two and one-half or three years and which is being planted in steeply sloped mine site areas where trees cannot be grown effectively. Although it has no commercial utility, it takes up a lot of water, grows fast, furnishes cover for small game and serves as food for small animals (who eat the berries) and deer (who eat the foliage). The foliage grows back quickly, and the berries are bright red; so bristly locust makes attractive cover. For these reasons, DER has sometimes recommended its use in moisture-heavy areas and on hillsides where there isn't any real possibility of turning the site into a tree-farm kind of operation.

Which brings us to the commercial aspects of mine site revegetation. Whether or not the operator (or landowner, if they're two separate people, as they often are) uses the site to grow a crop of some kind is up to him; DER is concerned only with improving the appearance of the site, preventing pollution and bringing the land back into condition so that it can serve some useful purpose. Some strip-mined land, particularly that in central Pennsylvania in the Indiana–Cambria County area, has been developed into Christmas tree plantations. This has worked well because Scotch pines are very acid-tolerant and you can count on roughly a foot of growth a year. This means that if, after you finish strip mining, you plant two-year-old seedlings, five years later you can be harvesting a crop of Christmas trees.

There's another species of tree that has a good chance of making a quick turnaround profit for the operator or landowner. It's called a hybrid poplar, and it will grow as much as two and a half to three feet a year, which means that it can grow into good-quality commercial pulpwood size in ten years—and in the forestry business this is very good indeed.

DER is also doing some research on the possible use of Ponderosa pine, a fast-growing timber tree (that is, one used to produce lumber for building) that grows in acid soil out in the Rocky Mountains. Cones have been imported from New Mexico, and DER will determine whether this species will flourish in Pennsylvania.

At present there is some discussion between agronomists and foresters as to when or whether DER should require the planting of both grasses and trees. When trees are planted, their first roots spread out horizontally in the topsoil; it is only after two or three years that they start sinking deeper roots to obtain moisture and food. Because of this early dependence on topsoil, some foresters claim that the young trees will suffer if grass has also been planted, since there will be more competition for nutrients and water. On the other hand, the grass does furnish "instant green" and it does prevent erosion. So at present DER is frequently recommending that operators put down strips of grass and trees, one alternating with the other in a pattern similar to contour plowing, and that this be done regardless of terrain. (The grass strips are planted immediately after completion of backfilling; the tree strips are left fallow until the next spring planting season.) This solution seems to work very well, combining the benefits of both types of ground cover.

There's another form of crop that can be grown on mine site land, and that's the agricultural crop that was grown before the mining took place. In certain counties, such as Somerset and Jefferson, where the land is relatively flat and we therefore find Pennsylvania's closest approximation to area strip mining, the mine operator frequently obtains the right to do strip mining by contracting with the landowner, who has been using the land to grow crops. As part of the royalty agreement between the two, the operator in the past often agreed to strip off the topsoil, stockpile it and then, after the mining has been done and the normal reclamation work completed, to redistribute the topsoil on top of the normal fill. (Under the new 1971 law operators are required to save and spread topsoil in all strip mining operations.) With this kind of procedure, the land can be back in crops a year after the strip mining is completed. One excellent example of this kind of restoration is a potato farm in Somerset County; on August 30-31, 1972, this farm will be the site of the state's annual "Agriculture Progress Days," which attract crowds of 25,000, including agricultural agents from all over the Commonwealth. On that one farm visitors will be able to see an exhibition which includes:

Spot A: potatoes growing before mining takes place
Spot B: a strip mine in operation
Spot C: restoration work underway
Spot D: growing potatoes again, and
Spot E: a potato-chip factory.

Of course, this kind of crop-after-strip is possible only where the area was previously farmland, but there are many parts of the state where this kind of restoration is possible and where it is being carried on today.

Let's go back again to the operator and his bond. Once he has done the soil testing, has put on any required soil amendments (lime and fertilizers) and has completed the required revegetation in grasses and trees, he can apply for release of his bond. DER regulations specify that, before the bond is released, the forester must determine only that the planting has been done "in a workmanlike manner"; there are no requirements that a specified percentage of trees or bushes must survive for a given period of time. DER officials report, however, that where the plantings pass the "workmanlike manner" test, survival rates one year after planting are usually in the neighborhood of 85 percent.

How does a mine site look after revegetation as compared with the way it looked before the mining was done?

Of course, this depends on the particular mine site—they vary somewhat. But generally most of the mining land in Pennsylvania was or is orginally in what we call scrub or tertiary woodland—woods that have been cleared a couple of times and then allowed to regrow naturally. After the mining is complete, a mixture of annual and perennial grass is planted, and the area soon looks like a grassy field. It will then follow the normal pattern of growth and coloration—fresh green grass each spring, brown dead grass during the winter. In some areas, volunteer or self-seeded trees may also appear on the restored site.

If trees are planted, it will be several years before they start showing to any significant degree. From then on, the area will look like a tree plantation, with young saplings of uniform size planted in uniform rows and growing at uniform rates.

Anthracite Strip Mining

So much for bituminous strip mining.

Anthracite strip mining is a very different matter, and in Pennsylvania the mine-site restoration being done is far less satisfactory. Most active strip-mine sites are as desolate as the surface of the moon, and restoration work is essentially just repair work—trying to make the best of a bad situation which has resulted from 50 or more years of virtually uncontrolled stripping.

Why is there such a difference between anthracite and bituminous mining in this regard? In the first place, anthracite beds or deposits are different. They aren't flat; they're steeply pitched, sometimes at 60- or 70-degree angles. And they're often thicker than bituminous seams, ranging from 6 to 20 feet or more in depth. As a result of these differences, anthracite stripping pits may be hundreds of feet deep, sometimes as much as 600 or 700 feet. In the second place, there is very little virgin land being strip mined in the anthracite area today. Most of it has already been deep mined and strip mined in earlier days, and most of the early stripping took place before the Commonwealth had effective legislation to insist upon proper restoration.

Let's examine the history of a typical anthracite mine to see how these two factors have combined to create special restoration problems.

In most anthracite mines, the first extraction of coal took place during the latter part of the 19th century, and it was done by deep mining. Shafts were dug more or less vertically from the surface down to tunnels which were carved out parallel to and below the coal seams. The coal, which was broken loose by blasting, was then carried through these tunnels to the surface (see Figure 9).

Later the same site was strip mined. As shown in Figure 10, the overburden was removed from the top seam and piled in a spoil bank (see the arrows), and then the top seam of coal was removed. When that was completed, the next overburden was removed, then the next coal seam, and so on to the maximum depth that the excavating equipment could operate. In most of these successive seams, the stripping operations removed coal left over from the earlier deep mining-pillars of coal and seam residues that had been left to support deep-mine walls.

Very frequently the same type of stripping operation also took place on the other side of the same coal seam, resulting in a post-stripping contour similar to that shown in Figure 11.

Where this kind of stripping takes place in several adjacent seams, the result is a landscape that is a series of deep cuts, spoil piles and ledges.

It's a gloomy landscape, and the environmentalists in Pennsylvania and elsewhere are rightfully concerned about it, particularly since many of the current operations (and frequently it's a third or fourth stripping that's

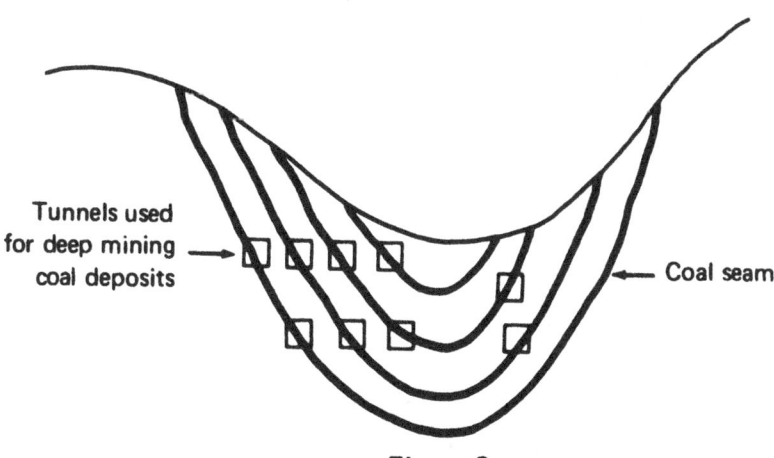

Figure 9
Deep Mining of Anthracite Deposits

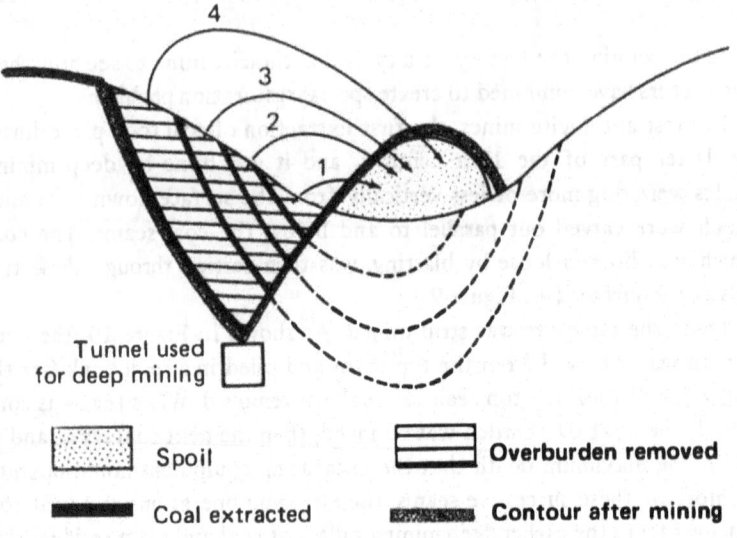

Figure 10
Initial Strip Mining of Anthracite

Spoil

Coal extracted

Overburden removed

Contour after mining

Tunnel used for deep mining

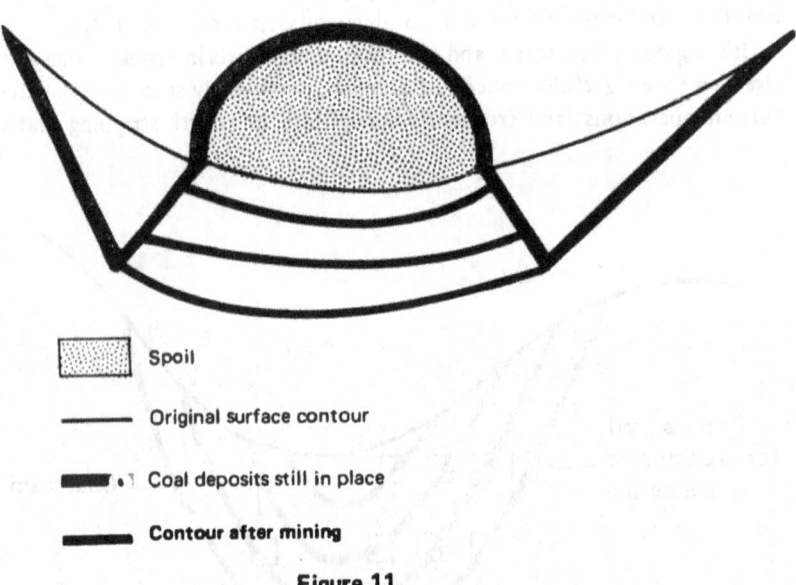

Spoil

Original surface contour

Coal deposits still in place

Contour after mining

Figure 11

**Contour after Strip Mining
Both Edges of Anthracite Coal Seams**

currently going on in a given site) are being performed by the same operators who did the original unregulated stripping.

Pennsylvania's approach to this situation can be summarized in one
maxim: *allow anthracite stripping only where it will result in eliminating
problems that wouldn't be solved otherwise.* In practice this means that
permits to strip virgin territory are not granted except in unusual circumstances—for example, where a small unstripped island is completely sur-
rounded by improperly restored sites. In this case, stripping of the island
can be made part and parcel of an effort to reclaim the entire area, and the
permit is granted only with a guarantee that this be done.

Pennsylvania's stripping maxim also means that when an operator is
granted a permit to restrip an old site, to go deeper and take out deposits
left behind during previous strippings, he is required to smooth the terrain,
to insure that there is no siltation as a result of the current operation and
to eliminate the old water-filled pits. These pits are sometimes 300 feet
deep, and they funnel water down into abandoned deep-mining tunnels,
where it becomes even more acid and then flows out to pollute the rivers—
sometimes to the tune of millions of gallons per day.

In other words, in order to restrip, the operator is required to clean up,
at least in part, the mess left from the previous excavations. This procedure is more a salvage job than ideal restoration, but it may generate some
useful real estate, and the site will look somewhat better than it did when
renewed stripping was begun. There are two alternatives to this kind of
controlled restripping. We can use public funds to restore the sites (and
this is a very expensive process—such projects have cost as much as $2,000
an acre in the anthracite area), or we can simply leave untouched the gray
lunar landscape created before the Commonwealth controlled stripping
activities.

Revegetation is no more difficult in anthracite then in bituminous
sites. In recontouring anthracite sites, the bottom layers of fill are made
up primarily of the large hunks and slabs of stone that usually surround
anthracite deposits and that are removed during anthracite stripping. But
the top layers of fill used in recontouring are made up of broken anthracite spoil, which, despite its sterile gray appearance, is not hostile to plant
growth. Quite the contrary. It has a lower sulfur content than bituminous
spoil and is therefore less apt to be or become acid. In addition, it doesn't
contain as much damaging iron and aluminum toxicity. In short, the basic
soil potential is there—so once earth-moving is complete, anthracite sites
can be revegetated successfully.

The problem with revegetating anthracite in the past has been that
anthracite operators have tended to wait until they have completed excavation of an entire area—which may be very large and include several

different stripping operations—before beginning the work of recontouring and revegetation. Under the new law passed in November 1971, new regulations will require that as a given section of an excavation is completed, the restoration work on that portion of the total area must begin immediately without waiting for the completion of adjacent excavations.

Another deterrent to effective reclamation of anthracite sites has been the fact that, until passage of the new law in November 1971, stripping of anthracite was covered under different legislation than bituminous. (Under the 1971 act, all surface mining in Pennsylvania—including bituminous, anthracite, limestone, sand and gravel—is covered under the same law, and enforcement will be carried on by the same agency.)

Before the new law was enacted, the regulations and restoration standards were supposed to be equally exacting for both kinds of mining. This was in theory; in practice it was not the case. Bituminous regulations were more rigorous and more vigorously enforced, primarily because of differences in the public's concern over and involvement in the whole issue.

Most people don't realize it, but in strip-mine regulation—just as in regular law enforcement work—enforcement agencies can do their work much more effectively if they have a broad base of public opinion and support. In the bituminous areas of western Pennsylvania, there has been that kind of support for many years. In the eastern Pennsylvania anthracite area, however, the general public has until recently been apathetic, and as a result the strip-mining laws and regulations were not utilized to their fullest potential.

Within the last few years, however, this has changed. Prodded by the press and a number of environmental groups, the general public has become aroused and involved in this issue, with the result that there has been increasing pressure on state officials for more effective control of anthracite stripping through stronger regulations and more rigorous use of the administrative tools at their disposal. This change in public attitude in the anthracite area was a powerful factor in the passage of the Surface Mining Conservation and Reclamation Act in 1971; many of the provisions of that act reflect this concern.

Increased public support is not, however, the only reason for increased emphasis on control of anthracite stripping. Another important reason is the fact that the present cost-price structure of anthracite coal makes it possible for strip-mine operations to do the required restoration and still make a profit. The cost of deep mining anthracite coal in Pennsylvania has been increasing rapidly and markedly, partially because of ever more stringent state and federal safety regulations and partially because of state-imposed requirements for the treatment of waters pumped from deep mines. Because of these increasing costs, companies who own strippable anthracite reserves are turning to strip mining because it is considerably cheaper. They now realize that, even though they are required to perform more extensive restoration in the future than they have in the past, the cost of strip-mined anthracite will still be less than that which has been deep mined.

The same kind of cost differential exists for bituminous, but the size of the cost gap is less because of lower costs of water removal. As discussed earlier, in anthracite there may be 40 tons of water that must be removed in the process of deep mining one ton of coal; in bituminous deep mining the ratio is only three to five tons of water to one of coal. Despite the smaller cost differential—and because of the public support discussed earlier—the bituminous miners have had several years of more intensive and extensive experience with strip-mine restoration. The fact that they have been able to absorb these costs and still operate a profitable business enterprise has not been lost on the anthracite operators, many of whom now recognize that the lessons learned in western Pennsylvania could be applied farther east.

Cost of Restoration and Revegetation

How much does all this cost?

The answer to this question must always begin with "It depends—" because the cost is so strongly affected by the kind of coal mined, the slope of the terrain, the depth of the coal deposit removed and the type of geological formations in and near the mine site. The cost of restoring a bituminous strip-mine site in Pennsylvania, for example, is far less on the average than the cost in mountainous coal-mining areas such as West Virginia and eastern Kentucky.

In Pennsylvania most of the restoration and revegetation work is done by the operators themselves, and for these projects no cost figures are available. We do know, however, how much it has cost the state of Pennsylvania to restore currently mined lands in cases where the operators have forfeited their bonds and DER has done the job: an average of $500 an acre for bituminous and $1,000 an acre for anthracite.

For several reasons, these cost figures are probably higher than those for the restoration jobs done by the operators. First and most important, restoration work can be done much more inexpensively if it is done immediately after the mining is completed, while machines and men are still on the spot and before time and weather have made the spoil harder to handle. Second, the restoration jobs performed by the state have frequently been the toughest ones; an operator is less apt to forfeit his bond when the required reclamation work is relatively simple and inexpensive.

And third, all state reclamation work is done by labor paid the prevailing wage rate as determined by the Pennsylvania Department of Labor and Industry, while the strip-mining industry itself does not operate under this requirement.

Enforcement

A strip-mine control law, however strict or carefully drawn, can be only as effective as its enforcement. As Pennsylvania's mine reclamation program has matured, the enforcement arm has developed such a reputation for integrity and efficiency that even its critics grant that it is among the finest—and may indeed be the best—in the entire nation.

An essential ingredient of an enforcement system's integrity is its ability to police itself. Pennsylvania's system proved itself in the spring of 1971, when a supervising inspector was suspected of extortion. As soon as there was clear evidence of his wrongdoing, he was suspended by DER and was later tried and convicted by a county court. The promptness and decisiveness of DER's action in this case was a major factor in establishing public confidence in this department and its work.

The ultimate success of an enforcement agency depends on what its personnel take pride in. With DER personnel, it's the appearance and utility of the strip-mine sites that remain when their work is done. Many of the inspectors are informally—but genuinely—competitive with each other, and will have heated arguments about who has the most attractive and useful projects in his district. All are proud of the fact that Pennsylvania's strip-mine reclamation work has been shown to visitors from all over the United States, and that its architects are called upon to give their advice to others who are wrestling with the same problems in other states and in Congress.

With its system of licenses, permits and bonding, Pennsylvania uses a combination of the carrot and the stick in motivating strip-mine operators to do proper restoration and revegetation. If an operator submits good restoration plans, follows proper practices while he is stripping and does the restoration and revegetation according to state standards, he will obtain his annual license without difficulty, will have his entire bond released and will continue to obtain permits to do further stripping when he applies. If he doesn't play by the rules, he'll be out of the game in the future, and his bond, or part of it, will be used by the state to restore and replant the site that he stripped.

If such a system is to work, of course, there must be a corps of honest and trained men who determine whether the operator meets these requirements; in Pennsylvania this job is done by state-employed inspectors and foresters, with technical advice from state-employed engineers.

The Commonwealth has a corps of inspectors and foresters who handle only strip-mining operations; deep-mine inspectors (who deal almost exclusively with enforcement of health and safety standards) are a separate group.

The roster of strip-mine inspectors and foresters is further subdivided into those dealing with bituminous and those covering anthracite operations. At present DER has 2 foresters and 22 inspectors assigned to bituminous strip-mine sites. Of the 22 inspectors, one is assigned to each of the 19 bituminous districts designated in the state, and 3 are at-large or roving inspectors, who visit all the sites to furnish guidance (to operators as well as to the regularly assigned inspectors) and to maintain uniformity of standards and reporting. Anthracite strip mining is covered by 1 forester and 8 inspectors, assigned and operating on a similar basis, in addition to 8 additional inspectors whose special responsibility is the control of blasting operations.

All strip-mine inspectors are required to file a report on every mine site every 30-45 days. The report forms are uniform throughout the state, and inspectors are carefully instructed as to what DER is particularly interested in and how it should be reported. In short, every attempt is made to make the inspection procedure as uniform as possible, so that all strip-mine operators in the state, regardless of where they are working, will be subject to the same kind of surveillance and enforcement.

The inspectors, who are responsible for all aspects of their assigned stripping operations (including enforcement of health and safety standards as well as site restoration) are state civil service employees. Until very recently, a man couldn't be eligible for an inspector's job unless he had had a minimum of five years of actual working experience in the type of strip mining he would inspect (i.e., bituminous or anthracite). However, the eligibility regulations are now being broadened to include men who have bachelor's degrees in mining engineering, agronomy, soil mechanics or geological engineering.

A candidate for the job of inspector may attend a series of ten classes offered by DER and covering such subjects as coal geology, state law, reclamation and revegetation rules and practices, handling of explosives and equipment maintenance. The lectures are given by DER engineers and inspectors, with occasional guest lectures by members of the Penn State faculty. These classes are offered whenever the state civil service register becomes depleted, that is, when there are only a few eligible candidates still waiting for appointments. To date the class series has been offered only once, since there is still an adequate number of qualified candidates on the register. The classes, which were widely advertised, were given in the evening so that men working in mines who were interested in becoming inspectors were able to attend. At the completion of the course, the men who passed the competitive written and oral state examination were put on the state civil service register, ranked in order of their performance on the exam. When a vacancy occurs, the position is filled from this register of qualified people. As long as an inspector performs his work satisfactorily, he may continue in his position until retirement, for which he is eligible at age 60.

There is stiff competition for the inspector jobs; the positions are attractive because salaries are good (beginning at $10,432 with regular annual increases of 5 percent), the job has prestige and a man's tenure is not affected by changes in the political complexion of the state government.

DER foresters and engineers, who are also state civil service employees, are required to hold bachelor's degrees in their fields; regulations controlling appointment, promotion and tenure are similar to those for inspectors.

Once he is assigned to a given district, a mine inspector is responsible for all aspects of all the strip-mine operations in that district. (At present inspectors in the bituminous districts may be assigned 20-30 different operators carrying on anywhere from 35 to 45 different stripping operations; in the 6 anthracite districts the numbers are smaller, ranging from 15 to 20 operators and 20 to 45 operations in a single district.)

An inspector's responsibility for a given operation begins when a coal operator applies for a strip-mining permit. After inspecting the area and the reclamation plan proposed by the operator, the inspector will make a recommendation to DER as to whether the permit should be issued. The recommendation, along with those of the engineers assigned to the DER central office in Harrisburg, weighs heavily in the DER decison, and an operator's past record—promptness and quality of the reclamation and revegetation work he has done and his record of compliance with health and safety and antipollution standards—is a major factor in the decision to grant or not to grant a permit to do more stripping.

After the permit is issued and stripping starts, the inspector is responsible for seeing that the state health and safety laws are enforced, that blasting requirements are met (a very sensitive issue with the public—seismographs are frequently set up to determine whether complaints about excessive tremors are valid) and that required pollution-abatement activities are carried on during the duration of the mining (which means seeing that diversion ditches are in place, that water is pumped out of the pit, that the pumpings are treated to reduce acidity and that any silt is settled out).

Once the stripping is finished and the operator has completed his reclamation work, the inspector is responsible for determining whether this restoration work has followed the operator's DER-approved plan; if so, most of the operator's bond is released by DER, and the inspector signs off on the operation, turning it over to the forester. This specialist, in turn, advises on, supervises and evaluates the revegetation work which the operator must carry out. As previously discussed, the forester must certify that the operator has done his planting in a workmanlike manner before the remainder of his bond can be released by DER.

How much does such a program of enforcement cost?

At present, 66 DER employees work in strip-mine enforcement: 1 director, 30 inspectors (22 for bituminous and 8 for anthracite), 3 foresters, 8 blasting control inspectors, the equivalent of 12 engineers full time, 12 overhead personnel such as clerks and stenographers and approximately 15 percent of the time of the two top men in DER. The estimated cost of these personnel, plus the equipment needed to do their jobs (including the four-wheel-drive jeeps used by the inspectors to drive from site to site) is approximately $600,000 a year.

Perhaps the most interesting thing about this cost figure is that most of it is borne by the strip-mining industry itself through payment by operators of license and permit fees; the amount of enforcement cost that must be appropriated by the legislature—and hence paid for out of tax revenues—is very small.

CURRENT UNDERGROUND-MINING OPERATIONS

So far this article has dealt exclusively with Pennsylvania's program to control current *strip-mining* operations in order to protect the environment; we turn now to the state's program in currently operated *underground mines*.

Most of Pennsylvania's programs in underground mines (or deep mines, as they are frequently called) emphasize the health and safety of miners, subjects not covered in this article. In the environmental field, the emphasis is on preventing four major types of problems: acid mine drainage, mine subsidence, refuse banks and mine fires. We shall discuss each of these in turn.

Acid Mine Drainage

In Pennsylvania, the law does not tell the operator specifically what he must do to prevent acid mine drainage. Rather it tells him what results his operation must have: either there must be no water coming from his mine, or, if there is, that water must meet the quality standards set by the Environmental Quality Board of DER. These standards are exactly the same for deep mining as for strip mining: not more than 7 milligrams of iron per liter of water, and a pH reading between 6 and 9. The mine-drainage permit procedure is also the same in deep mining as in strip mining; in applying for his drainage permit, the operator must describe in detail what his mining procedure will be and what his program for drainage control will be during and after the operation of the mine. The post-operation plans are particularly important, since under current Pennsylvania law if an operator has worked a deep mine during any year from 1966 on to the present, he is responsible for all drainage from that mine forever. Mines operated prior to 1966 may be sealed under Operation Scarlift.

In deep mines the drainage problem is complicated by the fact that there are several sources of water.

Above the coal seams there are frequently aquifers or layers of water-bearing rock. During or after the deep-mining procedure, the roof of the mine may buckle, causing what we call a subsidence; as a result of subsidence, the aquifers may be fractured, releasing water into the mine cavity.

Rainfall may also percolate from the surface down into the mine through these subsidence fissures.

Aquifers are also frequently found beneath the coal seams. After deep mining has been completed, the floor of the mine may heave up, breaking these underlying rock layers and permitting water to rise into the mine.

Strip mining on the surface may funnel water into the underground workings of the same coal seam.

Despite this multiplicity of water sources, a mine operator can meet DER's drainage water quality standards while the mining operation is going on; the technology to do the job has already been developed. The acid water is collected, treated to make it more alkaline (at a facility constructed near the mine) and then discharged into the nearest stream. (Since most deep mines operate for long periods of time, frequently over a period of 30-50 years, permanent construction of water treatment plants is financially feasible; for shorter-lived strip-mining operations, however, mobile facilities are frequently used.)

After his deep-mining operation is completed, however, the operator must take action to prevent future drainage, and this is a much tougher problem to solve. Usually he does this by sealing the mine. A great deal of research has been done to find out what method of mine sealing is most effective; different methods have been tried during various periods of the state's mining history. At present it appears that the best method is what we call the hydraulic or watertight seal. This method is essentially one of keeping the water in the mine by corking the mine entrance up tight with a concrete plug or stopper. If there are cracks in the mine roof which would allow water to enter into or flow out of the mine, the mine roof and nearby rock layers may be reinforced by what we call a grout curtain, which is an injection of cement that spreads laterally to create a strong area around the mine seal.

There are, however, some mines that it is simply not safe to seal. They are mostly older mines, and their layout and surrounding geological formations are such that even after sealing, water would continue to accumulate in the mine, creating dangerous pressure—and perhaps culminating in a watery explosion that would inundate the neighboring countryside—if the entrance were stopped up with concrete. For such a mine, the only answer is for the operator to attempt to reduce the flow of water into the mine, to partially seal the mine or, as a last resort, to continue treatment of the water.

Under current Pennsylvania law, operators may be required to post a bond to insure that they perform the required water treatment and mine sealing. (Operators with a long record of satisfactory performance may be exempted from this requirement.) The amount of the bond is determined by DER, and is based on DER's and the operator's estimate of the cost of measures that would be required to correct residual water pollution problems.

Mine Subsidence

When mine subsidence occurs—that is, when a mine caves in, causing the land above it to buckle and settle—there can be extensive damage. Since 1969, Pennsylvania has had a program to protect home owners in the bituminous areas of the state from this kind of damage during current mining operations; a similar program for anthracite areas has been discussed but not yet formulated.

In handling this problem, DER's basic approach has been the same as that used to handle the problem of acid mine drainage: the state does not tell the operator how to run his mine, but it does stipulate what the final results of his operation must be. In the case of mine subsidence, it's a question of what the results may *not* be—namely, that nobody's home may be damaged by the deep-mining operation and that, if such damage does occur, the operator is financially responsible for repairing it.

In current bituminous mining operations, the enforcement method used by DER is to require each operator to post a bond covering the value of the homes located above the area which he will be mining. (The bonding does not cover commercial establishments, outbuildings, barns, garages or the land itself. There is a bill now pending in the legislature, however, which would require the operator to compensate a property owner if his water supplies—such as wells—are affected by subsidence caused by current mining operations.)

Prior to the beginning of a new deep-mine operation, DER engineers make a house-by-house engineering inventory of the homes over the proposed site, noting in detail the condition of each home so that there can be no later question as to which damage was caused by the mining and which was already there when the operation began. The operator's bond is held by DER for a period of three years after the completion of the mining operation, at which time it is released if there has been no damage to homes over the mine site or if the operator has satisfactorily repaired any that have been damaged.

What do most operators do to prevent mine subsidence? Basically they usually extract less-than-maximum coal from a given mine, since the coal that is left, if there is enough of it and if it is located in the right places, can serve as effective surface support. So, for example, instead of mining 80 percent of the coal and leaving 20 percent behind in the form of roof support pillars—which might be enough residue to protect the miners from roof falls but not enough to prevent surface subsidence—the operator might mine only 50 percent of the coal and leave the other half.

In Europe, especially in Germany, another method has been developed to prevent mine subsidence, particularly in areas where the buildings on the surface would be expensive to replace. In this method, coal refuse (shale and slate which are separated from the coal in a coal preparation plant) is crushed and then restowed in the mine cavity. This procedure—essentially one of protecting the mine surface by stuffing the mine interior—performs the double function of refuse disposal and mine support. A bill requiring Pennsylvania operators to use this procedure will be introduced during the current session of the General Assembly.

Other methods of preventing mine subsidence are being investigated in engineering studies carried on at various universities throughout the United States. Some studies indicate that the probability and amount of surface subsidence may be related to the width of the mine entry and the coal-mining procedure used. As new engineering data are made available, the mining industry will be able to utilize this information to maximize coal production and minimize the problem of surface subsidence.

For example, there is a new mining technique that was brought to this country from Germany during the last decade and has only recently been put to full use in the bituminous mines. This technique uses a special machine known as a longwall miner, which removes all of the coal as the machine advances through the seam; a set of automatic advancing jacks holds up the mine roof immediately behind and parallel with the cutting bits. As the operation moves farther along the coal deposit, the jacks are also moved, leaving behind a completely mined-out area. The surface over this area settles, but because 100 percent of the coal has been removed, the settling or subsidence is uniform.

Uniform subsidence seldom causes damage to any surface structures which lie directly and entirely above the mining operation. The damage occurs when there are variations in the degree of subsidence—where, for example, one part of a house stays at the same level and the rest of the house drops several inches, in which case the masonry of the house can crack badly. Traditional mining techniques, with their coal pillars interspersed throughout the post-mining cavity, can and do cause this variable subsidence, the long-wall miner does not. In virtually all cases where the long-wall method has been used, little or no damage to surface structures has been reported, although there are frequently cases where wells run dry because of the fracturing of aquifers caused by the subsidence.

Formulas have been worked out to determine how much subsidence will occur as the result of a given long-wall mining operation. If a six-foot-deep deposit lying 100 feet beneath the surface is removed, for example, the overlying surface will sink six inches; if the deposit is thinner or lies deeper, the subsidence will be less.

Since there will be a drop-off of the land surface at the perimeter of each long-wall mining operation, care must be taken to plan the operation so that there is no surface structure sitting athwart this perimeter. For this reason, the new mining method is most suitable in areas which are rural and where the coal deposits occur in large blocks.

Refuse Banks

As the name implies, coal refuse piles are made up of the part of the coal that isn't used. Some of the refuse comes from the mine itself, and some is a discarded by-product of the coal preparation process. A coal refuse pile may contain a wide variety of materials, such as waste rock, small pieces of coal, shale, clay and slate—and it may also contain discarded mining supplies such as grease containers, rags and timber.

Coal refuse piles are a real environmental problem: they are unsightly, they frequently catch on fire and discharge dust (causing dangerous air pollution), they produce acid drainage and siltation and they are subject to slides.

In order to prevent these problems, Pennsylvania law requires that an operator must obtain a permit from DER before he may create a refuse pile. Before the operator is issued a permit, DER must be satisfied that the resulting pile will not cause air or water pollution and that it will be stable—*not only during but also after the completion of mining operations.*

Although DER furnishes guidelines and technical advice, the operator is responsible for working out the method used to meet the following standards.

Prevention of air pollution. The pile must be set down in such a way that it will not catch fire or be a source of wind-borne particulate matter. In order to prevent fires, various techniques are used to prevent the entrance of air into the pile: building one or more sides of the pile against a hill, depositing the refuse in well-compacted layers no more than a few feet thick, placing seals of noncombustible material between the compacted layers, compacting the exposed surfaces of the pile, sealing the sides and edges with earth or clay and allowing refuse to weather (be exposed to the atmosphere for a week or two) before depositing it in piles in order to cut down on its combustibility. Compacting, sealing and revegetation of the refuse piles will also help to keep dust from blowing out of the piles and polluting the air.

Prevention of water pollution. The operator must take the necessary precautions to insure that no acid- or silt-laden water will flow into the state's streams as a result of the refuse bank. Prevention measures include digging diversion ditches (to prevent rainfall from filtering through the pile) and constructing collection networks (so that any runoff can be collected and led to a treatment facility prior to its discharge into the streams of the state).

Stabilization of piles. Operators must apply the best principles of engineering and soil mechanics to insure that the refuse piles will be stable, that is, that they won't slip or slide.

In addition to the refuse pile permit described above, in Pennsylvania an operator is required to obtain a separate kind of permit if mine refuse is to be used to construct a dam (either formal or impromptu) which will impound water resulting from mining or coal-cleaning operations.

Mine Fires

A deep-mine fire may begin either in the mine itself (during current operations) or at the surface (in which case it usually spreads into an abandoned mine). Since in this portion of the article we are dealing with current mining operations, we will briefly discuss the former type of fire here.

A fire that begins in the mine may be caused by overloaded electric circuitry, by spontaneous combustion or (in years past) by the upset of miners' lamps. For example, the Laurel Run mine fire near Wilkes-Barre, which is now being extinguished at the cost of millions of dollars, is alleged to have started after World War II from an improperly placed miner's lamp.

Fire-prevention activities in the mines include:

- taking extensive safety precautions, such as keeping electrical circuitry in topnotch operating condition, using only equipment which has been authorized as "permissible" (that is, safe) by the Bureau of Mines, rock dusting to prevent gas explosions, cleaning up and removing fallen coal to prevent spontaneous combustion and carrying out elaborate systems of self-inspection by miners to be sure they carry no matches or smoking materials;
- using sophisticated monitoring systems which warn that fires have begun (such as devices which give instant warning if there is an increase in temperature of carbon monoxide anywhere in the mine);
- extinguishing fires promptly, using conventional equipment (such as portable fire engines which are mounted on tracked vehicles) and some new automatic devices, including a chemical canister which will burst and distribute fire-fighting chemicals when it is exposed to elevated temperatures.

Since deep-mine fires are dangerous to human life as well as to the physical environment, most of the activities described above are covered in the extensive federal and state health and safety legislation which has been passed to protect miners from the hazards of mine fires. Information on this legislation and on techniques used to prevent and control mine fires can be obtained by writing the Chief, Office of Mineral Information, Bureau of Mines, U.S. Department of Interior, Washington, D.C. 20240.

OPERATION SCARLIFT

So far this article has dealt with what is being done in Pennsylvania to save the environent while mining coal *now*.

But what about the damage done before Pennsylvania passed its landmark legislation in the 1960s? As a result of more than a hundred years of uncontrolled mining, the Commonwealth had a terrible legacy—thousands of acres of disturbed lands, or orphaned lands, as they are called, which badly needed attention and work, but which could not legally be made the responsibility of the current property owners.

Pennsylvania decided that the state government had a responsibility to help tidy up the mess, and serious state-financed reclamation efforts began in 1963.

In 1965, the legislature passed Appropriation Act 12-A, which provided $5.5 million of state money to be used as matching funds for mine restoration projects authorized in Section 205 of the Appalachian Regional Development Act.

In 1967 the legislature passed (in the required two consecutive annual sessions) a proposed Land and Water Conservation Amendment to the state constitution—an amendment which authorized a bond issue of $500 million to finance a wide range of conservation activities, including $200 million to correct the problems created by pre-1963 coal mining in the state.

In order to ratify an amendment to the state constitution in Pennsylvania, a majority of the voters must approve it in a referendum. Before the vote on the Land and Water Conservation Amendment, a bipartisan campaign was launched to educate the public about the issues involved. The press, conservation groups, sportmen's clubs and schools joined in an all-out effort to convince the electorate that it was important to vote "Yes."

The effort was successful. In the referendum of May 1967, the voters approved the amendment and the bond issue (which would be paid off with general tax revenues), and the amendment was implemented by Act 443, approved in January 1968. Operation Scarlift—the term coined to cover all mine-area reclamation efforts—was born.

Scarlift's program is just what the name implies: reclamation and restoration work to remove the five major types of scars left by a century of uncontrolled mining: acid mine drainage, burning mine refuse banks, underground mine fires, surface subsidence and abandoned strip mines which are unsightly and erosion prone.

These kinds of scars are expensive to remove; it is estimated that it would cost billions of dollars to repair the environmental damage caused by pre-1963 coal mining in Pennsylvania alone. In tackling a job of this size, the $200 million allocated to Operation Scarlift from the bond issue—plus the other state, local and federal funds designated to help with the job—can furnish only a beginning. But beginnings are important, and since 1963 Scarlift has done important work in removing all five kinds of scars.

Acid Mine Drainage

Drainage from abandoned deep and strip mines affects nearly 3,000 out of Pennsylvania's 50,000 miles of streams and rivers, polluting the state's natural waterways all the way from Pittsburgh to Philadelphia. Principally acid, this drainage prohibits aquatic life in the streams it affects and makes the water unfit for human consumption.

Between January 1, 1963, and August 31, 1971, 176 mine-drainage correction projects were completed. These projects have been of two main types: (1) operations to prevent and minimize the quantity of mine drainage and (2) construction of water treatment facilities. Among the latter is the Smith Township Water Authority plant at Burgettstown, the first plant in the United States designed specifically to transform mine-drainage water into an economic asset. Using an ion-exchange process to remove sediment and acidity, it creates a high-grade marketable water product that could put treatment of mine drainage on a pay-as-you-go basis.

In addition, 296 abandoned deep-mine openings were packed or sealed to prevent mine drainage, and 3 defective mine seals were repaired.

Mine Refuse Banks

As we have discussed earlier, refuse banks can cause several kinds of environmental problems; one of the most serious is the severe air pollution that results if the banks catch on fire. Several hundred burning banks dot the landscape in the coal regions of Pennsylvania; some are even located in the heart of metropolitan areas such as Scranton.

Refuse bank fires are extinguished by:

- smothering the fire by covering the bank with clay or other inert material which will keep out air;
- quenching the fire by using large high-pressure water nozzles called hydraulic cannons;
- quenching the fire by flooding the burning area or by loading out the burning material and placing it in water;
- excluding air from the burning material by injecting fine material into the bank to decrease its porosity.

Between the beginning of 1963 and the fall of 1971, 70 burning coal refuse bank fires were extinguished under Scarlift, and 3 projects to level and grade refuse banks were completed.

Underground Mine Fires

Dangerous underground mine fires, which emit noxious gasses and force wholesale evacuation of homes and businesses, frequently begin at the surface and then spread to abandoned underground mines. Some originate with refuse banks; it is believed, for example, that the Cedar Avenue mine fire in Scranton began when garbage was burned in a trash disposal site located on top of a pile of flammable mine refuse. The fire eventually burned through the refuse into the remaining coal in an underlying old stripping pit. Other deep-mine fires begin when garbage or trash is dumped in the bench left at an abandoned and unreclaimed strip-mine site; the dumped material may accidentally catch fire, and this in turn ignites the coal seam where it has been exposed by the earlier strip mining. The fire then runs underground, burning its way back along the seam. When the fire goes underground, it becomes much more difficult and expensive to extinguish.

This is certainly a case where an ounce of prevention is worth many pounds of cure. If burning refuse piles can be extinguished and if dumping of solid waste can be strictly controlled, most underground fires in abandoned mines can be prevented and many millions of dollars saved.

Once an underground fire has begun, however, there are several methods which may be used to extinguish it—all of them expensive and time consuming: loading out, flushing, surface sealing and isolation by trench barrier. These methods, all of which have been used in Scarlift, are described in "The Bureau of Mines Restoration Work in Appalachia," *Appalachia*, February 1971, p. 19.

By August 31, 1971, Scarlift had extinguished or brought under control 108 underground mine fires.

Surface Subsidence

Subsidence can and does occur whenever coal has been deep mined. The major metropolitan areas affected in Pennsylvania are Scranton, Wilkes-Barre and Pittsburgh, but there is also subsidence in less populated regions of the state.

The best way to minimize surface subsidence is to do the deep mining properly in the first place, as discussed earlier in this article.

When caving-in occurs in an abandoned mine, remedial action must be taken promptly to avoid damage to surface structures. The corrective work is basically that of filling up the voids with some substance that will be strong and solid enough to prevent further buckling of the surface. This is done by flushing the substance (frequently crushed solids prepared from coal refuse piles) through boreholes into the voids (see "The Bureau of Mines Restoration Work in Appalachia," *Appalachia*, February 1971, pp. 19-20, for details).

Between January 1, 1963, and August 31, 1971, Scarlift completed 39 flushing projects in abandoned underground mines, stabilizing hundreds of surface acres of land on which millions of dollars' worth of property are located.

Abandoned Strip Mines

In Pennsylvania there are approximately 200,000 acres of land which were strip mined before 1963, the date when stripping operations began to be regulated by state law. By September 1, 1971, Operation Scarlift had completed 98 projects to backfill abandoned open-pit coal mines in order to eliminate erosion, mine-drainage pollution and hazardous conditions or to provide sites for industrial development or public facilities. The largest single project was in what is now Moraine State Park in Butler County, where 178 acres were reclaimed in an area that had been extensively mined by both underground and surface techniques.

Scarlift work in Pennsylvania is now performed on both public and private lands. Act 57, passed in March 1970 to amend the 1968 legislation which created Scarlift, called for a board of realtors to evaluate private real estate before and after restoration work is performed by the state. After completion of the work, a lien is put on the real estate; this means that if the owner sells the property, he must pay the state either the cost of the restoration work or the increase in value of the property, which ever is less. This change in the law permits the state to restore more mining-damaged lands but prevents windfall profits to current property owners.

Accomplishments

From the beginning of 1963 through August of 1971, Scarlift completed a total of 797 projects costing $4.8 million. Most of this money has come from state bond issues and the 1965 appropriation act (Act 12-A), with additional contributions from the Appalachian Regional Commission and federal agencies such as the Bureau of Mines. Scarlift is Pennsylvania's first major step in reclaiming its past environmental losses.

CONCLUSION

This has been the story of Pennsylvania's reconciliation between coal mining and environment. It has covered current operations and those long since completed, anthracite and bituminous coal, strip mines and deep mines.

It is a story that tells us many things.

It tells us that when you take the coal out of the ground, however you do it, you have disturbed the environment—and that it costs time, money, and effort to repair the damage.

It tells us that it is better—and far cheaper—to mine carefully and selectively, minimizing the environmental trauma that must later be repaired.

It tells us that some coal deposits shouldn't be mined at all, because the damage is irreparable.

It tells us that although strong legislation is important in this crusade, no amount of laws and regulations will preserve our environmental heritage unless honest and dedicated men enforce them.

It tells us that we learn as we work, and that we must be willing to change our laws, our rules and our technology.

The stakes are high. If we can reconcile coal mining and environment, we'll have a more productive society and an environmental heritage we'll be proud to pass on to our children. If we don't, we'll have purchased our economic growth at a price beyond calculation—a portion of our planet gutted, scarred and eroded.

The choice is clear.

Appendix:
Types of Surface Mining

Contour strip mining is most commonly practiced where deposits occur in rolling or mountainous country. Basically, this method consists of removing the overburden above the bed by starting at the outcrop and proceeding along the hillside. After the deposit is exposed and removed by this first cut, additional cuts are made until the ratio of overburden to product brings the operation to a halt. This type of mining creates a shelf, or "bench," on the hillside. On the inside it is bordered by the highwall, which may range from a few to perhaps more than 100 feet in height, and on the opposite, or outer, side by a rim below which there is frequently a precipitous downslope that has been covered by spoil material cast down the hillside. Unless controlled or stabilized, this spoil material can cause severe erosion and landslides. Contour mining is practiced widely in the coal fields of Appalachia and western phosphate mining regions because of the generally rugged topography. "Rim-cutting" and "benching" are terms that are sometimes used locally to identify workbenches, or ledges, prepared for contour or auger operations.

Area strip mining usually is practiced on relatively flat terrain. A trench, or "boxcut," is made through the overburden to expose a portion of the deposit, which is then removed. The first cut may be extended to the limits of the property or the deposit. As each succeeding parallel cut is made, the spoil (overburden) is deposited in the cut previously excavated. The final cut leaves an open trench as deep as the thickness of the overburden plus the ore recovered, bounded on one side by the last spoil bank and on the other by the undisturbed highwall. Frequently this final cut may be a mile or more from the starting point of the operation. Thus, area stripping, unless graded or leveled, usually resembles the ridges of a gigantic washboard. Coal and Florida phosphate account for the major part of the acreage disturbed by this method, although brown iron ore, some clays, and other commodities are mined in a similar manner.

Anthracite strip mining in Pennsylvania is conducted on hillsides where the coal beds outcrop parallel with the mountain crests. Although most of the operations are conducted on natural slopes of less than 10 degrees, the beds themselves vary in pitch up to 90 degrees. Beds that are stripped are thicker than in the bituminous fields, most varying from 6 to 20 feet, and can be mined economically to much greater depths. Because of the angles at which the beds lie, the methods employed may not be correctly identified either as contour or area mining, but rather as a combination of both. In a few instances, the operations may resemble open pits and quarries, while others are long, deep, narrow canyons.

Auger mining is usually associated with contour strip mining. In coal fields, it is most commonly practiced to recover additional tonnages after the coal-overburden ratio has become such as to render further contour mining uneconomical. Augers are also used to extract coal near the outcrop that could not be recovered safely by earlier underground mining efforts. As the name implies, augering is a method of producing coal by boring horizontally into the seam, much as the carpenter bores a hole in wood. The coal is extracted in the same manner that shavings are produced by the carpenter's bit. Cutting heads of some coal augers are as large as seven feet in diameter. By adding sections behind the cutting head, holes may be drilled in excess of 200 feet. As augering generally is conducted after the strip-mining phase has been completed, little land disturbance can be directly attributed to it. However it may, to some extent, induce surface subsidence and disrupt water channels when underground workings are intersected.

Open pit mining is exemplified by quarries producing limestone, sandstone, marble and granite; sand and gravel pits; and large excavations opened to produce iron and copper. Usually, in open pit mining, the amount of overburden removed is proportionately small compared with the quantity of ore recovered. Another distinctive feature of open pit mining is the length of time that mining is conducted. In stone quarrying, and in open pit mining of iron ore and other metallics, large quantities of ore are obtained within a relatively small surface area because of the thickness of the deposits. Some open pits may be mined for many years—50 or more; in fact, a few have been in continuous operation for more than a century. However, since coal beds are comparatively thin—the United States average being about 5.1 feet for bituminous coal and lignite strip mined in 1960—the average surface coal mine has a relatively short life.

Dredging operations utilize a suction apparatus or various mechanical devices, such as ladder or chain buckets, clamshells and draglines mounted on floating barges. Dredges have been utilized extensively in placer gold mining. Tailing piles from gold dredging operations usually have a configuration that is similar to spoil piles left by area strip mining in coal.

Dredging is also used in the recovery of sand and gravel from stream beds and low-lying lands. In the sand and gravel industry most of the material (volume) produced is marketed, but in dredging for the higher-priced minerals virtually all of the mined material consists of waste that is left at the mine site. Some valuable minerals also are recovered by dredging techniques from beach sands and sedimentary deposits on the continental shelf.

In hydraulic mining a powerful jet of water is employed to wash down or erode a bank of earth or gravel that either is the overburden or contains the desired ore. The ore-bearing material is fed into sluices or other concentrating devices where the desired product is separated from the tailings, or waste, by differences in specific gravity. Hydraulic mining was extensively used in the past to produce gold and other precious metals, but is practiced only on a limited scale today. As both hydraulic mining and dredging create sedimentation problems in streams, some states exercise strict control over these techniques, either through mining or water-control regulations.

Part 3
Economic Impacts

10
The Economic Problem
of Coal Surface Mining

Robert L. Spore

At the present time, committees in both houses of Congress are considering proposed legislation for the control of surface mining, particularly surface mining for coal. The degree of concern over this problem is suggested not only by the number of published accounts of the adverse effects of surface mining,[1] but also by the success of anti-strip mining candidates in recent primary elections. That the control of surface mining should now be of national concern comes as no surprise to most persons, for surely the issue is consistent with the pattern of rising environmental awareness of recent years.

To the economist, it should come as no surprise that the issue is being debated and will no doubt be partially resolved in the absence of firm quantitative economic assessments of the dimensions of the problem. The same pattern of events occurred in the earlier controversies over the control of air and water pollution. At most, the role of the economist appears to be that of helping to predict what the repercussions or impacts of the proposed legislation will be on local or regional economies. But the primary issue of whether or not scarce resources should be devoted to the control of surface mining and, if so, to what extent and by what means or instruments of control, seems somehow to escape applied economic analysis. This state of affairs often tends to suggest, in turn, that perhaps surface mining is not really an "economic" problem, and that criteria for action other than efficiency in the use of scarce resources (e.g., that it is

From *Environmental Affairs*, vol. 2, no. 4, June 1973, pp. 685-693. Copyrighted by the Boston College Environmental Affairs Law Review. Reprinted by permission.

Robert L. Spore is an economist, Oak Ridge National Laboratory—National Science Foundation Environmental Program, Oak Ridge, Tennessee. Research sponsored by the National Science Foundation RANN Program under Union Carbide Corporation's contract with the U.S. Atomic Energy Commission. The comments and suggestions on an earlier draft of this article from E. A. Nephew and from Professors William Miernyk, Richard Gordon, and Bernard Booms are gratefully acknowledged.

"fair" or "equitable" that surface mining be controlled) should be employed in resolving this issue.

It is the purpose of this article to analyze the economic aspects of the problem of surface mining and to discuss the implications of that analysis in terms of the economist's conventional recommendations for corrective action. Such a review suggests that it has been the inability of economists to supply empirical or quantitative substance to their theoretical analyses which necessarily has led to a discussion and proposed resolution of the issue according to other criteria.

One theoretical aspect of the problem of surface mining is well understood by economists and is probably more or less obvious also to others. In general, the problem is one of external effects or spillovers: the act of surface mining results in environmental impacts and costs that are imposed on persons living downstream or downslope of the mining site but which are not accounted for in market transactions in that those imposing the costs are not required to pay for them. For example, downstream persons or firms must bear the costs (e.g., increased treatment expenses) associated with using water the quality of which has deteriorated due to acid drainage or sedimentation.[2] As a result, there is a discrepancy between the private and social costs of mining, and mine operators, by responding to private costs only, produce too large an output at too low a price. There occurs not only a misallocation or inefficiency in the use of resources—some of the resources devoted to mining could have been employed elsewhere to obtain a product of higher value—but also a problem of equity, for those persons suffering the environmental costs of mining are in fact unfairly subsidizing others who are able to enjoy lower prices on products produced from the commodity being mined.[3] The economist's suggested remedy in such a situation involves a justified intervention of government in the market economy, by corrective taxes or mining regulations, so that the external costs are "internalized" to the mining firm.[4] Economic principles can even be employed to define the preferred extent or degree of mining control and mined-area reclamation: society should continue to employ resources for the prevention and control of environmental impacts only up to the point where the value of the resources so employed just equals the value of external costs avoided.[5] Implementation of this analysis involves primarily the complex and difficult task of evaluating (by assigning dollar costs) the environmental or external impacts that occur under various degrees of surface mining control.

It is fair to question, however, whether the situation that would result following the introduction of such controls would really be acceptable, particularly in the case of coal contour surface mining in such mountainous areas as are found in Appalachia. Since many of the areas currently being mined are relatively uninhabited, and since the environmental impacts (other than perhaps the destruction of scenic views) often do not occur more than a few miles downstream or downslope, the magnitude of the external costs being suffered in the absence of control might not be sufficient to justify very extensive or complete mining regulation.[6] On the other hand, many persons would feel, at least intuitively, that society could devote still more resources to the control of surface mining and be better off. Such an intuitive feeling would be correct, for there are other aspects to the problem than the control of external effects.

Internalizing external costs amounts to making a correction to the market cost of mining some commodity. Presumably, the market cost of mining already includes some measure of the time stream of returns that could have been obtained from some alternative use of the land if it had not been mined. That is, the mining firm has had to pay some amount for the land it mines, either through direct purchase or lease of the site or through a royalty payment to the owner of the surface rights. The amount paid represents the market's evaluation (a discounted present value) of the stream of net returns that the land could yield.[7] It is necessary, however, to examine just what alternative uses and returns the market has considered.

For example, many areas of the Appalachian region underlaid by coal seams offer a wide variety of alternative and not mutually exclusive benefits to society. Some relatively uninhabited or primitive areas possess rare natural habitats and scenic views of potential value for various types of outdoor recreation. Less obviously, such areas also have value as pollution sinks and as genetic "banks."[8] A legitimate question is whether or not the market price of such land adequately reflects the value to society of these alternative uses. Unfortunately, the answer is that it does not, because of the particular nature of many of the services that land can provide that are "public goods."

A public good is a good of the type that, once it is supplied to one person, it is not possible or desirable to exclude others from its enjoyment.[9] For example, it is not possible to deny the benefits of a genetic "bank" to one person while providing them to others, and neither is it possible to assess him fairly for the benefit he gains. Up to a point, the additional costs of allowing one more person to make use of a recreation area are zero. Since efficiency in the use of resources requires that the price of an additional unit of some good or service be set equal to the additional cost of providing it, the price to be charged for an additional use of a recreation area should be set at zero; thus use should not be rationed by positive prices.[10] Many members of society would no doubt be willing to pay some amount for the preservation of natural habitats and scenic views regardless of whether they might ever actually visit and enjoy them.[11] But again, since the cost of providing such options to an additional person is zero, no price should be charged for their provision.

In all of these cases, where it is impossible to exclude users or where the efficient allocation of resources requires the charging of zero price, no private person or firm would provide such goods and services voluntarily, for it would be unable to gain sufficient revenue to cover the cost of provision. Thus, while the preservation of the land and natural habitats may provide a wide range of valuable alternative goods and services, the private market economy has no incentive to assure their availability. The market price of land reflects only the value of those services which a private firm can capture.[12] The efficient allocation of resources again requires the intervention of government, not in this case to internalize external costs, but to prevent an unwarranted conversion of natural habitats into mining sites.

Whether private land values accurately reflect social opportunity costs must be investigated on a case by case basis. The techniques which can be employed are again applications of cost-benefit analysis; in this instance, they involve explicit comparisons of the value of the goods and services flowing from alternative land uses.[13] If the analysis indicates that preservation of the natural environment is justified, then such preservation should be provided for through the public budget and paid for out of general revenues.[14]

It is important to realize the considerable difficulties involved when attempting to implement this decision-making framework. In particular, it is exceedingly difficult to evaluate the benefits that would accrue to society from the provision of a genetic "bank" or from the use (or option to use) a wilderness recreation area.[15] The desired measure of benefit is the individual's willingness to pay, but since no markets exist where such goods and services can be exchanged at equilibrium prices, methods other than the use of market prices for imputing benefits must be devised. Procedures must be found not only for estimating the demand for such goods and services but also for determining the appropriate dollar benefits that would be obtained from their provision. Although considerable progress has recently been made in the case of outdoor recreation,[16] much needed research remains. While in times past economists can perhaps be justifiably accused of having ignored this area of applied analysis, much of the lack of progress and development is directly related to the intractable nature of the problem.

Meanwhile, environmental awareness has continued to grow and demands for solutions to these problems have increased. Thus, it has been necessary to debate the issues and propose legislative solutions in the absence of the valuable cost-benefit information the economist might provide. Of necessity, then, criteria other than the efficient use of scarce resources have been applied to the problem. In this study, it is important to consider the implications of resolving these problems according to these other criteria.

Currently, much of the discussion concerning coal surface mining involves not whether some control might be justified, but rather what the resulting economic repercussions on employment, income, and regional economic development would be if surface mining were so strictly controlled as to make it unprofitable. This is also an economic problem, but one of income distribution or equity rather than resource allocation. Interestingly, concern over the possible adverse employment effects of the control of surface mining continues despite the fact that reclamation programs would require the use of additional labor and that, where surface mining is eliminated, the effect of a shift to deep mining would be an increase in total regional employment.[17] Thus, only in those areas where surface mining would be eliminated and deep mining would not be a feasible production alternative is concern over employment justified. In such areas, surface mining is presently providing employment while, on the other hand, preserving natural habitats and related public goods and services often are not types of economic activities which result in immediate local employment and economic development in the traditional sense of the word. Such is not necessarily the case in the long run, however. For example, the Appalachian Regional Commission has spent huge sums on the construction of highways designed to link the Appalachian states more closely to national markets and to remove the transportation handicap of the past. Since a major feature of several of these states is their wilderness appeal, various types of amenity-oriented economic activities will be attracted once these highways are completed. Continued surface mining will almost certainly prevent this development.[18]

It is important to recognize that it is not necessary to sacrifice the preservation of natural habitats and efficiency in resource allocation in the interest of providing a fair or just distribution of income. The two goals need not be at odds. If inequity in income distribution is the problem at hand—and surely it is in the depressed regions of Appalachia—then there are preferable means of meeting this problem. Employment dislocations can be met more effectively and efficiently through such devices as job re-training, improved educational systems, and the like.[19] Public funds are required for programs to solve this problem. We need not also sacrifice our natural environment and the irreplaceable services it can provide.

Another element entering the discussion is the possible effect on the energy supply if surface mining is too strictly controlled or eliminated.

While the issues surrounding the current energy "crisis" are complex, objectivity is required to put this particular aspect into perspective. Thus, the forecasts by mine operators and electric utilities of widespread fuel shortages if mountain surface mining were eliminated should be compared with the fact that of all fuels burned in the U.S. for electric power generation, only eleven percent comes from contour stripped coal, and much of this could easily be absorbed by increased deep mining.[20] Any resulting increase in coal prices will be paid by coal users, which is desirable in the interests of both resource allocation and equity.

In conclusion, while several issues cloud the discussion of surface mining, the basic problem is one of efficiency in the use of scarce resources. Since the private market fails to take into account the external costs of surface mining and the full opportunity costs of land use, scarce resources are currently not being used efficiently; therefore, governmental intervention to promote surface mining regulation and mined-area reclamation is justified. While economic analysis can provide important guidelines to aid government decision-making, much necessary research, particularly in the area of the valuation of public goods, must first be performed. Without this research, any surface mining controls that are implemented will be in accordance with other criteria, with no guarantee that the social welfare will in fact be improved.

Notes

1. See U.S. Department of the Interior, SURFACE MINING AND OUR ENVIRONMENT (Washington, D.C.: U.S. Government Printing Office, 1967); D. Brooks, *Strip Mining Reclamation and Economic Activity*, 6 NATURAL RESOURCES J. 13-44 (1966).

2. Other possible external effects include: landslides, increased flooding, destruction of scenic views, and deterioration of public roads by overweight coal trucks. D. Brooks, supra n.1, at 22-25.

3. The allocation and distribution repercussions mentioned are those that generally accompany production with external effects. See E. J. Mishan, *The Postwar Literature on Externalities: An Interpretative Essay*, 9 J. OF ECONOMIC LITERATURE 1-28 (1971). Such repercussions will in fact occur, however, only in those markets for the specific commodity, surface-mined coal. On the other hand, if the supply to a market consists of both surface- and deep-mined coal, if the average cost of surface-mined coal is lower than that of deep-mined coal, and if price equals minimum average cost of the deep mining producers, then the distribution consequences of external effects will be higher profits to the surface mining

operators rather than lower coal prices (and, thus, lower prices on products produced from coal). Evidence of the relatively high net returns that can accompany coal surface mining is presented in S. Brock, and D. Brooks, THE MYLES JOB MINE—A STUDY OF BENEFITS AND COSTS OF SURFACE MINING FOR COAL IN NORTHERN WEST VIRGINIA (Morgantown, W. Va.: Office of Research and Development, West Virginia University, 1968).

4. For an analysis of various alternative solutions to the problem of external effects, see O. Davis and M. Kamien, *Externalities, Information, and Alternative Collective Action,* in Joint Economic Committee, U.S. Congress, THE ANALYSIS AND EVALUATION OF PUBLIC EXPENDITURES: THE PPB SYSTEM, Vol. I (Washington, D.C.: U.S. Government Printing Office, 1969) 67-85.

It is appropriate to note that the effect of governmental intervention is to circumscribe somewhat the rights of property owners. Under differing institutional arrangements, e.g., where the rights to environmental quality are by law explicitly placed with the public, the imposition of government controls would not be as necessary. See R. Coase, *The Problem of Social Cost,* J. OF LAW AND ECONOMICS 1-44 (October 1960).

5. An example of this analysis as applied to the similar problem of air pollution control can be found in A. Freeman, THE ECONOMICS OF POLLUTION CONTROL AND ENVIRONMENTAL QUALITY (New York: General Learning Press, 1971) 9-10. If at the optimal level of control, damages still remain, they should be compensated for directly by the mining firm.

6. D. Brooks, supra n.1, at 27, estimates that it would require an expenditure of $50/acre to avoid the worst effects of surface mining, while a more complete restoration of mined land could cost $1800/acre to $3000/ acre.

7. Net returns are represented by land rents. Economics possesses differing theories of the determination of land rent. For example, rent may be conceptualized as representing a marginal revenue product or a differential between product revenues and factor payments (other than for land) for goods produced per acre. See R. Clower and J. Due, MICROECONOMICS (Homewood, Ill.: Richard D. Irwin, Inc., 1972) 327-336.

8. Pollution sinks can be defined as areas over which the atmosphere is able to cleanse itself of pollutant loadings through settling and dispersion, and where the proportions of normal atmospheric constituents (such as oxygen) can be re-established. Genetic banks can be defined as areas where the wide diversity of genetic evolution is allowed to continue uninterrupted so that these areas might provide a possible source of (1) natural enemies when biological rather than chemical control of obnoxious

or detrimental species (e.g. insect pests) is desired, and (2) new genetic material to aid in the development of disease resistant strains of species (e.g. for agricultural crops). For a discussion of the various potential benefits of preserving natural habitats, see J. V. Krutilla, *Conservation Reconsidered,* 57 AMERICAN ECONOMIC REV. 777-786.

9. See R. Musgrave, THE THEORY OF PUBLIC FINANCE (New York: McGraw-Hill, 1959) 6-14 for a classical discussion of public goods.

10. Bator, F., THE QUESTION OF GOVERNMENT SPENDING (New York: Macmillan, 1960) 97-102.

11. Such so-called "option" demands are discussed in Krutilla, J., et al., *Observations on the Economics of Irreplaceable Assets,* in A. Kneese and B. Bower, ENVIRONMENTAL QUALITY ANALYSIS (Baltimore: The Johns Hopkins Press, 1972) 95-111.

12. In addition to the inability to take account of public goods and services, there are other instances where the private market can fail to consider the full opportunity costs of alternative land use. For example, scientific forestry can be profitable only when practiced on a rather large scale, and such economies cannot be realized when land ownership is divided into many small holdings. In other instances, the division of land into arbitrary political jurisdictions, together with other similar institutional restraints, can limit the range of land use alternatives the private market can consider.

13. For example, see J. V. Krutilla and C. J. Cicchetti, *Evaluating Benefits of Environmental Resources with Special Application to the Hells Canyon,* 12 NATURAL RESOURCES J. 1-19 (Jan. 1972); R. L. Spore and E. A. Nephew, *Opportunity Costs of Land Use: The Case of Coal Surface Mining,* presented to the Conference on Energy: Demand, Conservation and Institutional Problems, Massachusetts Institute of Technology, February 12-14, 1973.

14. R. Musgrave, supra n.9, at 14-15.

15. Particularly troublesome are the difficulties of estimating the assumed future benefits. Not only can the value of benefits change over time, but the definition of society must be made clear. The question of what collection of persons, present and future, should be considered when scarce resources are being allocated is discussed by J. de V. Graaff, THEORETICAL WELFARE ECONOMICS (Cambridge, England: Cambridge University Press, 1967) Ch. VI.

16. See survey in M. Clawson, METHODS OF MEASURING THE DEMAND FOR AND VALUE OF OUTDOOR RECREATION (Washington, D.C.: Resources for the Future, Inc., 1959).

17. See W. Miernyk, *Environmental Management and Regional Economic Development,* presented to the Southern Economic Association

meeting, Miami Beach, Florida, Nov. 6, 1971.

18. I am indebted to Professor William Miernyk for suggesting this point.

19. R. Musgrave, supra n.9, at 18, suggests that, in general, it is preferable to meet objectives concerning the distribution of income through the tax and transfer mechanisms that exist, and to avoid interference in the allocation of resources as otherwise determined.

20. Environmental Policy Center Newsletter, August 31, 1972; R. C. Austin and P. Borrelli, THE STRIP MINING OF AMERICA (New York: Sierra Club, 1971) 54-58.

11
Strip Mine Reclamation and Economic Analysis

David B. Brooks

It was proved conclusively that the stripping had no near or substantial relationship to the public health, safety, morals or general welfare.

—Edwin R. Phelps[1]

With strip mining and its companion, the auger-mining process, the shades of darkness moved close indeed to the Cumberlands.

—Harry M. Caudill[2]

It has almost become a cliche to describe strip mining for coal as "rape of the land." Strip mining is a surface method in which large power shovels— some of them the largest in the world—"strip" off the soil and rock overlying coal beds, dump it to one side, and then load the underlying coal onto trucks.[3] An extremely productive method of mining,[4] it nevertheless evokes strong reactions because the unwanted soil and rock are turned into long, successive ridges of unsorted, ugly, and unproductive waste as "strip" after parallel "strip" of earth is mined. These man-made badlands extend over large areas, each ending in a deep pit, the last strip mined out, beside which is a cliff called the highwall. With "area stripping," used in relatively flat terrain, the entire surface area is turned into giant washboards. With "contour stripping," used in mountainous areas, the strips resemble looped shoestrings as they follow the sinuous outcrop of a coal seam, leaving a gash of one hundred feet or so in the hillside. Finally, with "auger mining," a relatively new technique, drills as large as seven feet in diameter bore into a seam (often into a high-wall left by stripping) from

Reprinted with permission from 6 *Natural Resources J.* (1966), published by the University of New Mexico School of Law, Albuquerque.

David B. Brooks is an economist, Resources for the Future, Inc., Washington, D.C. The author acknowledges with thanks the assistance of Robert K. Davis, Jack L. Knetsch, Allen V. Kneese and Edwin H. Montgomery, all of whom contributed to the paper through numerous discussions as well as by their comments on an earlier draft.

the surface, leaving it perforated by a series of holes from which the coal has been removed.[5] Any of these methods may cause extensive pollution and erosion damage downslope and downstream of the mine site unless the mine is carefully managed.[6]

Strip mining for coal in the United States will be one hundred years old in 1966, but during much of this time it was not an important method. Since the 1930s, however, strip mining has grown to account for one-half of all anthracite and nearly one-third of all bituminous coal and lignite mined in this country.[7] It has recently been estimated that operating and abandoned strip pits now occupy 500,000 acres in the Appalachian and Midwest coal fields.[8] Since the 1930s, coal strip mining has been attacked—and defended—in literally hundreds of emotional articles, speeches, and political campaigns. During the same period scientific knowledge about the effects of strip mining has been developed from a variety of sources. Both science and emotion are represented in current opinion and in the body of legislation that regulates strip mining in the important producing states.[9]

As yet, however, little effort has been devoted to subjecting these questions to economic analysis.[10] The purpose of this article is to indicate what economics has to say about coal strip mining and attendant efforts to protect other natural resources. More explicitly, I will argue that the private profit signals to which coal stripping firms must and should respond to maximize their profits are not adequate guides for maximizing social welfare. In many situations private market decisions can be relied upon to yield an approach to maximizing social welfare, but this is not the case whenever there is a divergence between private costs and social costs, like the situation presented here. The essence of the strip mining problem is that substantial costs resulting from the process of stripping are imposed on other individuals and are not reflected in the accounts of the coal mining firms.

It will be convenient to use the term "reclamation" to mean efforts devoted to controlling the use of land while it is being stripped as well as efforts devoted to bringing back to use land that was stripped in the past. The term "regulation" will refer to a legal enactment to accomplish one or both of these goals. In popular statements both reclamation and regulation are commonly called "conservation."

I. The Uniqueness of Coal Strip Mining

Why has coal strip mining attracted more attention than other mineral commodities mined in open pits? The answer lies in a combination of reasons. First, coal strip pits are common in the wooded and agricultural areas of the populous eastern half of the country, not in the remote and semi-arid West. Second, compared with other non-metallic minerals mined in large quantities in the East, coal stripping requires the production of much larger amounts of waste.[11] Third, compared with open pit metal mines, coal strip pits are very short-lived. The coal is mined from an area within a year or a few years, while iron or copper pits often remain in existence for half a century or more. Fourth, coal mines (not only strip mines) present certain problems not common to other mines. Coal, both in place and in dumps, is inflammable. Some 220 fires are burning in underground seams today and about 500 more burn in waste piles.[12] Many coal seams also carry iron sulfide minerals that react with air and water to form sulfuric acid, thus producing the widespread acid mine drainage that is toxic to fish and vegetation and which causes extensive corrosion damage.[13]

Finally, there are factors that are less definable. Coal mining is a symbol of the industrial revolution and carries with it a congeries of impressions for some people: impersonality, monopoly capitalism, absentee ownership, etc. To these, stripping adds the following: wholesale and rapid change in land use; serious deterioration in a familiar landscape; and extensive stream and valley pollution. It has also been suggested that stripping offends most seriously not by creating ugliness per se, but by creating ugliness in areas where one least expects to find it.

Given this complex of issues—partly rational, partly mystical, but always strongly felt—it is more apparent why individuals with otherwise diverse interests—sportsmen, farmers, conservationists, and even underground miners—could unite in their opposition to strip mining.[14] During the past several decades, therefore, strip mining has been generally and popularly regarded as an evil, mitigated only in part in its high productivity. But this was not the only dilemma that it posed. Conservationists looked with disgust upon the resulting landscape, yet they had to admit that strip mining recovered a greater proportion of the coal than did underground mining.[15] Agronomists emphasized the loss of arable land to strip pits, yet they had to admit that poor farming practices resulted in a far greater loss.[16] Social scientists worried about effects of stripping on local communities, yet they had to admit that stripping not only provided much needed employment in coal towns but also had a far better safety record than underground mining.[17]

Thus, to most people any judgment of the social value of coal strip mining has always been a matter of balance. And it is just this kind of balancing, of choosing among alternatives when there are real and difficult conflicts, that economic analysis is designed to handle.[18] Economic analysis does this by providing a *rational and operational* set of rules for determining whether the benefits from any action outweigh the costs. Moreover, in situations like strip mining, where private costs are not equal to social costs, *all* costs can, at least in principle, be incorporated so that the general goal of public policy, to maximize net social benefits, can be pursued.

The remainder of this paper is divided into three sections. The first is a review of how approaches to strip mine reclamation have changed during the past several decades. The second is a series of conclusions pertinent to economic analysis that I have drawn from the literature, from interviews, and from field observations. Then, in the third section tentative suggestions are made about the application of economic concepts to policy problems.

A final note before proceeding. The emphasis in this paper is on the effects of strip mining on natural resources. There is reason to think that the more immediate problems may relate to the people who live in and move out of strip mined areas. Indeed, a large proportion of strip coal comes from the poverty-stricken region defined as Appalachia.[19] Human resources and natural resources are related, of course, and I could not disagree if it were stated that the first emphasis in these areas should be placed on education rather than on reclamation.[20]

II. Changing Approaches to the Problem

Beyond noting the few articles in economics journals, the purpose of this section is not to review the extensive literature on strip-mine reclamation and regulation.[21] Rather, it is to point out the decided change in both tone and content discernible in serious considerations of the subject.

A. Agriculture and Agricultural Journals

Scattered articles on the effects of strip mining and on the minor reclamation efforts of the time began to appear in the 1920s.[22] Discussion warmed considerably in the following decade but focused less on the ill-effects themselves than on the amount of land that was taken, probably permanently, out of agricultural production. The arguments were not well supported and tended to reflect agrarian values.

During the late 1930s, two forces initiated a change in the tenor of discussion. The first was the research interest that state agricultural experiment stations and the Central States Forest Experiment Station of the United States Forest Service began to show in strip mine reclamation. (In the case of acid mine drainage, state engineering experiment stations and the United States Public Health Service served in a similar relationship.) The experiment stations viewed problems created by mining like they did those problems created by farming: they saw damages; they analyzed their nature; and they sought ways of coping with them.[23] Moreover, they financed or inspired studies by individuals in related fields—ecologists, fish and wildlife biologists, hydrologists—so that many disciplines have contributed to our present knowledge of strip pits.

Strip mine legislation was the second force. West Virginia passed the first regulatory law in 1939,[24] and other states followed suit. As state agencies were established to administer the law and carry out reclamation activities, a demand was created not only for researchers but for foresters and agronomists who could put findings into practice over large areas. But perhaps the main contribution of the state laws was a shift of emphasis from cure to prevention, from post-mining reclamation to regulation designed to avoid damages. Moreover, as the postwar agricultural revolution muted the argument that stripped land was needed for food production, the public-oriented perspective of state agencies encouraged them to further shift their emphasis toward recreational use of stripped land.

B. Mining Industry and Journals

For the most part during the prewar years the strip mining industry denied legal or moral responsibility for the effects of stripping. However, as the first results of reclamation research became available, a few companies did experiment with reforestation. Also, several statewide associations of strip mining firms—usually the larger ones—were formed to carry out reclamation programs.[25] Gradually the prevailing attitude shifted from do-nothing to one that could be called "industry *oblige*." But so long as voluntary reclamation was held to be the appropriate policy, strippers fought every state law.[26] Organized efforts were devoted to opposing bills introduced in state legislatures and, when passed, fighting them in the courts. Nevertheless, some laws were passed and, with the exception of a poorly drafted Illinois statute, upheld by the courts as a legitimate use of the police power to protect the general welfare.[27]

Today state regulation is no longer opposed by the strip mining industry as a whole. Indeed, one often hears a call for stricter enforcement.[28] There remains some opposition to extending legislation to states which do not now regulate strip mining,[29] but the more broadly supported industry position is to oppose: (1) federal investigation of any kind,[30] and (2) state laws placing responsibility on the industry for lands stripped and abandoned before existing legislation went into effect.[31]

Articles on strip mine reclamation have appeared regularly in the mining press since about 1946.[32] Most articles have been written by officials of the now very active reclamation associations set up by the strippers. These organizations, staffed by foresters and agronomists, were better equipped to utilize the techniques developed by the experiment stations than were mining companies. Their professional attitude is probably the source of the most recent shift in the industry attitude. The goal of "industry *oblige*" was to reduce opposition to stripping, much as institutional advertising might improve the public image. But agronomists and foresters, like miners, are interested in production; they shifted the emphasis from public relations to gaining income from mined-out land through commercial forestry, grazing, or (increasingly) charging user fees for recreational use.[33]

It is surprising that during three decades of widespread interest only four articles on strip mining have appeared in economics journals.[34] Of these, only one considers strip mining in a framework explicitly separating private and social values.[35] Another essentially proposes application of a social rate of discount to strippable farm land to retain it in agriculture.[36] A third presents a useful critique of strip mine legislation.[37] And the fourth, written by a geographer, describes the effects of strip mining in a semi-arid region.[38] As a matter of fact, the work of several geographers deserves substantial credit for today's more rational climate of opinion and comes close to providing, albeit qualitatively, the kind of analysis urged in this paper.[39]

The comments above should not be taken to imply that economic considerations are absent in other studies, for information on reclamation cost is given in many articles. However, the data presented are typically very general or very specific. More important, cost is reported as if reclamation were a production process in which private costs could be simply tabulated against private returns. In short, economic data have sometimes been reported, but economics has not been used as a decision framework incorporating social as well as private values.

III. Economic Observations

To formulate public policy for strip mining with the objective of increasing the net benefits to society, the place of strip mining in our socioeconomic system must be described. The following conclusions, drawn from a variety of sources, seem relevant to an analysis of strip mining in this context.

1. The day of depletion of the coal minable by surface methods is not at hand, as some have suggested. Technologic advances, manifested in the pit by mammoth shovels, are making it possible to move larger and larger amounts of overburden to reach underlying coal. Furthermore, in thermal generation of electricity, the most important use of coal today, the lower quality coal usually produced at strip mines can be burned as efficiently as the more expensive, higher quality coal produced at underground mines.

2. Under existing economic arrangements coal strip mining is the highest use of most land stripped or sought by strippers. That is, the present value of the time stream of private net revenues from coal production is greater, usually considerably greater, than the market price of that land for any other use.[40] Not only are the per acre returns from coal higher than from other commodities, but they accrue within such a short time that their present value is not greatly diminished by discounting the future. The difference in capital values is indicated by the active market existing for strippable land.

In other words, both strip mining firms and land owners appear to be making appropriate decisions in terms of the private costs and returns that each must consider.[41] In this framework the long standing argument whether or not strip mines consume land of good, average, or marginal agricultural quality is irrelevant.[42] The same analysis applies whatever the quality of land is involved, though coal companies will presumably have to pay more for higher quality.

3. By private standards the strip mining industry is acting in an efficient manner. Like the exploitation of many other natural resources, the difficulty with coal strip mining is that private standards are not sufficient to define social efficiency. This market failure results because the decisions of strip miners impinge upon other individuals in the economy and affect the miners' production and consumption decisions in ways that are not reflected in their cost calculations. These effects are what economists call technical externalities or external costs. They are of interest not only because they are tangible or intangible costs imposed on others by the mining operation, but more importantly because there is no compensation for such costs and, therefore, no need for the coal operator to control them. They are outside his market calculations—hence the name, external costs—even though they are significant costs to society.

Through the years an almost endless number of ill-effects have been attributed to strip and auger mining. Upon closer examination many of these accusations have been found to be untrue. Other damages, those affecting the sales value of land held by coal companies, should come to be reflected in private decisions. But there are external costs that are real enough, and they form the heart of the strip mining problem. Inasmuch as these costs have been the subject of most of the nontechnical articles about stripping, they need not be discussed here in any detail,[43] but they should be reviewed briefly.

(a) Air pollution is a relatively minor problem, confined to dust at some pits and to smoke from burning waste piles or coal seams.[44]

(b) Water pollution, resulting from acid drainage or sedimentation, or both, is much more serious than air pollution. Acid drainage (actually a greater problem with deep mining) occurs as direct runoff from pits and as seepage from auger holes. It is responsible for caking in boilers and for corrosion of boats and bridges at considerable distances downstream from its point of origin. Acid drainage is also responsible for long reaches of some streams that are permanently devoid of fishlife or vegetation and for occasional fish kills in other reaches. Sedimentation, a more serious problem with contour stripping, results from the erosion of spoil banks, denuded hillsides, and access roads. Sediment in streams destroys fish habitat, erodes bridges and roadways, clogs culverts, and aids in undercutting stream banks. It shortens the life of flood control and water storage projects. Both acid drainage and sediment contribute to increased treatment costs for downstream users.

(c) Land problems go hand-in-hand with those of watercourses. The land downslope or downstream of a strip mine may receive eroded material from the mine area. It may become devegetated. In some cases sediment and coal fines have choked stream valleys until the fields become swampy and useless for agriculture. There is some evidence that choked stream beds and the bursting of sediment-built dams are responsible for increased flood damages.[45] Forest development is often altered and wildlife habitat destroyed; stagnant pools commonly develop in old strip pits, and there are cases in which coal fires have set forest fires.

(d) Intangible or less measurable effects derive from aesthetic and cultural values that are not directly tied to markets. Important aesthetic effects result from the loss of a natural environment, whatever its original character. Other aesthetic effects result from the absence of vegetation for years on some spoil banks and from the debris remaining after mining. Aesthetically speaking, the small proportion of land actually consumed by strip pits is of less importance than the much larger area over which its effects are visible.[46] Such intangible costs are imposed not only on residents but on visitors traveling through the area. Equally important are the effects on communities near stripping areas. The character of many may be adversely affected by the transient nature of coal strip mining. Tax burdens for those who remain in the area may rise while the level of, or access to, public services declines because people move away or routes of communication are disrupted. Finally, the high-wall itself presents a safety problem near built-up areas.

Some of the external costs discussed above are incurred directly by existing producers of products other than coal and by consumers. The remainder are represented by local income lost because additional productive opportunities are reduced by stripping.[47] There is no question that income from fishing, tourism, and other recreational activities is reduced while stripping is in progress, and that such income may remain low for years after abandonment of the mine. More questionable are the effects of strip mining on potential industrial development. It is considered important by the Area Redevelopment Administration,[48] and at least one power company has engaged in a reclamation program in the hope of increasing industrial development within its market area.[49]

4. Less widely recognized than the external costs of coal strip mining are certain external benefits. That is, in some cases stripping confers benefits on individuals or on the community at large for which the coal company is not recompensed. For example, it has been claimed that men employed in strip mines learn skills more widely used in other industries than are those learned in timbering or in underground mining.[50] Other effects are more tangible. When stripping occurs over old underground mines, the process often collapses the roofs and seals openings so that the flow of acid mine water from the deep mines is reduced or eliminated.[51] It has already been noted that some flood control benefits are claimed. In other cases, strip mining can be an effective way of extinguishing fires in coal seams.[52]

5. It is now rather widely held that technologic problems associated with reclaiming strip-mined land have been solved, and that today's problems relate to managing land and making it more productive.[53] As a general statement, this is no doubt true. However, there are areas in which further technical research would probably significantly lower the cost of reclamation. Most of our reclamation knowledge pertains to the relatively flat terrain stripped in Indiana, Illinois, western Kentucky, and elsewhere. Smaller but still large amounts of strip coal come from the contour mines in the hills of West Virginia, Pennsylvania, eastern Kentucky, and eastern Ohio. These areas are also the home of the auger mine. But there is little research and still less experience to guide reclamation efforts in mountainous terrain.[54]

Additionally, only a small part of the research on reclamation has treated the method of mining as a variable. It has been shown that the tandem system—a method in which a dragline on the edge of the pit removes and segregates the soil and overburden while a shovel in the pit digs the coal— produces better reclamation results but raises the direct cost of mining.[55] However, there have been no systematic studies of the relationships existing between mining methods, reclamation results, and total costs. This probably results from thinking of mining and reclamation as separate stages of production. In contrast, German coal operators have for years incorporated reclamation practices directly into their mining methods.[56] The same approach is being followed at phosphate mines in Florida.[57] In both cases substantial costs savings are claimed over procedures that divorce reclamation from mining.

6. Useful information on the cost of strip mine reclamation and control of acid drainage is not readily available. What has been published is often of little meaning because there is no indication of what is included in the cost figures. Such reported "costs of reclamation" may include anything from piles of spoil bulldozed against the highwall to the development of fields and forests. Moreover, costs vary with the nature of the terrain, with local employment conditions, and with the purpose for which the land is being reclaimed. Grading costs, perhaps the major variable, are reported to range from 1 1/4 cents per ton to 43 cents per ton (over $1000 per acre).[58] Nor is it always clear whether "per acre" figures refer to acres actually stripped or acres affected in other ways. Finally, it is impossible to dissociate costs of mining from costs of reclamation in many reported instances.

Despite the problems of generalizing about reclamation costs, it is nevertheless useful to have some idea of the magnitude of the costs involved. The most frequently cited cost figure is fifty dollars per acre. This amount is supposed to include a very little grading, some soil preparation, simple erosion control, and planting of tree seedlings; it presumes reasonably flat terrain. In rougher terrain the same figure may be used with the understanding that no grading or soil preparation is included but that greater precautions are taken to ensure correct drainage. Reclamation for purposes other than reforestation is generally more expensive.

It is likely that the figure of fifty dollars per acre represents a minimum program serving to avoid the worst effects, rather than an average cost of reclamation. The other extreme is represented in the estimates prepared by a special committee appointed by the Secretary of Agriculture when it was proposed to open a wooded, mountainous area of a national forest to stripping. The committee estimated that the cost of "restoring" mined land to something like its original contour and original forest cover would be $1800 to $3000 per acre, plus $800 to $1500 per acre for land that was disturbed but not actually mined.[59]

The minimum figure can apparently be borne by the coal industry, but the higher figure—assuming the full costs are to be paid by the coal company—would preclude mining. Between these extremes one can find cited almost any cost figure that he considers more representative. My own impressions are that costs of $50 to $250 per acre are appropriate for reforestation and pollution control on relatively level land; and that costs in the mountains are unlikely to be less than several hundred dollars per acre, despite claims to the contrary.[60]

7. Although time has provided considerable experience, it does not appear that strip mine reclamation has been privately profitable.[61] In the majority of cases the net monetary return to a coal company would

be greater if the company could avoid performing any reclamation activities at all. This does not mean that the returns (from harvesting timber, leasing, charging user fees, etc.) are insufficient to recoup the direct costs of maintaining and paying taxes on the land. But it does mean that the private returns are insufficient to recoup these costs plus the initial investment in reclamation if any reasonable interest rate is charged for the funds. In short, granting that for one reason or another coal companies have decided to reclaim land, they have made the best of the situation;[62] but the costs and returns are not usually such that an outside investor would look at strip mine reclamation as an attractive venture.

This is in contrast to the position of the reclamation associations and the large coal companies that reclamation is privately profitable.[63] No doubt in special sets of circumstances it is profitable. However, most statements about the "profits" are found on closer examination to include only a comparison of revenue and direct cost, not revenue and total cost. In other cases hidden subsidies are involved, as when a company "loans" the use of its earthmoving equipment to the reclamation project or charges off costs for the replacement of soil as an expense of mining. Rarely is reclamation recognized as an investment process on which discounted net returns should amount to at least a normal profit if reclamation is to be regarded as privately profitable. In the few cases for which there are sufficient data to roughly compute and discount net returns, the results run to less than three percent per year.[64]

8. Although it is likely that the net private returns from strip mine reclamation are less than a firm could earn from other investments, there is good evidence that over some range the net social returns are high. Social returns include all the benefits from some action, no matter to whom they accrue, whether or not they can be marketed (as social costs include all the costs of some action, no matter who pays them or whether there is a market for them). To restate my conclusion, the direct returns from reclamation, which could be collected by a public body rather than by a private one, *plus* the tangible and the intangible returns accruing to others will often considerably exceed the costs of reclamation. Because these latter, non-direct returns—largely but not entirely represented by external costs avoided—are not collectible in the ordinary sense, strip mine reclamation can be socially, but not privately profitable. However, like private investment, social investment must be justified in incremental amounts. It is not enough to know merely that investment in strip mine reclamation is worthwhile in an overall sense. The benefits and the costs of reclamation vary from place to place—and not always in the same direction. Before investing, one should also know where and in what amount investment will yield the greatest net return. The problem presented by

comparison of the social benefits of reclamation with the social costs of reclamation is discussed in the next section.

IV. The Application of Economics

A. Benefit-Cost Analysis

The main burden of this paper is that benefit-cost analysis offers the most useful framework for making decisions about strip mine reclamation. Benefit-cost analysis is essentially the same sort of decision-making process that is used in ordinary market calculations. However, it can be used in situations in which for one reason or another private market calculations do not produce good results, e.g., external costs in strip or auger mining. In either benefit-cost or private market calculations a comparison is made, in monetary measures, between (1) the gains to be realized if some action is taken, and (2) the things that have to be given up in order to take that action. The action is justified if the benefits exceed the costs or, more accurately, if the benefits exceed the costs by a greater amount than for any alternative action.

The same benefit-cost principles apply whether operating strip mines are being regulated or abandoned pits are being reclaimed. However, it is simpler to illustrate the latter case. Consider a limited budget of, say $1000 available for recreational development at three pits. Pit A is near a city; pit B is on rolling farmland well out from the city; and pit C is in the mountains. Because of differences in the availability of construction equipment, in terrain, and in the types of development proposed (playgrounds in the city park, trails in the mountains, etc.), the costs of reclamation, assumed constant at each pit, vary among the pits as follows:

Pit A—$200/acre, Pit B—$100/acre, Pit C—$300/acre.

Benefits do not remain constant but vary with the amount of land developed. Ignoring for the moment how gross benefits are determined, assume that for three successive acres in each case they are:

	Pit A	Pit B	Pit C
1st acre	$600	$250	$600
2nd acre	550	150	400
3rd acre	300	100	200

By subtracting the per acre reclamation cost for each of the three acres at each pit, the net benefits are:

	Pit A	*Pit B*	*Pit C*
1st acre	$400	$150	$300
2nd acre	350	50	100
3rd acre	100	0	-100

Costs are lowest for pit *B*. The "three-acre benefit-cost ratio" is highest for pit *A*. Neither is a sufficient criterion for optimizing investment. The greatest net social gain can be won by developing the first acre at pit *A*, the second also at pit *A*, the third at pit *C*, and so forth. Thus, some pits may receive extra reclamation funds while others are not reclaimed at all.

What are the benefits, and what are the costs of strip mine reclamation? As emphasized above, the main benefits of both regulation and post-mining reclamation are represented by external costs avoided. When corrosion of boats or silting of ponds and streams can be reduced, this is a benefit. In addition, there are benefits from making the land productive. Represented by profits from grazing or tree harvesting, and in recent years, from orchards, homesite construction, or recreation fees, these benefits have often been captured by private owners. Other productive uses are likely to lie within the public sector. Use of strip pits for sanitary dumps is among these.[65] Also with the public sector are certain recreational uses and the production of fish and wildlife, particularly when they are treated as primary products of reclamation, rather than by-products.[66] It has even been suggested that strip pits themselves be used as tourist attractions.[67]

The costs of strip mine reclamation appear in two stages. Some are incurred after mining is completed and are clearly associated with the reclamation program. When abandoned pits are being reclaimed, all costs are of this type. But operating pits also incur costs because of strip mine regulations and anticipated reclamation activities. Such hidden but, nevertheless, additional costs must also be counted against the benefits of strip mine reclamation.

By moving directly into illustrations of benefits and costs, an important step has been omitted. It has been implicitly assumed that by evaluating social benefits and social costs in terms of dollars, the social value of proposed actions may be approximated. It is not possible to justify this step here. It is sufficient to say that there is broad agreement that market prices or information on willingness-to-pay (which may consist of surrogate measures in the absence of markets) are socially valid indications of the desires of the members of a community for certain quantities of goods and services.[68] Moreover, prices and willingness-to-pay data provide rational and operational guidelines for investment decisions that will maximize society's gain from the use of its resources. By the same token, public intervention in the market is justified when something interferes with the maximization process. This implies that intervention is costless, which is of course not true; however, in the case of strip mining the costs are probably not excessive when compared with the costs imposed by unregulated market operation. As reflected in benefit-cost analysis, prices provide the tools for making public decisions about strip and auger mining that cannot be provided by such nonoperational slogans as "full reclamation."

B. The Role of Public Policy

The first requirement for the systematic use of benefit-cost analysis in public policy toward strip mining is an explicit statement of the social optimum being sought. The appropriate criterion for a social optimum involving strip mining activities is that all costs associated with an optimum level of mining be minimized. This criterion will not be satisfied whenever strip mining imposes costs that are not included in the coal operator's calculations, nor will it be satisfied if cheaper solutions to some problems are feasible but are not open to individual operators. These two conditions, external costs and economies of scale, to use the economist's terms, are the most important general rationales for public intervention.[69]

Given the criterion for a social optimum, what is the role of public policy when there are uncompensated externalities? Its main role is redistribution of costs in a manner ensuring that those who are responsible for external costs have an incentive to take them into account. Only when costs can no longer be shifted to others in the economy will private costs correspond with social costs and the social optimum be realized. For example, in many areas strip mine operators have no incentive to prevent mine wastes from being picked up and carried off by streams. Because the miner has free use of the water, a valuable resource, his costs are understated. Simultaneously, a farmer downstream has lower profit from his land because acid and sediment are in the stream. Hence, the farmer's costs are overstated. If the downstream losses are greater than the costs of control at the strip mine, there is a net social loss and society is receiving less from the use of its resources than it could. But if costs are redistributed so that the mine operator must pay compensation to the farmer for damages, this net loss cannot occur.[70] The operator will have an incentive to control the release of sediment and acid to the point at which the added benefits from further control are no longer worth the added expense. If damages remain, it will be cheaper (and socially appropriate) for him simply to compensate the farmer. Once again, net social returns are being maximized. Moreover, they are being maximized by the normal market process in which a private resource owner attempts to minimize his costs. The only difference is that social costs are now made equal and are reflected in his private costs.[71]

There are several things about this process of cost redistribution that deserve further attention. First, not all external effects are eliminated. To do so would be as much a waste of society's resources as controlling none of them. The social costs of moderate control measures plus some damages will usually be less than the social costs of eliminating all external effects. Similarly, there will be some abandoned pits for which the external costs avoided plus the potential net returns with reclamation will amount to less than the cost of reclamation at that location, and such pits would not optimally be reclaimed. On the other hand, with cost redistribution the scale of mining activities, the "optimum level of mining," will also differ from what it would be with an unregulated market. There are some lands that can be strip mined profitably now because certain costs need not be considered by miners. If the miners of these lands had to bear all the costs of strip mining, the operation would not be profitable and the land would probably remain in its natural state. Finally, social benefits and costs must be computed in net terms. In the example above the social cost of crops lost by pollution is the profit expected from those crops, not their gross value. Similarly, the social benefits of a reclamation program include the profit from the crops saved plus any profit that can be earned from the reclaimed land itself.

Redistribution of costs is the major role that public policy can play in the strip mining problem, but that is not the only role of public policy. It also has a role whenever regional or multipurpose approaches to reclamation can capture economies of scale and thus yield cheaper solutions than could be obtained with mine-by-mine approaches. For instance, it has been shown that large multipurpose dams often achieve a significant reduction in damages from acid drainage through dilution of the acid,[72] though it is an open question whether this method is preferable to mine-by-mine methods. Again, better reclamation results can often be achieved by coordinated work in larger parcels of land than may be controlled by one operator. The importance of such economies of scale is indicated by the success of coal operators' conservation associations and local soil conservation districts in West Virginia, where the strip mine law permits the miner to contract with them to do his required reclamation.[73] Regional or multipurpose projects introduce additional questions about sharing the costs of the program. For example, it is not obvious how the costs of a regional program for replanting strip land in a depressed area should be distributed among mining firms, direct beneficiaries, and the general public.

Finally, the time dimension of strip mine reclamation deserves mention. Many of the damages from strip mining are temporary. An important aspect of benefit-cost analysis is to determine when the costs imposed by temporary losses or temporary ugliness are greater than benefits that may become negligible in a fairly short time. If a strip mine will reforest itself in five or ten years, it would no longer be correct to assign benefits to the reclamation program after that time. Should reclamation be left to nature in such a case? In some cases this might be appropriate action, but if this were the only area near a city for fishing or hiking, then even a temporary loss might impose large costs. Acid mine drainage presents a particular problem in this regard because its effects are so persistent. It has been reported that a stream may require thirty months for restoration after concentrated acid has flowed for barely one hour.[74] That is, the damages are much less reversible than are damages from other pollutants. Consequently, the importance of keeping acid out of streams or of maintaining adequate dilution flows at all seasons of the year becomes critical.

The reclamation program can also be designed to serve varying purposes during the passage of time. It has been persuasively argued, for example, that too much emphasis has been placed on reclaiming land in ways that lead directly to marketable products. A socially preferable procedure may be to make the initial goal one of obtaining cover on the bare soil and eliminating the ugliest aspects of the scar. Later phases of the program may then be devoted to commercial forestry or other profitable pursuits.[75] In any event, the sequence of reclamation activities is another variant in the search for the optimal reclamation program.

C. Evaluation

Thus far statements about benefits and costs have been made as though it were possible to evaluate them simply and accurately. This is, of course, far from the truth. They can be exceedingly difficult to evaluate. However, there are many benefits and costs whose market prices can be directly incorporated into the analysis. Value of timber produced, cost of seedlings, and fees collected are a few of those regularly used in evaluating government projects. There are other benefits and costs that can be evaluated indirectly, though no market exists for the particular benefit or cost in question.[76] In these cases values can be imputed by substituting market prices that do exist. For example, in a Public Health Service study, the amount of money spent each year because of mine acid-induced corrosion of boats and marine structures, caking of boilers, and added treatment by industries downstream was calculated. The annual value imputed to acid drainage control was then the amount of these costs that would be avoided each year.[77] Flood damages, erosion damages, and other costs imposed by strip mining could be evaluated in the same way. Moreover, there are still other costs and benefits, once thought to be unmeasurable, that are proving at least partly tractable to analysis. Recreation is the most important of these.[78] It would seem entirely feasible today to use one of these techniques and the information available on the costs of different types of recreational sites to make a benefit-cost calculation of the net benefits of reclaiming strip land for recreational use.

There will remain, however, benefits and costs that are presently unmeasurable, and whose absolute values may be in principle unmeasurable. But this does not mean that these effects must be completely excluded from benefit-cost analysis. Kneese has suggested that the best way of handling "socially valid goals for which for one or another reason there are no values commensurable with the values pertaining to other elements of the system" is to treat them as explicit requirements in any proposed program.[79] Referring to water pollution control programs, he states:

> This can be done by initially treating these goals, expressed in physical terms, as limits or constraints upon the cost minimization objective. . . . Conceivably this would require a very different combination of units with different operating procedures than a system designed without the constraints. Presuming the constraints are effective, i.e., not automatically met if costs are minimized, they would result in a higher cost system than could otherwise have been achieved. The extra cost represents the limitation which the constraint places upon the objective.[80]

For example, it might be decided that for aesthetic reasons stripped land will remain denuded for no longer than one year. To accomplish this it may be necessary to save and replace topsoil, to do more soil preparation, or to avoid mining in certain sites. All of these procedures would increase the cost of the mining-reclamation process.

This method of making social goals explicit has the further advantage that it permits us to calculate their minimum value. It has been stated by Kneese:

> One useful way of stating the results of variation of constraints which represent goals . . . not valued directly by, or imputable from, the market . . . is in terms of what they must 'at least be worth.' . . . [By] comparing the optimum system with and without the constraint, it is possible to indicate what the *least* value is that must be attached to the increment of pleasure in order to make that level of control procedures worth while.[81]

In short, we are in fact putting a monetary valuation on aesthetic or social goals whether or not we like to think of it that way.

Actually this point is quite general and worth emphasizing. Any restriction or regulation that is placed on the processes of strip and auger mining (or anything else) implies an evaluation. Each has an economic cost that can be made explicit, and one must be able to argue that the social benefits to be gained by imposition of the requirement are worth *at least* this much.

D. Methods and Techniques

In the two preceding sections some principles of benefit-cost analysis and its application to strip and auger mining have been discussed in general terms. The final step in this preliminary assessment of the role of economics is to offer suggestions about how one might actually base decisions on benefits and costs. At this point it becomes convenient to separate the problem of regulating existing strip mines from that of reclaiming abandoned ones.

What methods are available for making benefit-cost calculations for reclamation of abandoned strip pits? The most promising approach is the method now coming into use for determining the social value of soil conservation projects.[82] These techniques require careful estimation of expected returns over time and clear recognition of the principle that reclamation must be justified on investment criteria. The data needed, but not presently available, to make these analyses include expected returns from different types and different sequences of reclamation activities on strip mined land of varying qualities and different locations. (Changes in land values may be a clue here.) Additionally, it would be essential to systematically collect data on the external costs of strip mining and to estimate the present value of future damages avoided. Some information of this type may come out of the cooperative study on acid drainage in several river basins in the northern coal fields of West Virginia recently begun by the United States Public Health Service and the West Virginia Bureau of Mines. In the same project various methods of coping with acid drainage will be compared, careful cost accounts being kept for each. The same approach could be fruitfully applied to an area in which the whole set of problems associated with strip mining is at issue.

With such data in hand it would be possible to adapt the techniques applied in soil conservation projects (which already include both direct returns and external costs avoided as benefits) to strip mine reclamation proposals. The problem is essentially no different. Moreover, the method is flexible. It would be possible to use a lower rate of interest for funds loaned in a depressed area; in areas where aesthetic values are high, limits on the depth or location of strip mining could be imposed as constraints on cost minimization.

When one turns to the more difficult problem of regulating existing mines, he finds that none of the seven state laws presently in force are adequate to handle the range of problems presented by strip and auger mining.[83] Most laws do not recognize that conditions vary, hence that external costs vary, with the state. Nor do these laws recognize that both reclamation costs and potential benefits vary with location and terrain conditions. The differences between area stripping and contour stripping are usually ignored. Regulations are applied across-the-board. For example, almost all of the laws impose a single standard for the grading of stripped land and spoil piles. Actually, the appropriate kind and degree of grading depends upon the terrain, adjacent land use, and proposed use of the reclaimed land. Professors Deasy and Griess have specifically urged laws designed to foster

> selective and local modification of the terrain, without major remodeling of the entire surface, [thereby permitting] development, at reasonable cost, of the widest variety, frequently aesthetically most pleasing, and on the whole economically most profitable types of highly specialized land usage—recreation, education, water conservation and waste disposal.[84]

On the other hand, a useful feature of some laws is their provision for substitution of land. Rather than reclaim land now being mined, an operator may elect to reclaim an equal number of acres of land not previously reclaimed. Although open to possible abuse, substitution does permit the reclamation effort to be concentrated on land that will return greater net benefits. It is not difficult to think of other techniques for concentrating the effort, possibly making it more efficient physically as well as economically.[85]

There is no need to belabor the point. The few instances cited indicate that much could be done to make existing strip mine legislation and its enforcement a more effective tool for reducing social costs by requiring certain practices of strip miners and by creating conditions under which socially more profitable reclamation procedures can be followed.

Perhaps there has been altogether too much reliance on control of strip and auger mining by legislative regulations. For existing operations other techniques may be applicable. In the field of water quality management, techniques such as zoning, effluent standards, and effluent charges have been successfully used to redistribute external costs.[86] Effluent standards are implied by Pennsylvania's "Experimental Rules and Regulations for the Operation and Maintenance of Strip Mines."[87] The rules provide that acid in drainage shall be reduced as close to zero as possible in the outflow and that the iron content shall not be so high that it precipitates as "yellow boy" on the stream bottom. The rules also suggest that hillsides be zoned so that certain areas, notably water courses, be left unstripped.

Similarly, the *Stearns* case decision, in which a specially convened board refused to permit stripping in Cumberland National Forest, was a zoning decision.[88] The *Stearns* decision was based not on the fact that the land was public land, but upon the hilly and forested character of that land. It was pointed out that the social costs of stripping would be much greater than the net value of the coal produced. And there was no reason to think that the coal under this land was of any greater value than coal that could be mined without such large social costs. Thus, this decision is not in conflict with other decisions permitting stripping in other national forests where conditions differ. Although rather broadbrush zoning to prevent stripping has been held unconstitutional,[89] there is no reason to think that zoning based on an evaluation of social costs would be so held.

The bonding system, common to all seven state laws, shows great promise as a device to redistribute costs to bring private and social costs in line. These bonds are required of strip miners before they begin operations and are released upon the completion of specified reclamation activities. Unfortunately, there is little evidence that the bonding system is being used as a device to direct reclamation along the socially most efficient path. Rather it is viewed only as a club over the heads of the operators, and with lax administration it need not be a very heavy club. First, the amount of the bond is usually fixed by law. It is not varied with the character of the land, the proposed method of mining, the nature of the reclamation problem, or the past performance of the coal operator. Second, there is no attempt to use the bonds as a device to gather blocks of land into planned reclamation areas. One strip pit could be reclaimed for forest, the adjacent pit for meadow. Third, in many cases the bonds are set so low that it is cheaper to forfeit than to perform any reclamation. Finally, the bond is usually returned on the basis of certain activities, not on the basis of certain accomplishments. The fact of seeding, not of growth, is sufficient to have the bond released. In short, there is little economic rationale for the amount of the bonds or for their terms as they are used today. If, instead, the bonds were set according to some benefit-cost guidelines taking into account the nature and beauty of the terrain, proximity to urban areas, the time required for natural revegetation, and alternative uses of the land, among other things, the net benefits to society from the whole strip mining process would be significantly increased.

All of the methods suggested to achieve a socially preferable allocation of resources would require more complex administrative procedures than do the current across-the-board rules. Public intervention is never costless. The justification for the added administrative costs lies in the social gains that can justifiably be expected from the application of economic concepts to the problems created by strip and auger mining. Finally, there is every reason to think that the strip mining industry could accommodate itself to a new regime. It is a remarkably resilient industry that has taken many other problems in its stride. With further research it may appear that the socially optimal position is not so privately expensive after all.

Conclusion

This is an opportune time to review the problems associated with strip and auger mining for coal. Public concern is high, and this concern derives increasingly from the desires of numerous individuals and groups residing in urban areas, rather than from those few whose interests are directly affected. The legislatures of a number of states have considered new or amended strip mine laws in the past year or two. Such legislative proposals invariably generate even broader public interest. The chairman of the mineral law section of the Pennsylvania Bar Association noted that the 1963 amendments to the strip mine laws of that state were "the single, most controversial piece of legislation of the past decade."[90]

With passage of the Appalachian Region Development Act of 1965,[91] strip mining has become for the first time an explicit concern of the federal government. This federal concern is potentially the most important development for solution of the problems of strip and auger mining. The act authorizes the spending of funds to cope with the effects of both surface and deep mining. Indeed, President Johnson, upon the insistence of Governor Scranton, agreed to substantially more funds than had originally been proposed.[92] Additionally, Senator Lausche's perennial bill to authorize a federal study of strip mining, which in other years had aroused vehement opposition and had never passed out of committee, was incorporated almost completely in the act as one of the few sections made applicable to the entire country rather than solely to Appalachia.[93] The importance of this new federal involvement in relating the problems of local and Appalachia is reflected in the strong positions taken outside of government. At one end are those who see an expanding coal industry, largely by means of strip and auger mining, as the key to Appalachian redevelopment. At the other end are those, like Harry Caudill, who claim that the social costs of strip mining in the mountains are so high that it should be completely prohibited.[94]

With interest focused on strip mining from a number of sources, there is danger that the desire for action will foster uneconomic or inconsistent programs. The problem is basically one of allocation of resources. An unregulated market will not produce a socially optimal allocation because of technical externalities. The purpose of this paper has been to indicate that a rational public approach to this problem, based on benefit-cost analysis, is within the capabilities of our analytic methods. Three different but interrelated goals have been implied, goals that may complement or oppose the others.[95] The first goal is national productivity, maximization of the net value of output from the resources that society puts into production. This goal is presumably approached by firms operating through the market system in response to free consumer choice. However, it is also this goal that requires government intervention to minimize the total of all costs associated with an optimum level of strip and auger mining whenever (1) costs associated with mining need not be considered by miners, or (2) a regional or multipurpose reclamation program would be more efficient than a mine-by-mine approach.

The second goal includes cultural and aesthetic values that cannot, for one reason or another, be put directly into the cost minimizing calculation. They are represented by constraints on the system forcing it away from the minimum cost point. The implication is that added benefits received are greater than added costs incurred. The importance of this quality of the environment goal is increasing. A substantial proportion of the public seems to be opting for beauty, or at least for the absence of ugliness.

The third goal for programs of strip mine reclamation is redistribution of income to individuals in stripped areas whenever these areas fall within the scope of the poverty program. For our purposes, this goal can be narrowed to local employment. If local employment in certain areas is accepted as a benefit, it follows that the "cost" of certain reclamation projects will be reduced to the extent that men who would otherwise be unemployed will secure jobs. On the other hand, if action directed to other goals reduces the amount of mining in some areas, an important conflict that must be resolved develops among the goals. Again, this goal of redistribution of income is real because the public seems to be opting for the elevation of poverty stricken areas.

The three goals of national productivity, quality of the environment, and local employment together represent a rationale for public policy on strip and auger mining. By the use of benefit-cost analysis conducted under constraints, an explicit and flexible framework becomes available for considering both regulation of existing surface mines and reclamation of the "orphan pits" abandoned in earlier years. Unfortunately, we are still a long way from having the data necessary to make such analyses. If the primary purpose of this paper is to emphasize the applicability of economic analysis to strip and auger mining, its secondary purpose is to indicate the lack of appropriate data and to stimulate the collection of it.

Notes

1. E. R. Phelps, *Current Practices of Strip Mining Coal*, in Proceedings of Symposium on Surface Mining Practices 8 (Univ. of Ariz. College of Mines 1960).

2. H. M. Caudill, Night Comes to the Cumberlands 305 (Atlantic-Little, Brown 1963).

3. The word "strip" is used both as a verb indicating the removal of overburden and as a noun describing the long, thin plan of the areas mined out in each stage of advance. Many discussions of strip mining are available: O. E. Kiessling, F. G. Tryon & L. Mann, The Economics of Strip-Coal Mining (Economic Paper No. 11, U.S. Bureau of Mines 1931); H. D. Graham, The Economics of Strip Coal Mining (Bull. No. 66, Bureau of Economic and Business Research, Univ. of Ill. 1948); University of Ariz. College of Mines, Proceedings of Symposium on Surface Mining Practices (1960), especially E. R. Phelps, *Current Practices of Strip Mining Coal*, id. at 1.

4. In 1962 the average productivity at bituminous coal and lignite strip mines in the United States was nearly 27 tons per man per day. The average at underground mines was 12 tons. The absolute difference between the two rates has been increasing. 2 U.S. Bureau of Mines, Minerals Yearbook, Fuels 71, 86 (1962) [hereinafter cited as Minerals Yearbook, Fuels].

5. Minerals Yearbook, Fuels 98-101. See also W. A. Haley & J. J. Dowd, *The Use of Augers in Surface Mining of Bituminous Coal*, in Report of Investigations 5325 (U.S. Bureau of Mines 1957). The average productivity at auger mines in 1962 was about 35 tons per man per day; cf. note 4 supra.

6. Among the many descriptions of the effects of strip mining on landforms, watercourses, and land use, the following, which do not agree in all respects, were found particularly useful: G. S. Bergoffen, A Digest: Strip-Mine Reclamation (U.S. Forest Service 1962); C. R. Collier et al., Influences of Strip Mining on the Hydrologic Environment of Parts of Beaver Creek Basin, Kentucky, 1955-1959 (Professional Paper No. 427-B, U.S. Geological Survey 1964); G. F. Deasy & P. R. Griess, *Coal Strip Pits in the Northern Appalachian Landscape*, J. Geography, Feb. 1959, p. 72; A. Doerr & L. Guernsey, *Man as a Geomorphological Agent: The Example of Coal Mining*, Annals of the A. of American Geographers, June 1956, p. 197; L. Guernsey, *Strip Coal Mining: A Problem in Conservation*, J. Geography, April 1955, p. 174; G. A. Limstrom, Forestation of Strip-Mined Land in the Central States (Agricultural Handbook No. 166, Central States Forest Experiment Station 1960); Tenn. Dep't of Conservation and

Commerce, Conditions Resulting From Strip Mining for Coal in Tennessee (1960); TVA, An Appraisal of Coal Strip Mining (1963).

7. Minerals Yearbook, Fuels 84-86, 172. A brief history of surface mining is presented by J. W. Feiss, Surface Mining—Minerals, Men, and Divots, Paper Delivered to the Conference on Surface Mining conducted by the Council of State Governments, Roanoke, Va., April 13, 1964, p. 6 (mimeo.).

8. TVA, op. cit. supra note 6, at 4. An estimate made five years ago concluded that, "on the average, some two acres of every square mile in the Northern Appalachian Coal Fields, and a slightly smaller acreage of every square mile in the Eastern Interior Coal Fields, consist of strip pits." Local concentrations are, of course, much higher. Deasy & Griess, supra note 6.

9. In 1962 bituminous coal and lignite stripping was practiced in 22 states. However, six states—Illinois, Indiana, Kentucky, Ohio, Pennsylvania, and West Virginia—accounted for about 85% of the tonnage produced. These states and Maryland have laws regulating strip mining. R. G. Meiners, *Strip Mining Legislation,* 3 Natural Resources J. 442, 443 (1964).

10. Bergoffen, op. cit. supra note 6, at iii.

11. Typically the ratio of waste to coal is 12:1. The ratio of waste to usable product is much higher in low-grade metal mines, but the great bulk of the waste is not produced at the mine but at mills and smelters where it can more easily be handled. Feiss presents an outline comparing the physiographic effects of different mining methods; op. cit. supra note 7, at Fig. 1.

12. U.S. Dep't of the Interior, Annual Report of the Secretary for the Fiscal Year 397 (1962); R. W. Stahl, Survey of Burning Coal-Mine Refuse Banks 1 (Information Circ. No. 8209, U.S. Bureau of Mines 1964).

13. The following are useful introductions to the acid mine drainage problem: G. P. Hanna et al., *Acid Mine Drainage Research Potentialities,* 35 J. Water Pollution Control Federation 275 (1963); G. D. Beal, Common Fallacies About Acid Mine Water (Sanitary Water Bd., Pa. Dep't of Health 1953) (mimeo.); and any of the papers by S. A. Braley appearing in mining journals during the 1950s.

14. Indications of both open and hidden attacks by underground miners on the lower cost strippers can be found scattered through the mining literature. Rather more surprising is the fact that the TVA, once the delight of conservationists, is being cast by them in the villain's role for allegedly ignoring the effects of strip mining to purchase cheap coal for low-cost thermal power.

15. Strip mines recover 90% or more of the coal in place whereas under-

ground mines seldom recover more than 50%. This conflict is typified in an article by W. C. Bramble, *Strip Mining: Waste or Conservation?*, American Forest, June 1949, pp. 24-25.

16. See, e.g., H. R. Moore & R. C. Headington, Agricultural Land Use as Affected by Strip Mining of Coal in Eastern Ohio 34 (Bull. No. 135, Ohio State Univ. Agricultural Experiment Station 1940) (mimeo.).

17. In 1959 the accident frequency rates at underground bituminous mines were 1.02 fatal and 42.71 nonfatal accidents per million man-hours. The rates at strip mines were respectively 0.46 and 20.69. At auger mines the rates were 0 and 21.20. D. Drury, The Accident Records in Coal Mines of the United States 96-97 (Dep't of Economics, Univ. of Ind. 1964).

18. P. A. Samuelson, Economics: An Introductory Analysis 1-7 (5th ed., McGraw-Hill 1961); R. A. Dahl & C. E. Lindbloom, Politics, Economics and Welfare 18-28 passim (Harper & Bros. 1953).

19. President's Appalachian Regional Comm'n, Appalachia 42-44 (1964).

20. This is surely a major theme of Harry Caudill's book, *Night Comes to the Cumberlands* (Atlantic-Little Brown 1963), especially pp. 305-24. It is also the principal conclusion in M. J. Bowman & W. W. Haynes, Resources and People in East Kentucky 244-46 (Johns Hopkins Press for Resources for the Future, Inc. 1963). These two books should be acknowledged as the source of my interest in these problems.

21. The pamphlets by Limstrom and Bergoffen, note 6 supra, include reviews of the literature. Three bibliographies have been prepared: G. A. Limstrom, A Bibliography of Strip-Mine Reclamation (Misc. Release No. 8, Central States Forest Experiment Station 1953); K. L. Bowden, A Bibliography of Strip-Mine Reclamation 1953-1960 (Dep't of Conservation, Univ. of Mich. 1961) (mimeo.); D. T. Funk, A Revised Bibliography of Strip-Mine Reclamation (Misc. Release No. 35, Central States Forest Experiment Station 1962).

22. These early activities were usually reported in the *Journal of Forestry*.

23. Much of this work was published in the *Proceedings* of the state academies of science rather than in an official publication.

24. W. Va. Acts 1939, ch. 84.

25. A. L. Toenges, *Reclamation of Stripped Coal Land*, in Report of Investigations 3440 (U.S. Bureau of Mines 1939); L. E. Sawyer, *Reclamation and Conservation of Stripped-Over Lands: Indiana*, Mining Congress J., July 1946, pp. 26-28.

26. For a recent statement that reflects the earlier opposition to any compulsory reclamation, see W. H. Schoewe, *Land Reclamation*, Mining Congress J., Sept. 1960, pp. 92-97, and Oct. 1960, pp. 69-73.

27. Meiners, supra note 9, at 445. G. D. Sullivan, Presentation Before the Mineral and Natural Resources Law Section, American Bar Association, Chicago, Aug. 12, 1963 (mimeo.).

28. A. E. Lamm, *Surface Mine Reclamation—Why and How*, Mining Congress J., March 1964, p. 25; D. Jackson, *Strip Mining, Reclamation, and the Public*, Coal Age, May 1963, p. 94. Interstate groups like ORSAN-CO are also favored over federal regulation; see W. A. Raleigh, *Acid-Drainage Curbs Are Here*, Coal Age, April 1960, pp. 80-84. There are two "ulterior purposes" that are at times alleged to be of influence in the call for stricter enforcement: (1) an attempt to take the steam out of efforts to strengthen existing laws, and (2) an attempt to force the smaller stripping concerns out of business.

29. Schoewe, supra note 26.

30. Lamm, supra note 28; Meiners, supra note 9, at 460. This position is somewhat inconsistent with complaints that reclamation requirements in one state are more expensive than those in another.

31. West Virginia is alone in having a fund into which strip miners pay a fee for reclamation of land mined in the past. Meiners, supra note 9, at 458. The ORSANCO rules for control of acid drainage define no responsibility for abandoned mines.

32. Most of these articles appear in *Coal Age* or *Mining Congress Journal.*

33. L. Cook, *A New Approach to Strip Land Reclamation*, Mining Congress J., Aug. 1963, p. 68, and *Reclaiming Land for Profit*, Coal Age, Oct. 1963, p. 94; Jackson, supra note 28. In 1963 a national organization, the Mined-Land Conservation Conference, was formed in Washington, D.C., to coordinate and publicize the work of state associations. The "Voluntary Industry Program for Surface Mined-Land Conservation" of the Conference would be ideal if it were actually practiced. See Mined-Land Conservation Conference, Surface Mine Land Conservation 1-4 (undated) (mimeo.).

34. In addition, strip mining in the context of establishing "safe minimum standards" for conservation practice has been discussed by S. V. Ciriacy-Wantrup, Resource Conservation: Economics and Policies 264-65 (Univ. of Cal. 1952).

35. H. W. Hannah & B. Vandervliet, *Effects of Strip Mining on Agricultural Areas in Illinois and Suggested Remedial Measures*, 15 J. Land & P.U. Econ. 296 (1939).

36. C. L. Stewart, *Strategy in Protecting the Public's Interest in Land with Special Reference to Strip Mining*, id. at 312.

37. Meiners, supra note 9.

38. A. H. Doerr, *Coal Mining and Changing Land Patterns in Okla-*

homa, 38 Land Econ. 51 (1962).

39. See especially G. F. Deasy & P. R. Griess, *Coal Strip Mine Reclamation,* Mineral Industries, Oct. 1963, p. 1; Guernsey, *Strip Coal Mining: A Problem in Conservation,* supra note 6.

40. Graham, op. cit. supra note 3, at 29-31, 46-51; Guernsey, *Strip Coal Mining: A Problem in Conservation,* supra note 6, at 178.

41. This is not to say the market is working in ideal fashion. First, the bargaining advantage lies with the coal companies because they have the drilling records. Graham, op. cit. supra note 3, at 50; Guernsey, *Strip Coal Mining: A Problem in Conservation,* supra note 6, at 178. Moreover, while some farmers may welcome stripping as a way to get their capital out of the farm, others who would prefer to continue farming may be forced to sell because the area loses economies, perhaps in marketing or in the supply of factors, when too much land is withdrawn from farming. Fear of such diseconomies could set up a chain reaction that in effect lowers property values. Guernsey casts some light on these possibilities; id. at 179-81. See also G. H. Walter, *Agriculture and Strip Coal Mining,* Agricultural Economics Research, Jan. 1949, pp. 26-28.

42. Coal operators have generally held that the land stripped was of marginal quality, whereas others have held that it was of higher quality. Evidence indicates that land stripped is neither largely good nor largely poor land for agricultural purposes. Graham, op. cit. supra note 3, at 43-44; TVA, op. cit. supra note 6, at 5.

43. For varying appraisals of the importance of these costs, see references cited note 6 supra; also Hannah & Vandervliet, supra note 35. Graham, op. cit. supra note 3, at 52-61, emphasizes the effect of strip mining on tax collections. Several admittedly biased but nevertheless vivid pictorial reviews have also been published. See, e.g., *Kentucky's Ravaged Land,* Louisville Courier-Journal, Jan. 5, 1964 (special supplement).

44. E. Hall, *Air Pollution From Coal Refuse Piles,* Mining Congress J., Dec. 1962, p. 37.

45. Collier, op. cit. supra note 6, at B-1, B-18. However, W. G. Jones argues that presently used methods of backfilling after strip mining contribute to flood control. He claims that the strip pits themselves act as terraces to prevent rapid runoff and that the backfill is more porous than natural soils and holds more water. Jones, *Land Conservation in Pennsylvania Open Pit Mines,* Mining Congress J., Oct. 1963, p. 53.

46. The point that stripping consumes a small proportion of the total land surface was relevant when the community was worried about the destruction of agricultural land. It obviously has no relevance when the effects in question occur away from the site of mining. And it is almost equally irrelevant when many recreational uses of land are considered.

47. It is not necessarily true that local income losses are net losses to the economy. They may simply be transfers from one region to another. However, given the depressed conditions in many strip mining areas, a case can be made for considering them as net losses.

48. The same approach is implicit in the Appalachia program. The less optimistic side of the argument is carefully presented by Bowman & Haynes, op. cit. supra note 20, at 135-59.

49. *Program Drawn To Enhance Landscape*, Electical World, Sept. 17, 1962, p. 94. No doubt this motive also underlies in part the TVA's recent interest in strip mine reclamation.

50. Graham, op. cit. supra note 3, at 41-42.

51. Jones, supra note 45, at 54, states that strip mining in areas once mined by underground methods has been the greatest single factor in controlling acid drainage in Pennsylvania; see also Jackson, supra note 28, at 89.

52. The Carbondale, Pennsylvania, program is the best known example of controlling a fire by strip mining. However, this case does not qualify as an external benefit because the purpose of fire control was fully recognized in the contract signed between the city and the coal companies. *Towns Built Over a Furnace*, Business Week, May 4, 1963, p. 98.

53. G. S. Bergoffen, A Digest: Strip-Mine Reclamation 22 (U.S. Forest Service 1962); R. F. May, *Surface-Mine Reclamation: Continuing Research Challenge*, Coal Age, March 1964, p. 98.

54. Bergoffen, op. cit. supra note 53, at iv, 12; Feiss, op. cit. supra note 7, at 9. Actually, much the same statement might be made about reclamation in semi-arid areas, which is not a problem today, though it might become one if lignite is ever mined in large amounts. See Doerr, supra note 38.

55. Bergoffen, op. cit. supra note 53, at 26; Limstrom, Forestation of Strip-Minded Land in the Central States 26 (Agricultural Handbook No. 166, Central States Forest Experiment Station 1960).

56. W. Knabe, *Methods and Results of Strip-Mine Reclamation in Germany*, 64 Ohio J. Science 75 (1964).

57. U. K. Custred, *New Mining Methods Rehabilitate Florida's Strip Mines*, Mining Engineering, April 1963, p. 50; *Land Reclaimers Plan for '68*, Chemical Week, Nov. 14, 1964, p. 55. Of course, reclamation in level and semi-tropical Florida is simpler than in the Appalachian or Midwest coal fields.

58. TVA, An Appraisal of Coal Strip Mining 9 (1963). Cost figures for strip mine reclamation are usually reported in terms of cents per ton or in terms of dollars per acre. One can be converted to the other by assuming that coal weighs 75 pounds per cubic foot, so that one acre of coal one

foot thick (one acre-foot) contains 1600 short tons of coal. If a stripping seam is 3 feet thick, a reclamation cost of $50 per acre is roughly equivalent to 1 cent per ton. Typically divergent views on costs in relatively flat terrain can be found in L. Guernsey, *The Reclamation of Strip Mined Lands in Western Kentucky*, J. Geography, Jan. 1960, p. 11, and in J. Hyslop, *Some Present Day Reclamation Problems: An Industrialist's Viewpoint*, 64 Ohio J. Science 157, 159-64 (1964).

59. S. T. Dana, *The Stearns Case: An Analysis*, American Forests, Sept. 1955, p. 44.

60. This impression is corroborated by experiments carried out in Pennsylvania. See H. B. Montgomery, *Conscientious Coal Stripping*, Coal Age, July 1962, p. 87. Additional evidence is found in the fact that costs of establishing timber stands in California after burns or harvesting run close to $100 per acre. See J. R. McGuire, *What Are All the Costs of Stand Establishment?*, in Economics of Reforestation 3 (Proceedings of the Annual Meeting of the Western Reforestation Coordinating Comm. 1963). The costs reported by the TVA are much lower, but there seems to be an inconsistency between the amount of coal produced and the acreage mined. TVA, op. cit. supra note 58, at 10.

61. As a generalization this conclusion is not common. However, it is supported by many studies on particular projects: G. H. Deitschman & R. D. Land, *How Strip-Mined Lands Grow Trees Profitably*, Coal Age, Dec. 1951, p. 95; P. N. Seastrom, *United Electric Coal Companies Land-Use Program*, Mining Congress J., Dec. 1963, p. 27; H. Kohnke, *The Reclamation of Coal Mine Spoils*, in Advances in Agronomy, vol. 2, at 341 (1950); *Symposium of Strip-Mine Reclamation*, 64 Ohio J. Science 98, 146 passim (1964).

62. Thus, recognizing that coal strip mining is a land use generally incompatible with farming, the companies have turned in most instances to commercial forestry or commercial grazing. In England, where a very different land situation exists, reclamation of open pit mines has been directed toward the production of cereals. See the series of three articles by W. M. Davies, *Bringing Back the Acres*, Agriculture, March, April, May 1963.

63. See Mined-Land Conservation Conference, op. cit. supra note 33, at 3. In support of the industry position, it is often pointed out that reclaimed strip land is worth more, or is more productive, than adjacent non-stripped land. Such statements are evidence of successful physical reclamation but are irrelevant economically because considerable money was spent on the stripped parcel of land, whereas none was spent on the other parcel. Therefore, the time stream of costs as well as of returns is different, and it is not immediately obvious that the stripped land is the more profitable.

64. But one much-quoted figure of $3.71 profit per year from reforestation implies a return of 6 or 6½%. The figure was apparently estimated by Professor L. A. Holmes and first published in Strip Mine Investigation Comm'n, Report to the 63rd General Assembly of Illinois 24 (1942).

65. G. F. Deasy & P. R. Griess, *Strip Pits and the Sanitary Landfill Process*, Mineral Industries, Nov. 1960, p. 1.

66. The best example of the use of strip mined land for public recreation is Kickapoo State Park in Indiana, part of which was built on strip land. (Indeed, almost every brochure on strip mine reclamation carries a picture of people fishing at Kickapoo Park.) Charles V. Riley of Kent State University has conducted pioneering studies on the use of strip land for wildlife production.

67. P. R. Griess & G. F. Deasy, *Economic Impact of a New Pennsylvania Tourist Facility*, 40 Land Econ. 213 (1964); K. L. Bowden & R. L. Meier, *Should We Design New "Badlands"?*, Landscape Architecture, July 1961, p. 226. Use of the unique character of pits is contemplated in Sweden where architects are making long range redevelopment plans for the iron mines; id. at 228. Similar proposals have been made but never implemented for the Lake Superior iron district of the United States.

68. An extended discussion of the theory underlying benefit-cost analysis can be found in J. V. Krutilla & O. Eckstein, Multiple Purpose River Development 3-77 (Johns Hopkins Press for Resources for the Future, Inc. 1958). A shorter treatment is presented by Allen V. Kneese, Water Pollution: Economic Aspects and Research Needs 18-20 (Resources for the Future, Inc. 1962). R. K. Davis offers a useful discussion of some "conceptual weeds," such as the notion that economic valuation implies commercialization, which can readily be expanded from recreation planning to strip mine reclamation. Davis, *Recreation Planning as an Economic Problem*, 3 Natural Resources J. 239, 241-44 (1963).

69. F. M. Bator, The Question of Government Spending 76-120 (Harper & Bros. 1960); see also Kneese, op. cit. supra note 68, at 29-32.

70. In some states, notably Kentucky, there are legal qualifications to the responsibility of coal operators to pay for damages. *Kentucky's Ravaged Land*, supra note 43, at 8-9; H. M. Caudill, Night Comes to the Cumberlands 74-75, 305-09 (Atlantic-Little, Brown 1963). These qualifications, upheld by the courts, derive from the contracts by which coal companies obtained mineral rights to the land around the turn of the century. This legal principle does not invalidate the economic principle stated in the text.

71. This process of "internalizing" external effects is discussed at greater length and with more attention to the theoretical underpinnings by Kneese, op. cit. supra note 68, at 20-27.

72. C. S. Clark, Mine Acid Formation and Mine Acid Pollution Control,

Paper Delivered to the Fifth Annual Symposium on Industrial Waste Control, Frostburg State College, Frostburg, Md., May 7, 1964 (to be published in the *Proceedings* of the Symposium).

73. E. Leadbetter, *There Oughta Be a Law*, Soil Conservation, Sept. 1957, p. 36.

74. G. D. Beal, Common Fallacies About Acid Mine Water 4 (Sanitary Water Bd., Pa. Dep't of Health 1953) (mimeo.).

75. Bergoffen, op. cit. supra note 53, at 21-22; F. W. Collins, Triple-Phase Strip-Mine Reclamation (Div. of Strip Mine Reclamation, Ky. Dep't of Conservation) (undated).

76. A. V. Kneese, *Socio-Economic Aspects of Water Quality Management*, 36 J. Water Pollution Control Federation 257 (1964).

77. U.S. Public Health Service, *Acid Mine Drainage Studies*, in Ohio River Pollution Control 973-1023 (Supplement C to Part II, 1944).

78. J. L. Knetsch, *Outdoor Recreation Demands and Benefits*, 39 Land Econ. 387 (1963); Davis, supra note 68.

79. Kneese, *Socio-Economic Aspects of Water Quality Management*, supra note 76, at 258.

80. Kneese, Water Pollution: Economic Aspects and Research Needs, op. cit. supra note 68, at 32-33, 42-44.

81. Id. at 34-35.

82. A. J. Coutu, W. W. McPherson & L. R. Martin, Methods for an Economic Evaluation of Soil Conservation Practices (Tech. Bull. No. 137, N.C. Agricultural Experiment Station 1959); R.N.S. Harris, G. S. Tolley & A. J. Coutu, Cropland Reversion in the South 61-69 (Agricultural Economics Information Series No. 100, N.C. State College 1963). See also certain of the papers in *Economics of Reforestation*, op. cit. supra note 60.

83. Detailed comment on these laws is given by Meiners, *Strip Mining Legislation*, 3 Natural Resources J. 442 (1964), and a summary of their provisions is given by Bergoffen, op. cit. supra note 53, at 26-42. The laws of individual states are generally reviewed in detail in law journals shortly after passage or amendment.

84. Deasy & Griess, *Coal Strip Mine Reclamation*, supra note 39, at 1. On the other hand, Meiners, supra note 83, at 449 passim, attacks the laws for being too flexible. He seems to view every permissible relaxation of regulation as an unwarranted gift to the strip miner. But in economic terms rigid restrictions, rigidly enforced, may have no more to offer than administrative simplicity. However, Meiners is certainly correct when he argues that whatever the flexibility permitted by law, it is poor practice to allow the mining company alone to determine the degree to which the law will be applied, as is done in some states.

85. The West Virginia practice of allowing soil conservation districts to contract with coal operators to perform required reclamation is one such technique.

86. A. V. Kneese, *Water Quality Management by Regional Authorities in the Ruhr Area, with Special Emphasis on the Role of Cost Assessment,* in Proceedings of the 1962 Meeting of the Regional Science Association (in press). See also other papers by Kneese for elaboration on the use of these techniques.

87. Sanitary Water Bd., Pa. Dep't of Health, Experimental Rules and Regulations for the Operation and Maintenance of Strip Mines to Prevent Pollution of Waters of the Commonwealth (1952) (mimeo.). The ORSAN-CO acid drainage control program is similar; see Raleigh, *Acid-Drainage Curbs Are Here,* Coal Age, April 1960, p. 80.

88. Dana, supra note 59. There was an additional legal question in this case involving mineral rights reserved when the land was taken into the national forest. However, the board was instructed not to consider this question but only to evaluate the long term public interest.

89. G. D. Sullivan, Presentation before the Mineral and Natural Resources Law Section, American Bar Association, Chicago, Aug. 12, 1963, pp. 11-12 (mimeo.).

90. D. B. Dixon, *Report of the Mineral Law Section,* 34 Pa. Bar Ass'n Q. 456, 457 (1963).

91. 79 Stat. 5 (1965) [Pub. L. No. 89-4, 89th Cong., 1st Sess. (March 9, 1965)].

92. Washington Post, April 23, 1964, p. A-6, col. 1, and April 29, 1964, p. A-11, col. 1.

93. Pub. L. No. 89-4, Sec. 205(c), U.S. Code Cong. & Ad. News, March 20, 1965, pp. 100-01.

Struck by the inconsistency of proceeding simultaneously with reclamation and with a study of how best to go about it, Senator Lausche succeeded in having the Appalachian Bill amended to provide that no federal funds be spent to restore privately owned strip land, pending completion of the study. Immediate reclamation of public land is permitted. Pub. L. No. 89-4, Sec. 205(d), id. at 102. (This amendment to the bill was one of two passed on the floor of Congress. Washington Post, Feb. 2, 1965, p. A-4, col. 2.)

Some work on the application of benefit-cost analysis had already been started in the Department of the Interior. Interview With E. H. Montgomery, Resources Program Staff, Dep't of the Interior, June 3, 1964.

94. H. M. Caudill, *Appalachia: Path From Disaster,* The Nation, March 9, 1964, p. 240. The special supplement to *The Courier-Journal* stated

that such a prohibition would be ideal, but that it was unattainable. *Kentucky's Ravaged Land*, Louisville Courier-Journal, Jan. 5, 1964, p. 13 (special supplement). See also Knabe, supra note 56, at 141-42.

95. Bowman and Haynes outline policy criteria for eastern Kentucky in terms of a set of goals, and I have drawn upon their formulation. Bowman & Haynes, Resources and People in East Kentucky 259-66 (Johns Hopkins Press for Resources for the Future, Inc., 1963).

12
The Cost of Coal

George E. Dails
Elizabeth C. Moore

What coal mining does to people and the environment depends on which kind of mining is being discussed. In this article mining will be divided into two traditional categories: underground mining (or deep mining, as it is commonly called) and surface mining (which includes both strip and auger mining).

In both discussions, costs will be analyzed over a specific 30-year period, the years 1970-2000, and the analysis will be based consistently on the following assumptions:

1. *The cumulative production of coal during the period 1970-2000 will be 20 billion tons.*[1] Although this estimate represents an extrapolation of past production history, it may well be conservative because of the escalating world demand for energy and the shortages or high prices of alternative fuels such as oil and natural gas.

2. *During the 30-year period, an average of 60 percent of the coal (12 billion tons) will be surface-mined, and 40 percent (8 billion tons) will be deep-mined.* As shown in Figure 1, there has been a dramatic increase in the percentage of coal that comes from surface mines in the U.S.—up from less than 2 percent in 1920 to 48.9 percent in 1972. According to L. W. Westerstrom of the Bureau of Mines, the figure for 1973 will probably be about 51 percent when data for that year are complete.[2] With the increased amount of mining in the West, where surface mining is more prevalent, this upward trend is expected to continue.

3. *There will be no significant breakthroughs in mining or reclamation technology.* There will therefore be no marked increase in productivity in the coal-mining industry, and no marked change in effectiveness of methods used to prevent or repair environmental damage.

From *Appalachia*, vol. 8, no. 2, Oct.-Nov. 1974, pp. 1-29. Reprinted by permission.

The notation system in this chapter is as it appeared in the original article.

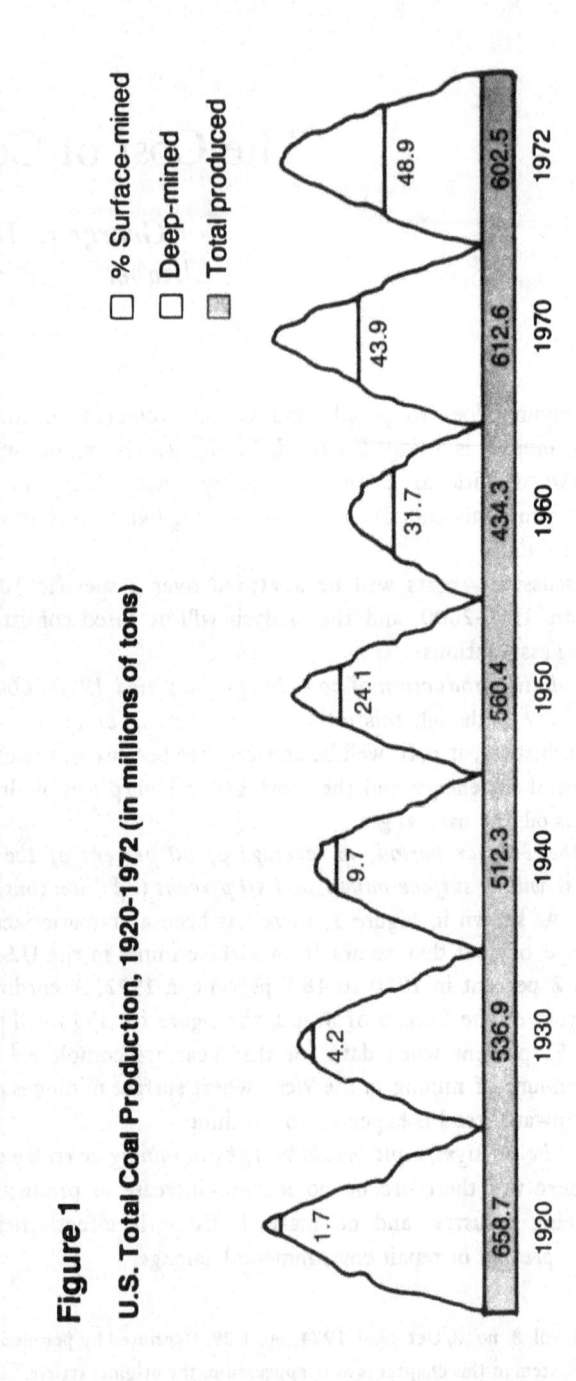

Figure 1

U.S. Total Coal Production, 1920-1972 (in millions of tons)

☐ % Surface-mined
☐ Deep-mined
☐ Total produced

1920	1930	1940	1950	1960	1970	1972
1.7	4.2	9.7	24.1	31.7	43.9	48.9
658.7	536.9	512.3	560.4	434.3	612.6	602.5

Source: Figure 1 is based on combined production totals for bituminous, lignite, and anthracite coal. Data on bituminous and lignite were obtained from the *Minerals Yearbook, 1965,* p 83 (for years 1915-1960), *Minerals Yearbook, 1970,* p. 2 (for 1970), and a preliminary release of information pending publication by the Bureau of Mines of the *Minerals Yearbook, 1972* (for the year 1972). Data on anthracite were furnished by Dorothy Federoff of the Bureau of Mines.

4. *There will be no restrictive legislation that will affect the type of mining practices now permitted.*

History may prove that some of these assumptions are way off base. The authors hope, for example, that the third and fourth assumptions will be proved wrong, and soon. But in estimating the social costs of coal, past experience must be leaned on heavily, and that means holding some important variables—notably technology and legislation—constant during the analysis.

One important point must always be kept in mind by the reader; the social costs calculated in this article will be only those resulting from coal production during the 1970-2000 time period, and will *not* include any cost for damage in the past.

The costs for past damage, such as payments to present black-lung victims and the costs of reclaiming strip-mined "orphan" lands, are large; many people feel that they should be paid by the current consumers of coal, since so many of them are the same people and groups who consumed the coal when the damage was done. Others argue that these compensation and reclamation costs should be paid by the federal government out of general revenues—which really means that they should be paid by all the taxpayers in the nation.

This is an important argument, and one that generates much heat. But it is beyond the scope of this article. In the analysis that follows, we shall frequently indicate the scope of these damages in the past, and past experience will be used as a basis for estimating costs during the 1970-2000 period. But we shall not confuse the two; scrupulous care has been taken to separate past from future in developing the cost figures.

The following social costs of coal will be analyzed in this article for deep-mined coal: mine fires, disposal of mine wastes, control of acid mine drainage, prevention of subsidence, loss of potential production due to accidents, black lung disease, and other costs (gaps in restitution to miners and their families after accidents, lowered property value near mine sites, damage to roads and highways). Analyzed for surface-mined coal will be: land reclamation, control of acid mine drainage, sedimentation, diminished recreational value of land, loss of potential production due to accidents, and other costs (same as those for deep-mined coal).

DEEP MINING

Mine Fires

There are two types of fires that occur during deep mining: underground mine fires and fires in coal refuse piles. Both are damaging and expensive to control. In the 1971 Senate subcommittee hearings on surface mining, John Quarles, then general counsel of the Environmental Protection Agency (EPA), reported that, taken together, the two kinds of fires had caused the deaths of 50 people and the destruction of property valued at more than $2 billion.[3]

Today, however, coal operators use effective techniques to prevent fires in deep mine and refuse banks—and the money spent on fire prevention is routinely included in production costs, and hence in the selling price of coal. Mandated by the Mine Health and Safety Act of 1969, use of these techniques means that new outbreaks are expected to be relatively rare in the future. For this reason, assessment of the social costs of coal mined from 1970-2000 includes no charge for control of mine fires.

Disposal of Mine Wastes

Some mine waste in deep mining comes from the mine itself, and some of it is a discarded by-product of the coal preparation process. Most of this waste is simply put into piles, creating a real environmental problem. Mine refuse piles are unsightly; they discharge particulates and noxious gases; they result in acid drainage and sedimentation; and they are subject to massive slides.[4]

The magnitude and cost of these wastes are enormous. Bureau of Mines studies in Kentucky and Alabama show that one-quarter of a ton of waste is produced for each ton of coal mined, and that disposal of these wastes costs an average of $.07 per ton of coal mined.[5]

At this rate, the 8 billion tons of deep-mined coal projected for the period 1970-2000 will result in 2 billion tons of waste, and disposal of the waste would cost $560 million.

An intensive search is underway to find constructive and profitable ways to use these mine wastes—in the manufacturing of lightweight aggregates for concrete or cinder blocks, as a fill for flat coastal areas, in the manufacture of bricks, in the construction of secondary roads in the form of burned refuse, and in generation of electricity. In mining areas, these wastes are sometimes used to backfill both active and abandoned underground mines.[6]

The happy day when mine waste can be turned into profit cannot come too soon, but until it does, it seems reasonable to use the $.07-per-ton figure from the Bureau of Mines studies as the average cost of mine waste disposal.

Acid Mine Drainage

The acid water which drains from both deep and surface mines is justly infamous. Frequently orange-red in color, it kills the aquatic life in streams and makes the water unfit for human consumption.

The total amount of drainage each year is tremendous. It is estimated that back in 1966, when coal production was considerably less than it is now, mining operations in the Appalachian region alone discharged 4 million tons (nearly 1 billion gallons) of acid each year into 5,700 miles of streams, resulting in continuous and serious deterioration of water quality. In the U.S. as a whole, nearly 11,000 miles of streams are affected by acid mine drainage on either a continuous or intermittent basis.[7]

The effects of acid mine drainage are expensive. It has been estimated that it adds about $3.5 million *each year* to the cost of industrial water users, municipal water suppliers, navigation, and public facilities.[8]

There are a number of methods that can be used to control acid mine drainage, as shown in Table 1,[9] such as neutralization, distillation, reverse osmosis, ion exchange and so on. Some of them involve treating the water to make it nonacid, and others reduce the amount of drainage that takes place at the mine.

Which methods are used depends on many variables—the type of mine (surface, underground below drainage level, or underground above drainage level), whether it is active or inactive, and what the geological and

Table 1

Cost and Effectiveness of Techniques for Controlling Acid Mine Drainage

Technique	% of Effectiveness	Cost (in dollars)
Treatment		
Neutralization	80-90	$.10-1.30/1,000 gal.
Distillation	97-99	.40-3.25/1,000 gal.
Reverse osmosis	90-97	.68-2.57/1,000 gal.
Ion exchange	90-92	.61-2.53/1,000 gal.
Freezing	90-99	.67-3.23/1,000 gal.
Electrodialysis	25-95	.58-2.52/1,000 gal.
At the source		
Water diversion	25-75	300-2,000/acre
Mine sealing	10-80	1,000-20,000/seal
Surface restoration	25-75	300-3,000/acre
Revegetation	5-25	70-350/acre

hydrological conditions are. Costs, as can be seen in Table 1, vary widely—as does the effectiveness of the various methods.

What will be the cost of acid mine drainage caused by the deep mining of 8 billion tons of coal between 1970 and 2000, and how much would this cost be on the average for each ton of coal mined? (Costs of acid mine drainage in surface mining are discussed later.)

The answer does not seem to lie in a detailed analysis of specific control methods. There are too many different methods, and the differences in costs and effectiveness are too great to allow us to calculate averages on any sensible basis. Instead we will take a look at our past national experience with this problem, and then use that information to estimate the future.

In 1970, the Department of the Interior estimated that it would cost $6.6 billion to clean up all the acid mine drainage in the nation—that is, to treat all the acid water so that it would be usable and nontoxic.[10] This $6.6 billion figure deals with past damage; it represents the total result of all the coal mining done in the U.S. from the beginning of our history up to the year 1970. It can also be used as a basis for estimating what the future costs of acid mine drainage will be.

But before using the $6.6 billion figure for this purpose, we must recognize that it includes the cost of not only the acid mine drainage that originated in the mines themselves (both deep and surface), but also of the acid mine drainage that originated in refuse or waste piles—from rainwater leaching down through the piles and turning acid in the process. Remember that we have just discussed the importance of disposing of mine wastes—and have developed a figure ($.07 per ton of coal) for the cost of doing so. In our calculations of the per-ton cost of acid mine drainage, therefore, we must ignore—indeed, we must be careful to eliminate—any costs that are the result of acid mine drainage from waste or refuse piles. Otherwise we will be double charging, and our cost figures will be accordingly inflated.

So the first step is to estimate how much of the $6.6 billion cleanup cost should be subtracted for this reason. There are no precise figures available, but David Maneval of the Appalachian Regional Commission estimates that somewhere between 10 and 15 percent of the pre-1970 acid mine drainage originated in waste piles.[11] Using a midpoint estimate of 12.5 percent and subtracting it from the total cost, it would cost $5.78 billion to clean up the acid mine drainage coming from mines.

Of this $5.78 billion worth of drainage, over 70 percent,[12] or about $4.05 billion, originated in deep mines, which produced a total of 69 billion tons of coal during this period of time.[13] The cost therefore averages out to $.0587 per ton. In other words, if all the acid mine drainage produced by deep mines up to 1970 were cleaned up, it would cost an average of $.0587 for each ton of deep-mined coal produced during this long period.

Recent experience indicates that $.0587 per ton is a very conservative figure. On the basis of experience in treating acid mine water at four typical bituminous deep mines in Pennsylvania, Maneval has estimated that the average cost of treatment was $.20 per ton of coal mined.[14]

So although we do not include the costs of past damage in calculating the real cost of coal during the period 1970-2000, we can use past experience as a basis of estimating what future costs will be. In this case, we conservatively estimate that it will cost an average of $.0587 per ton to clean up the acid mine drainage caused by deep mining 8 billion tons of coal during the last 30 years of the twentieth century. The total cost of abatement would be $470 million.

Prevention of Subsidence

Mine subsidence occurs when a mine caves in, causing the land above it to buckle and settle. It can cause extensive and expensive property damage, particularly in urban areas—as citizens of Scranton and Wilkes-Barre, Pennsylvania, can testify.

Most subsidence can be traced directly to the type of mining techniques used. In the room-and-pillar technique, which has been popular in the U.S., the columns or pillars of coal that are left in place to furnish support represent lost production—coal that cannot be sold. So the pillars are often of only intermediate size and strength, and the unsupported sections of the mine frequently subside. It is this nonuniform subsidence that causes the most damage on the surface.

New mining techniques, particularly the longwall method brought to the U.S. from Germany during the last decade, can help alleviate this problem. A special machine, known as a longwall miner, removes all of the coal as the machine advances through the seam; a set of automatic jacks holds up the mine roof immediately behind and parallel with the cutting bits. As the operation moves farther along the coal deposit, the jacks are also moved, leaving behind a completely mined-out area. The surface over this area settles, but because 100 percent of the coal has been removed, the settling or subsidence is uniform—and hence not usually damaging to surface structures. Until the longwall method is more widely used in the U.S., however, the threat of subsidence remains a real problem. At the Senate hearings in late 1971, the EPA's John Quarles estimated that there was a threat of subsidence or cave-ins in more than 250 communities in 28 states.[3]

When coal is mined with the room-and-pillar technique, the most common method used to prevent subsidence is to backfill the abandoned mine—that is, to fill up the voids with some strong, solid substance that will prevent buckling of the surface. At the same 1971 Senate hearings, Assistant Secretary of the Interior Hollis M. Dole estimated the cost of backfilling abandoned mines at $10,000 to $15,000 per undermined acre. He based his estimate on Bureau of Mines experience in subsidence control projects conducted in the anthracite region of Pennsylvania under the Appalachian Regional Development Act of 1965.[15]

The next question, then, is how many acres will have to be backfilled. According to calculations by Harry Perry, approximately 0.0001 of an acre of land was undermined in the U.S. for each ton of coal extracted from deep mines from the beginning of mining through the year 1969.[1] This means that 800,000 acres of land will be undermined to produce the 8 billion tons of coal projected to be deep-mined during this period of 1970-2000.

However, not all of the undermined acres will require backfilling to prevent subsidence. Perry estimates that of the acres undermined up to 1969, about one-third have subsided.[1] It is difficult to predict just which areas will be in jeopardy, but David Maneval reports that research now being funded by the Appalachian Regional Commission deals with this problem, and that by 1975 a thorough and reliable method of measuring subsidence-proneness should be available.[11]

With this tool at hand, then, we can assume that of the 800,000 acres undermined between 1970 and 2000, only about 267,000 will subside if they are not taken care of—and that these areas can be spotted ahead of time and backfilled at a cost of $15,000 per acre. The total cost of backfilling will be approximately $4 billion, or about $.50 per ton of deep-mined coal.

Deep-Mining Accidents

Measuring the cost of damage to human beings—the cost of accidents—is far more complex than measuring costs to the environment, and it may in some ways be impossible. Who can put a price tag on pain, or a wife's worry, or the way a child feels when his father is killed in a mine explosion? Can anybody pretend that any sum of money will really make up for what it means to go through life minus an arm, or paralyzed from the waist down?

Granting all this, however, part of the social cost of an accident can be given a price tag. When a man in the prime of his life is cut down by a mine cave-in, society is poorer in many ways that can be expressed in dollars and cents.

Three types of accident costs—the costs of administration, prevention, and restitution—are already being paid by the mine operators, and are then passed on to the people who buy and use coal.

Accidents create certain administrative costs. The accident must be investigated. If production levels are to be maintained, substitute workers must be recruited and trained, and their wages must be paid. Productivity probably sags, for a while at least, since substitute workers are not usually as efficient and productive as the men who were injured or killed in the accidents. These costs are paid by coal operators, and hence by coal consumers.

In order to prevent accidents, and to meet standards set under the 1969 Mine Health and Safety Act, operators install and operate mine-safety equipment, and many operators also organize and operate programs in mine-safety education. These expenses are part of the cost of mining coal and are passed on to those who buy and use it.

In the case of coal-mining accidents (as well as accidents in other industries), restitution is accomplished primarily through benefits provided by by workmen's compensation programs in all 50 states of the union. These benefits include: burial allowances and survivor's benefits to widows and dependent children of miners killed at work; weekly income payments to a miner and his family during the period when he is unable to work because of a disabling accident; medical benefits to help cover doctor bills and hospital expenses; and the cost of rehabilitation programs, including those which retrain injured miners to prepare them for new jobs.

The cost of workmen's compensation programs in the coal industry is paid by the coal operators, usually in the form of insurance premiums paid either to a special state workmen's compensation fund or to private insurance companies that furnish this kind of coverage. (In some states, coal operators may qualify as self-insurers, and large coal companies sometimes choose this approach.) In any case, directly or indirectly, it is the coal operators who pay the cost of workmen's compensation programs for their employees, and the operators in turn pass on the cost to coal consumers.

There are, however, significant gaps in the coverage provided by workmen's compensation. Each state has its own program, and there are sizable variations in the amount of coverage provided. Of the 24 coal-mining states, for example, 16 pay all medical bills for injured miners—but Wyoming, at the other end of the spectrum, pays only a maximum $880. Income-replacement benefits paid to a totally disabled miner and his family are $175 a week in Alaska and $150 a week in Arizona—but in eight of the coal states the maximum payment is below the poverty level set by the U.S. Department of Labor in 1973 ($82.69 a week for a family of four). Even the most generous state programs have limits on the dollar amount of weekly benefits that will be paid to injured miners or their widows and families.

Other types of financial help are available in case of mining accidents. Depending on the duration of the disability and the age and number of dependents, some miners and/or wives and families, are eligible for Social Security, union benefits, or both, in case of accidental injury or death. Even so, the restitution for coal-mining accidents is not usually complete. James O'Brien, assistant director of AFL–CIO's Department of Social Security, puts it this way:[16] "However all the various payments are put together—workmen's compensation, Social Security, union payments— the combination usually does *not* replace the income lost, and it frequently does *not* pay all of the special expenses, such as burial or medical expenses, that are incurred as a result of the accident."

Family, friends, or organized charity, public or private, must pitch in and help—and the dollar value of this needed help is a social cost of coal that must not be ignored. It is, however, a cost which at present cannot be translated into a dollars-and-cents figure for the coal industry as a whole. We know it is there, but we do not know how big it is. So it is listed as one of the "other costs" in Table 3, and no dollar figure is posted for it. As long as the present situation continues, this part of the cost of coal is an inequity to those who must pay it.

To summarize, many of the costs of accidents—the costs of administration and prevention and some (though by no means all) of the costs of restitution—are paid by coal operators, and are therefore already included in the price the consumer pays for coal.

Loss of Potential Production

There is, however, still another kind of cost which is incurred when a coal miner is injured or killed in an accident—and this cost is levied, not on the miner or his family or the coal operator or the coal consumers, but on the U.S. economy as a whole. When a person of working age is killed in an accident, the national economy is poorer because it loses whatever that worker could have produced during the rest of his or her working life (or during the period of disability, in case of injury). We have all seen magazine articles and health insurance ads that estimate the annual cost to the U.S. of heart disease, cancer, or alcoholism. In these articles, the costs are usually measured in terms of what happens to the productivity of working-age people who are disabled or killed by the disease in question, and these cost estimates are frequently—and successfully—used to justify spending large sums of money to help prevent, detect, and treat the illness. The same reasoning is valid in reckoning the cost of industrial accidents, including those that happen to coal miners. Our supply of manpower is one of our greatest national assets, and anything which diminishes or damages that supply levies a real cost on the nation.

This concept is easy to grasp when studying full-employment industries—including today's coal industry, in which current shortages of skilled manpower frequently mean that an injured or killed miner cannot be replaced except by hiring someone from another mining operation. In these cases, a mining accident has immediate and clear results: the loss of coal desperately needed in our energy-short economy. In industries where there is a pool of unemployed workers, an accident does not cause this immediate loss of current production—but the loss in *potential production* is just as real, and its result is just as serious for the long-run health of the U.S. economy.

One analyst, estimating the economic value of a human life lost in an airplane crash, included not only what the man (or woman) would have earned during the remainder of his working life, but also the value of the time he would have contributed to community service had he lived. Some economists feel that the loss of a person's contribution to family life is also an important cost of a fatal accident, and should be assigned a dollar figure.

In estimating the loss in potential production due to coal-mining accidents during the period 1970-2000, the authors have chosen to include *only* the amount of wages that would have been earned by dead and injured coal miners if no mining accidents had taken place. We have not included estimates of the value of a miner's services to his family and community. The result is a *very* conservative estimate of what our nation will lose in potential production as a result of coal-mining accidents during the last 30 years of the century.

The costs that are developed should be levied on the consumers of coal—and the levied money should promptly be spent by public agencies to cut down the accident rates in coal mines. There is much research that needs to be done on the causes and prevention of accidents, including the design of better ways of selecting miners, better ways of training them to do the job, and better equipment with which to do the job. Such research, including demonstration projects in different kinds of mines throughout the nation, could be a boon to miners and operators alike—and it could, if it is effective, soon lower this type of cost to the coal consumer. This is, in short, a cost which will eliminate itself if the funds are wisely spent.

Calculating Loss of Earnings

The next question immediately arises: How does one calculate the earnings that are lost as the result of accidents in the coal mines—and then convert this figure to a cost per ton of coal?

First, it should be noted that in making the cost calculations we have been forced to generalize. In real life, each mining accident is different, and the earnings that are lost will depend on the age of the miner, exactly what happened to him, what kind of work he does, how much he is paid—or how long he would have lived if the fatal accident had not occured. But since data are not available on all these variables, we must be satisfied with averages.

Second, in this analysis we do recognize that a mining accident can mean anything from a slight cut requiring just a bandage to an explosion or cave-in that causes death. In this discussion we shall exclude completely what are called nondisabling accidents—those in which the employee loses less than one day of work—because no detailed information is available on them. All other mining accidents will be divided into two major categories—nonfatal disabling and fatal—and separate costs will be calculated for each.

The method used here to develop per-ton figures is essentially a straightforward one that can be represented by the following fraction:

$$\frac{(1) \times (2) \times (3)}{8 \text{ billion tons}} = \text{Per ton cost of accidents in deep coal mines}$$

where (1) is the total number of accidents, 1970-2000; (2) is the average number of workdays lost per accident; (3) is the average daily earnings of coal miners; and 8 billion tons is the amount of coal that is projected to be deep-mined during 1970-2000.

As the reader follows the step-by-step computations that follow, first for nonfatal disabling accidents and then for fatal accidents, it may be helpful to refer back to this fraction occasionally to see exactly what is being done and why.

Loss of Earnings Due to Nonfatal Disabling Accidents—Item (1). What is the number of nonfatal disabling accidents that can be expected to take place in deep mines in the period 1970-2000? The latest complete accident data available from the Bureau of Mines are for the year 1971, when there were 52.81 nonfatal disabling accidents per one million man-hours in underground mines.[17]

Since the 52.81 rate is quoted for one million man-hours, the next step is to find out how many man-hours will be required to mine the 8 billion tons of deep-mined coal. According to Bureau of Mines statistics for 1971, the average productivity in underground mining was about 1.5 tons per man-hour;[18] at that rate it would take 5.333 billion (or 5,333 million) man-hours to mine the 8 billion tons.

We can now complete the calculations: 5,333 million man-hours times 52.81 accidents per million man-hours equals 281,636 nonfatal disabling accidents in deep mines during the period 1970-2000. For ease in subsequent calculation, this number will be rounded to 281,500. Thus, item (1) in the fraction is 281,500.

Item (2). What is the average number of days lost from work per accident? According to the latest (1970) figures available from the Mining Enforcement and Safety Administration (MESA), formerly a part of the Bureau of Mines, an average of 52 days was lost or charged for each nonfatal disabling accident in the mining industry.

The "lost or charged" terminology used by MESA has important meaning, since the term "nonfatal injuries" covers a lot of territory, ranging all the way from a bruised leg (which would be what is called a temporary disability, and would be charged for the actual number of days off from work)[19] through the loss of the first joint of a toe (a permanent partial disability that is charged at 35 days)[20] to the loss of both legs (a permanent total disability charged at 6,000 days).[21] (Some fifty to sixty assigned time charges were developed by the American National Standards Committee Z16 to measure the comparative severity of various kinds of industrial injuries. This committee is made up of representatives from industrial associations, labor unions, insurance associations, and departments of state and federal governments. The time charges that were developed were based on the committee's combined experience with industrial accidents of all kinds.)

Because there is such a range of severity in nonfatal accidents, a weighted average of 52 days was derived, as shown in Table 2.[22] Thus, item (2) in the fraction—the average number of days lost or charged as a result of nonfatal accidents—is 52.

Item (3). What are the average daily earnings in the coal-mining in-

Table 2

Average Number of Days Lost or Charged as Result of Nonfatal Disabling Accidents in Coal Mining, 1970

Type of Disability	Number of Injuries	Total Number of Days Lost or Charged	Average Number of Days Lost or Charged
Temporary	11,310	376,696	33
Permanent partial	234	176,810	756
Permanent total	8	48,000	6,000
Total nonfatal disabling	11,552	601,506	52

dustry? Preliminary Bureau of Labor Statistics data indicate that as of June 1973 the average gross earnings of miners in the coal industry were $232.93 per week,[23] or $46.59 per day on the basis of a five-day week. (The average work week on that date was 41.3 hours, or an average of 8.26 hours per day on a five-day-week basis.) For ease in calculation, the average earnings figure has been rounded to $47 per day. Thus, item (3) in the fraction is $47.

The formula can now be put together as shown below.

If the total cost of $688,000,000 is divided by the 8 billion tons of coal projected to be deep-mined during the period, we find that the average loss of earnings due to nonfatal disabling accidents in deep mines would be $.086 per ton of coal mined (at 1973 wage rates). This estimate of lost earnings can be used as a measure of the loss of potential production caused by such accidents.

Loss of Earnings Due to Fatal Accidents. Using the same formula, the per-ton loss of earnings as a result of fatal accidents in underground mines can be calculated as follows.

Item (1). Number of fatal accidents predicted for deep mines between 1970 and 2000. Bureau of Mines figures for 1971 show that during that year there were .84 fatal accidents per million man-hours, an improvement over the 1970 figure of 1.20.[17] Since mining 8 billion tons of coal will require 5,333 million man-hours, the projected number of fatal accidents during the 30-year period would be 5,333 times .84 or 4,480.

Item (2). Average number of days lost or charged per fatal accident. According to MESA, a fatal accident is charged as 6,000 days.[24] (This figure was also developed by American National Standards Committee Z16.)

Item (3). Average daily earnings in the coal-mining industry are $47 per day, as described earlier.

If this total cost of $1,263,360,000 is divided by the 8 billion tons of coal projected to be deep-mined during the 1970-2000 period, we find

Deep Mining

(1) Total Number of Accidents, 1970-2000		(2) Average Number of Days Lost or Charged/Accident		(3) Average Daily Earnings (1973 Wage Rates)		Total Loss of Earnings Due to Nonfatal and Fatal Disabling Accidents
Nonfatal 281,500	x	52	x	$47	=	$ 688,000,000
Fatal 4,480	x	6,000	x	$47	=	$1,263,360,000

that the average loss of earnings due to fatal accidents in deep mines would be $.158 per ton of coal mined (at 1973 wage rates). This loss of earnings can be used to measure the loss of potential production due to fatal accidents.

The combined loss of potential production (as measured by lost earnings) in deep mines during the 1970-2000 period is as follows:

Due to nonfatal disabling accidents	$.086 per ton
Due to fatal accidents	$.158 per ton
Total loss of potential production due to accidents in deep mines	$.244 per ton

Further Deep-Mining Costs

As of May 1973, black lung payments to miners and their widows had cost $1 billion, and were expected to top $8 billion by 1980. These are costs, however, that are the result of the way deep mining was done in the past. Presumably the future is going to be different—and better. Federal standards under the 1969 Mine Health and Safety Law required that by the end of 1972 the level of respirable dust could not exceed 2 milligrams per cubic meter of air; the Bureau of Mines reported in the summer of 1973 that coal dust had been reduced to acceptable levels in 93 percent of the mines in the nation.[25]

The standards in the 1969 law were set after consultations with experts and presumably they are rigorous enough to ensure that new miners coming into the industry will not contract the disease. If these standards are rigorous enough, and if they *are* enforced, the social cost of coal deep-mined during the last 30 years of the twentieth century will not include any costs attributable to black lung. If the standards are not rigorous or not enforced, it will be quite another story.

In this analysis we will assume the best, and therefore not include in the cost table a charge for black lung disease. But this is one entry that merits the closest scrutiny as more experience and data are accumulated on miners' health under the new law.

We have already discussed one of the costs of coal that is very real but which cannot be translated into national dollar figures: the gaps in restitution (medical and income benefits) to miners and their families after accidents. These gaps exist in both deep mining and surface mining.

There are other costs that fall in this same unquantifiable category: the lowered property values near mine sites (because of noise, dust, and general unsightliness) and the damage to roads and highways caused by heavy equipment and coal trucks. These latter two social costs are usually greater in surface mining than in deep mining, but they exist in both. They appear in Table 3 with question marks in the cost columns.

The estimated social costs of deep-mined coal are summarized in Table 3. They add up to $.873 per ton, or a total of $6.982 billion for the 8 billion tons projected to be deep-mined during the period 1970-2000.

SURFACE MINING

Land Reclamation

Without proper reclamation, thousands of square miles of surface-mined "moonscape" will be left as a legacy to future generations; large portions of some Appalachian states already have this appearance. Thanks to the work of many fine photographers, we all now know what surface mining does to the land, and most people in the U.S. would agree without hesitation that if there is going to be surface mining there should also be good reclamation afterward. There is great disagreement, however, as to what "good reclamation" means and how much it costs.

In 1971, Assistant Secretary of the Interior Hollis Dole estimated that it would cost an average of *$500 per acre* for what he called "basic reclamation" (grading, revegetation, and drainage control) of the approximately 2.3 million acres of unrestored surface-mined land that existed in 1970. In the same testimony, he pointed out that Bureau of Mines data on reclamation work in the Appalachian states showed an average cost of *$1,100 per acre* to return surface mined land to productive use, with some special projects running as high as *$15,000 per acre.*[26]

In 1971, William Guckert, director of the Mine Reclamation Division in Pennsylvania's Bureau of Land Protection and Reclamation, reported that the cost of reclamation performed by mine operators in Pennsylvania averaged from $250-$500 per acre, with some costs of $750 per acre in high terrain.[27]

In a U.S. Bureau of Mines project in Pennsylvania in 1966, the cost of grading and filling a strip-mined area ranged from $912 to $2,770 per acre depending on the methods used. According to estimates by the Center of Science in the Public Interest, reforestation or planting of forage for grazing could add *another $250 per acre.*[28]

In its 1972 calculation of the total cost of cleaning up our national environment, the Council on Environmental Quality used a figure of *$2,000 per acre* as the average cost of reclaiming land disturbed by surface mining.[29]

Ernest Preate, president of Help Eliminate Life's Pollutants (HELP) a citizen conservation group in Scranton, Pennsylvania, reported costs of $2,000 an acre (for backfilling, grading, and landscaping 120 acres of surface-mined land near Scranton), $5,730 an acre (for rough grading,

Table 3

Estimated Social Costs of Coal, 1970-2000

	Costs/Ton (in dollars)	Total Cost (in billions of dollars)
Deep-mined coal **(8 billion tons)**		
Mine fires	—	—
Disposal of mine wastes	$ .0700	$ .560
Control of acid mine drainage	.0587	.470
Prevention of subsidence	.5000	4.000
Loss of potential production due to accidents	.2440	1.952
Black lung disease	—	—
Other costs		
Gaps in restitution to miners and their families after accidents	?	?
Lowered property values near mine sites	?	?
Damage to roads and highways	?	?
Total	.8727	6.982
Surface-mined coal **(12 billion tons)**		
Land reclamation	.5000	6.000
Control of acid mine drainage	.3932 max.	4.718
Control of sedimentation	—	—
Diminished recreational value of land	.0676	.810
Loss of earnings due to accidents	.0420	.504
Other Costs		
Gaps in restitution to miners and their families after accidents	?	?
Lowered property values near mine sites	?	?
Damage to roads and highways	?	?
Total	1.0028	12.032

compacting, and controlled filling of a 20-acre tract to be used as the site of a vocational-technical school in Lackawanna County, Pennsylvania), and $8,000 an acre (for site preparation work on 27.4 acres of surface-mined land in Norton, Virginia which were used as the location of an elementary school).[30]

The Soil Conservation Service reports costs of $2,000 an acre for reclamation work performed by an independent mine operator near Catoosa, Oklahoma, where the land is as flat as a tabletop.[31]

The Environmental Protection Agency reports that the per-acre cost of regrading and revegetating surface-mined land includes the following: $500 to $2,500 per acre for regrading,[32] $50 for fertilizer,[33] $60 for liming to reduce soil acidity,[33] and $200 to $800 for seeding.[34] No costs were estimated for stockpiling topsoil,[35] for surface preparation (raking, reburying toxic and dark-colored materials, and breaking up compacted surfaces)[36] or for irrigating dry areas,[37] but these costs would be high in many mining operations.

Cost of reclamation is obviously a slippery statistic. It not only depends on all the obvious variables—the topography, the soil and the geology of the area affected, the depth of the pit, or the size of the cut. It also depends on what standards must be met. Does "good reclamation" mean just restoring the land to useful condition—any use at all? Is it enough for example, to restore former cropland to a condition such that it is suitable only for pasture? And what attention do we pay to the aesthetics of the situation? Some of the mountains of Appalachia, for example, had no specific economic use before they were surface-mined—but they were an important part of the scenery, and many people enjoyed them.

In calculating the social costs of coal, the standards of acceptable reclamation are beyond dispute: after reclamation is complete, the surface-mined land must be at least as economically productive and attractive to the eye as it was before mining began.

On this basis, of course, there is some land that should never be surface-mined at all, since it can never be fully restored, no matter how much money is spent. That is why several of the surface-mining bills that have been proposed in Congress have prohibited surface mining in steepslope areas, such as those common in Central Appalachia. In the western part of the U.S., where the coal seams are often very thick, there are other problems. "If you remove a 30-foot thickness of earth," says David Maneval, "surface mine a 100-foot-thick seam of coal, and then just put the earth back, as is frequently done in parts of the West, you finish up with an enormous hole."[11] This is not acceptable reclamation by any standard.

The authors hope that federal legislation will soon prohibit surface mining in areas where good reclamation is impossible. If this happens, the average per-acre cost of reclamation in the U.S. will go down. As of now, however, the average must include what it costs to do everything possible to restore these lands, even though the final results will not be satisfactory. Difficulty must not be used as an excuse for doing nothing.

On this basis, then, how do we decide which reclamation cost figure to use? The U.S. averages that are generally available are not satisfactory. They are based heavily on experience in Pennsylvania, where most of the surface-mined land is flat or rolling and rainfall is adequate—two great advantages in reclamation work. Perhaps even more important, the quality of reclamation varies widely from one part of the U.S. to another. Since there has to date been no quality control on a national level, averaging out reclamation costs in the U.S. today is almost a meaningless exercise.

So in developing an average per-acre reclamation cost for this article, we have chosen to go another route: to find a place where large-scale reclamation work has been done over a long period of time, where the caliber of the work has been high enough to return surface-mined land to truly productive use—and then to find out how much that kind of reclamation costs.

Germany's Example

The place is West Germany, where fuel needs were and are such that "it is not possible to consider seriously the luxury of banning the surface mining of brown coal. Instead, methods of mining and land restoration had to be developed which would permit continued production of brown coal without incurring serious environmental damage."[38]

German reclamation in the brown-coal fields of the Rhineland begins before the mining starts. Detailed plans are made for the evacuation and relocation of populated settlements and for the restoration of land after the mining is over. Land-use patterns are proposed and approved far in advance, and the post-mining landscape is planned accordingly. With this kind of advance planning, the mining operations can be tailored to fit the land restoration work that is to follow—instead of vice versa, as is so often the case in the U.S. During the mining, huge wheel excavators selectively strip off and save the top layer of loess, a form of topsoil—the extremely fertile loam which lies at the surface. The remaining sand, gravel, and clay overburden is then removed, and the coal is extracted. After mining is completed, mammoth spreaders are used to fill the overburden back into the mined-out pits, which are then leveled off with bulldozers. The loess, which has been combined with water to make a slurry, is then reapplied to the surface. The slurry dries out after several months, leaving behind a three-to-six-foot-thick layer of fertile topsoil on which almost anything can grow. Grain and hay, for example, thrive on land restored less than five years earlier.[39]

The institutional arrangements used by the Germans for planning, regulating and enforcing this program are of particular interest, and although there is not enough space to discuss them here, we strongly recommend that they be studied by any person or group interested in reducing the environmental damage we now suffer from surface mining.[40]

The German program has been operating slightly more than twenty years. Under its aegis, 33,000 acres of land have already been restored for forestry, agriculture and recreational uses. Officials there report that the cost of restoring mined-out lands to full agricultural productivity ranges from $3,000 to $4,500 per acre.[41]

Experience in Great Britain, where strip-mining regulations are very strict and reclamation is first-rate, bears out the German cost figures. A Sierra Club study shows that British costs range between $1,350 and $7,542, for a rough average of about $4,400 per acre—clearly in the same range as the West German costs.[42]

On the basis of this German and British experience, therefore, a reclamation cost of $4,000 per acre will be used in this analysis. Of course there are differences between the work done in Europe and that carried out in the U.S. Few U.S. reclamation projects, for example, would involve transplanting whole towns, as was sometimes the case in West Germany; so in this regard the American work might be done more cheaply. On the other hand, however, a great deal of surface mining in the U.S. takes place in mountainous areas, where reclamation is far more expensive than in either Germany or Britain; strip-mining in those countries is permitted only on relatively flat terrain.

On balance, the $4,000-per-acre figure is reasonable—assuming that we insist on first-class reclamation of our surface-mined lands. If we try to do it for less, we will get just what we pay for: lands that are far less productive and frequently far less beautiful than they were before mining took place.

Per-Ton Cost of Reclamation

To calculate the total cost of reclamation, we must know not only the cost per acre, but how many acres will be affected. The Bureau of Mines has estimated that the average thickness of bituminous and lignite coal seams surface-mined in 1970 was 5.7 feet,[43] no corresponding figure is available for anthracite. If the average density of surface-mined coal is 1,800 tons per acre-foot,[44] this means that the coal content of surface-mined seams would average 10,260 tons per acre (1,800 tons per acre-foot times 5.7 feet). With an 80 percent recovery factor (about 20 percent of the coal in a seam is lost or wasted in the mining process),[45] the actual yield would average 8,200 tons of coal per acre.

This means that approximately 1.5 million acres will have to be stripped to obtain the 12 billion tons of surface-mined coal projected for the 1970-2000 period. Using the reclamation cost of $4,000 per acre, the total cost of reclamation will be $6.0 billion, or $.50 per ton.

How will this per-ton cost estimate be affected by today's coal rush to the American West, where thick seams of low-sulfur coal (frequently yielding from 40,000 to 80,000 tons per acre) lie relatively close to the surface? It is too soon to know what the per-ton cost of reclamation will be in these surface-mining operations. In a recently published report by the National Academy of Sciences (NAS) to the Energy Policy Project of the Ford Foundation, a panel of scientists flashed some warning signals. The major findings of the report were summarized in a magazine article by Thadis W. Box of Utah State University, chairman of the study committee:[46]

"In the western coal areas, restoration (which the panel defines as duplicating the pre-mining conditions of the site) is rarely, if ever, possible. Rehabilitation, however, should be performed for all stages of mining."

Note that the definition of "rehabilitation" used by the panel contains some rather elastic terms, as indicated by authors' italics: " 'Rehabilitation' as used in this report implies that the land will be returned to a form and productivity *in conformity with a prior land use plan* including a stable ecological state that does not contribute *substantially* to environmental deterioration and *is consistent with* surrounding aesthetic values."

Even if one accepts this "rehabilitation" as an appropriate standard, there is no guarantee of success. "Those areas receiving ten inches or more of annual rainfall can usually be rehabilitated," writes Professor Box, "provided that evapo-transpiration is not excessive, if the landscapes are properly shaped and if techniques that have been demonstrated successful in rehabilitating disturbed rangeland are applied. . . . The drier areas, those receiving less than ten inches of annual rainfall or with high evapo-transpiration rates, pose a more difficult problem. . . . Where natural revegetation of a disturbed site may occur in five to twenty years on a high-rainfall eastern U.S. site, it may take decades or even centuries for natural revegetation to occur in a desert. . . . Arid ecosystems are delicately balanced and, once disturbed, may never reach natural equilibrium on a time scale acceptable to modern man. . . . Revegetation of these (arid) areas probably can be accomplished only with major, sustained inputs of water, fertilizer and management."

But this is not the whole story. "Rehabilitation of mined lands requires more than achieving a stable growth of plants," Professor Box continues. "In these other aspects—wildlife, aesthetics, erosion control and water quality—pertinent data for rehabilitating mined land are virtually nonexistent. The necessary research has barely begun. . . . In the West the absence of a climatic 'safety factor' means that even supplying all the requirements will not guarantee success. Instead, successful rehabilitation requires a commitment to the proper application of proven techniques at all the critical times."

In short, even the somewhat fuzzy goal of "rehabilitation" is not going to be easy to attain in the West, and its cost and success are still uncertain.

There is still a third standard defined in the NAS report: "reclamation," which is something less than complete restoration but something more than mere rehabilitation. It is defined as "making the site habitable to organisms that were originally present, or others that approximate the original inhabitants," a standard of acceptability very similar to that used by the authors of this article. If we insist on this standard for surface-mined land in the West, cost and success are even greater question marks—and we need to start finding the answers before we heedlessly allow any more of the landscape to be damaged by strip-mining.

Control of Acid Mine Drainage

In surface mining, most of the acid mine drainage originates in the spoil banks—the piles of dirt, rock, and shale which are stripped from the surface to get at the coal underneath. These spoil banks are potent generators of acid for a number of reasons:[47]

1. In surface mining there is great exposure of the layers of shale, which are heavy producers of sulfuric acid when they are exposed to air and water. Since these shale layers are usually found directly above and below the coal seam, they are customarily the last to be stripped off, and hence they usually finish up at the top of the spoil pile, where they can do the greatest damage.

2. Even if the shale is not at the top of the pile, spoil banks encourage formation of acid mine drainage because the loose material in them allows both air and water to percolate down ten feet or more.

3. Higher temperature produces more acid water, and during the summer months the temperature in spoil piles is very high.

4. It is hard to control the acid produced by surface mines before the acid water from spoil banks can flow in all directions, down off and down through the spoil.

In 1970 the total cost of treating all the then-existing acid mine drainage was estimated by the Department of the Interior to be $6.6 billion,[10] of which $5.78 billion was attributed to the mines themselves, as calculated earlier in this article. Of this cost, 30 percent,[12] or about $1.73 billion, was due to surface mining. From the beginning of coal mining until

1970, 4.4 billion tons of coal had been surface-mined.[48] The average cost of abating acid drainage would therefore be $1.73 billion divided by 4.4 billion tons, or $.3932 for each ton of coal surface-mined during that long period.

If proper land reclamation is carried out, as we have already discussed in this article, will the amount of acid mine drainage in surface mining be reduced to zero? The answer to this question depends on the kind of reclamation being talked about. The work done in West Germany did not include any special provisions for preventing acid mine drainage since there are no sulfur-bearing minerals in the brown-coal fields of the Rhineland.[49] In areas where the sulfur-bearing shale layers do exist, the problem is unavoidable. In the words of David Maneval:[50]

"In strip mining, the union between coal, air, and water is almost inevitable. Coal seams are exposed to the air during stripping operations, and rainfall furnishes the water, which frequently runs down the hillside into the area where the stripping is going on."

Special actions are therefore required to prevent formation of acid mine drainage. First, the sulfur-bearing strata must be segregated, and this requires great care since they are usually the layers which lie nearest the coal seams. Then this sulfur-bearing material must be put at the bottom of the strip pit, and the top layers of the piles must be compacted tightly to prevent air and water seepage.[47] Simultaneously, all water must be diverted from the mining operations, and if any water accumulates in the pit as the result of direct rainfall, it must be pumped out and treated immediately.[51]

In short, reclamation must include some very specific practices if it is to prevent acid mine drainage altogether, and most of these practices were not included in the German work because the problem did not exist there.

There is no question, however, that even without these specific practices, the amount of acid mine drainage will be cut down markedly by what is called basic reclamation work—putting the spoil back after mining is completed, restoring the topsoil, and revegetating the area. So if any kind of decent reclamation work is done, the cost of controlling acid mine drainage will be considerably less than the $.393 per ton estimated cost for past mining—precisely how much less is not known at present.

For these reasons, $.393 a ton will be the maximum cost of acid mine drainage for surface-mining coal, and the cost figure will be so labeled.

Control of Sedimentation

Another major water pollution problem that results from strip-mining is sedimentation. Strip-mined areas continuously erode, filling streams and

rivers with sediment which impedes the flow of water, fills the stream channels and promotes flooding, coats stream bottoms and prevents the growth
of aquatic plant and animal life, fills reservoirs and impoundments, clogs
public water systems and transmits pathogenic organisms. According to
studies by the U.S. Geological Survey, erosion and sedimentation rates on
strip-mined land are 500 times as great as those on neighboring land that
has not been stripped.

Furthermore, erosion also loads water with large quantities of other
minerals, such as manganese, aluminum, ammonium, magnesium, calcium,
potassium, and sodium. These minerals are needed in the stripped soil to
encourage revegetation; without them the soil is not hospitable to new
growth. Maybe even worse, when the minerals collect in runoff water in
toxic concentrations, they destroy fish and plant life wherever these
waters flow.

Unfortunately there have been no specific and detailed analyses of what
it costs to correct this particular impact of surface mining. Removing sediment from streams—or preventing it from going there in the first place—
have been viewed as integral parts of land reclamation, and it has been
assumed that the cost of reclamation covers the cost of coping with sedimentation as well.

This is probably a safe assumption if the reclamation is really first-rate.
If it is not then specific actions must be taken to prevent erosion and remove sediment from mine water, and those actions cost money. Since the
reclamation analyzed and recommended in this article is of high quality
and includes a high degree of concern for these problems, no separate cost
will be assessed for control of sedimentation.

Diminished Recreational Value

One cost of surface mining that is frequently forgotten is the diminished
recreational value of disfigured land and polluted water. Hunters, campers,
hikers, fishermen, and swimmers spend money in pursuit of their sports,
and this expenditure is lost or greatly diminished when areas are ravaged
by surface mining.

A study published in 1968 by the Bureau of Sport Fisheries and Wildlife in the Department of the Interior stated that approximately 3.2 million acres of land had been surface-mined (for all minerals, not just coal)
in the U.S. from the beginning of mining through 1967. Two million of
these acres (or about two-thirds) were fish and wildlife habitat damaged
by the mining.[52] The study then estimated that, if there were "minimal
or basic reclamation that would control silt, sediment, and acid pollution"
on these 2 million acres, two changes would occur: the number of man-

days of use of the land by sportsmen would increase from 7.5 million to 40 million per year, and the recreational value of the land would therefore increase from $10 million to $58 million per year.[53]

In other words, reclaiming 2 million acres of fish and game habitat would give a payoff (in terms of increased use by sportsmen) of $48 million per year, or $24 per acre per year. Or, worded still another way, whenever an acre of surface-mined fish and game habitat is unreclaimed, it costs society $24 per year in terms of its diminished recreational value.

How do we translate these figures into a per-ton cost for surface-mined coal? Taking it step by step:

1. We have already shown earlier in this article that an estimated 1.5 million acres will have to be stripped to obtain the 12 billion tons of surface-mined coal projected for the period 1970-2000. On the basis of past experience, this would mean that roughly two-thirds of this acreage, or 1 million acres, would be fish and game habitat.

2. More than just the 1 million acres of habitat will be affected by the stripping however, since use of adjacent land is also often impaired. The Bureau of Sport Fisheries and Wildlife has reported that for every acre of land disturbed in Appalachia by surface mining, use of an additional 1.25 acres has been impaired[54]—denied to hunters, campers, hikers, and fishermen because of highwalls, piles of mine refuse and tearing up of the landscape by machinery and trucks. So the surface mining of 1 million acres of fish and game habitat during the 30-year period would result in the impairment of an estimated total of 2.25 million acres.

3. If each unreclaimed acre of habitat costs society $24 per year in reduced recreational use, the total cost of 2.25 million impaired acres would be $54 million ($24 times 2.25 million) for each year these acres are being mined or are not yet reclaimed after mining is completed.

4. The next question is: How many years, on the average, will these surface-mined acres be out of commission as far as hunters and fishermen are concerned? Harry Perry estimates that an average of three to five years elapses from the time mining begins until reclamation can be started.[55] But this is only the beginning. Some time is required to do the reclamation itself, although concurrent reclamation—reclaiming as you go during the mining operation—is more and more common. Even more significant, a long period of time must pass before the ecological balance is restored so that fish and game are again available to sportsmen. Many things must happen, and some take longer than others. Fish populations must be replenished by introducing native species into the streams or ponds. There must be a regrowth of ground cover dense enough to attract native animals back to the mined areas—and this means time to grow bushes and small trees as well as grass. These things do not happen quickly. Willard M.

Spaulding, Jr., deputy regional director of the U.S. Fish and Wildlife Service, estimates that after first-rate reclamation work has been completed, it still takes an additional ten years, on the average, to develop an ecosystem that will allow sportsmen to harvest fish and game.[56] Thomas Flynn of the Division of Environment, Bureau of Mines, confirms this estimate.[57]

On the basis of this testimony, we assume in our analysis that it takes a total of fifteen years, on the average, to surface-mine habitat land, to do first-class reclamation, and then to restore the fish and animal populations so that sportsmen will again find the mined areas attractive.

We are now ready to calculate the per-ton social cost that results from the diminished recreational value of fish and game habitat. Remember that the total loss of recreational value has been estimated to be $54 million for each year the 2.25 million surface-mined acres were not available to sportsmen. If we multiply this $54-million annual cost by fifteen years (the average "out-of-commission" time), the result is a total cost of $810 million, or $.0676 per ton for the 12 billion tons of surface-mined coal projected for the period 1970-2000.

Loss of Potential Production

In calculating the loss of potential production due to accidents in surface mining, the method used is the same as that used for deep mining. First, the loss will be calculated for nonfatal disabling accidents.

Item (1). Number of nonfatal disabling accidents predicted for surface mines between 1970-2000. Bureau of Mines figures for 1971 show that there were 25.86[17] nonfatal disabling accidents per million man-hours in surface mines. The Bureau of Mines also reported in 1971 the average productivity in surface mining was approximately 4.5 tons per man-hour.[18] At that rate, it would take 2.667 billion (or 2,667 million) man-hours to mine the 12 billion tons of surface-mined coal projected for that period.

The number of nonfatal disabling accidents would therefore be 2,667 times 25.86, or 68,970. For ease in calculation this number will be rounded to 69,000. Thus, item (1) in the formula is 69,000.

Item (2). The average number of days lost from work per accident is 52.[22]

Item (3). The average daily earnings in coal mining are $47.[23]

Calculation of the formula is as shown on page 223.

If the total cost of $168,000,000 is divided by the 12 billion tons of coal projected to be surface-mined during the 1970-2000 period, we find that the average loss of earnings due to nonfatal disabling accidents in surface mines would be $.014 per ton of coal mined (at 1973 wage rates). As was true for deep mining, this estimate can be used as a measure of the loss of potential production caused by such accidents.

The loss of earnings from fatal accidents is calculated as follows.

Item (1). Number of fatal accidents predicted for surface mines between 1970-2000. Bureau of Mines figures for 1971 show .45 fatal accidents per million man-hours.[17] Since surface mining 12 billion tons of coal will require 2,667 million man-hours, the total number of fatal accidents would be 2,667 times .45, or 1,200. Item (1) in the fraction is 1,200.

Item (2). The average number of days lost or charged per fatal accident is 6,000.[24]

Item (3). The average daily earnings in coal mining are $47.[23]

Calculation of the formula is as shown below.

If the total cost of $338,400,000 is divided by the 12 billion tons of coal projected to be surface-mined during the 1970-2000 period, we find that the average loss of earnings due to fatal accidents in surface mines would be $.028 per ton of coal mined (at 1973 wage rates). This loss of earnings can be used to measure the loss of potential production due to fatal accidents.

The combined loss of potential production (as measured by lost earnings) in surface mines during the 1970-2000 period is as follows:

Due to nonfatal disabling accidents	$.014 per ton
Due to fatal accidents	$.028 per ton
Total loss of potential production due to accidents in surface mines	$.042 per ton

The estimated social costs of surface-mined coal are summarized in Table 3. They add up to $1.033 per ton, or a total of $12.032 billion for the 12 billion tons projected to be deep-mined during the period 1970-2000.

Surface Mining

(1) Total Number of Accidents, 1970-2000		(2) Average Number of Days Lost or Charged/Accident		(3) Average Daily Earnings (1973 Wage Rates)		Total Loss of Earnings Due to Nonfatal and Fatal Disabling Accidents
Nonfatal 69,000	x	52	x	$47	=	$ 168,000,000
Fatal 1,200	x	6,000	x	$47	=	$ 338,400,000

EFFECT ON COST AND PRICE

How must would the cost of producing coal increase if the social costs listed in Table 3 were added? It is impossible to answer this question, for the simple reason that it is impossible to get any authoritative information as to what production costs are in the coal industry. Coal companies consider production costs to be confidential information, and the Bureau of Mines does not gather data on the subject.

We can, however, indicate what the effect would be if social costs were added to average selling price, and this effect is shown in Table 4: an increase of roughly 9 percent in the price of deep-mined coal and an almost 18 percent increase for surface-mined coal.

Even here, we are limited by the data available. The calculations in Table 4 are based on average U.S. selling price (f.o.b. the mine) during 1972, the latest year for which average figures are available from the Bureau of Mines.[2] Domestic coal prices have subsequently skyrocketed; Harry Perry reports prices as high as $25 to $35 a ton for low-sulfur coal during early 1974.[55] Of course, as selling prices rise, the burden of paying social costs becomes relatively less. If deep-mined coal is selling for $9.70 per ton, as in 1972, and you add $.88 in social costs, the price increase is 9 percent. If the selling price goes up to $35 per ton, however, adding the $.88 will up the price by only about 2 percent.

A WORD OF CAUTION

The figures in Tables 3 and 4 must be thoroughly understood and cautiously used. In the first place, the various costs are not completely independent of each other, as we have emphasized several times. For example,

Table 4

Effect of Adding Estimated Social Costs to the 1972 Selling Price of Coal

Type of Coal	Average U.S. Selling Price/Ton*	Social Cost/ Ton	Total Selling Price Including Social Costs	Percentage Increase In Price
Deep-mined	$9.70	$.881	$10.57	9.0%
Surface-mined	5.65	1.006	6.65	17.8%

*Source: Personal conversation with L. W. Westerstrom, Bureau of Mines, U.S. Dept. of the Interior.

effective backfilling of deep mines to prevent subsidence would also affect the amount of acid mine drainage from those mines. Adequate reclamation of a strip-mined area would reduce the acid mine drainage, the sedimentation and the loss of income from recreation. To make a proper analysis of these interrelationships would require more cost data and an extremely detailed and complicated econometric model, both of which we hope will soon be available.

In the second place, all of the estimated social costs are based on certain assumptions, and some of these costs might be changed markedly if reality does not correspond with these assumptions during the last 30 years of this century. Some examples:

A change in deep-mining technology—increased use of the longwall method, for instance—would probably result in a marked decrease in the per-ton cost of preventing subsidence, the largest single social cost for deep-mined coal. Many U.S. coal companies have already made this switch, and the rate of conversion is expected to increase.

Increased productivity in deep mining, surface mining, or both, would decrease the per-ton costs of accidents. If a given number of tons of coal can be produced with fewer man-hours, and if the number of accidents per million man-hours remains the same, there will be fewer accidents—and a smaller social cost.

If the new regulations controlling the permissable levels of respirable mine dust are not properly enforced, black lung costs (which are not given a dollar figure in Table 3) would have to be added into the total and that might cause a sizable increase in the social cost of deep-mined coal.

In the third place, the average per ton cost of strip-mine reclamation in the U.S. may be affected if surface mining continues its phenomenal rate of growth in the West—but no one knows for sure just what the effect will be. Other things being equal, the thick seams and resulting high yields would result in lower per-ton reclamation costs. But other things are not equal. There are some special reclamation problems in the West: those big holes that are left when the deep seams are mined out, the limited moisture, and the very fragile ecosystems that may take decades or even centuries to restore. As a result, the cost and success of western reclamation are still question marks.

In the fourth place, some of the coal-mining states have already taken action to pay some of the social costs of coal. They do it by collecting taxes and fees of various kinds from the coal operators and then spending these revenues—and sometimes other revenues as well—to prevent or repair the human and environmental damage caused by the mining. The four most common sources of state revenue are: permit fees (flat sums paid to the state for all new mining operations), acreage fees (so much for each acre of land disturbed by mining), severance taxes (so many cents for each ton of coal mined), and taxes on gross or net income from coal mining.

Unfortunately, most of the states have been much more interested in collecting the revenues from coal mining than in repairing the damage caused by it. As a general rule, the fees and taxes collected from coal operators have been put into the states' general revenue funds, where the money is used to pay for the regular functions of state government— many of which have little or nothing to do with the land or the people directly affected by the mining.

There are exceptions. Both West Virginia and Maryland, for example, collect acreage fees and earmark the funds for use in reclaiming abandoned or "orphan" lands damaged by mining. The Pennsylvania legislature has appropriated state funds and authorized state bond issues to pay for mine restoration projects. Several states use the acreage or permit fees they collect to pay for the enforcement agency that monitors the coal industry in the state.

Despite many inquiries, the authors were not able to obtain even an estimate of how much state revenue comes from coal mining nationwide— let alone any hard facts about how much of that revenue is used to pay social costs. From the skimpy information that is available, one thing is certain: since the social costs of coal will be an estimated $600 million per year during the last 30 years of this century, the current level of state spending for reclamation and enforcement is only a very small drop in a very large bucket.

On balance, the social costs summarized in Table 3 are probably too low. They are based on some assumptions that may very well not work out quite that way—the assumptions that health and safety standards will always be rigorously enforced and that no new mine fires will break out, for example.

Even more important, the cost figures are conservative because of the impact of inflation. They are based primarily on past experience—which means that many of them are already understatements of what these same costs would be today. By the end of the century, if current inflationary trends continue, the costs shown in Table 3 will have increased substantially.

With all the ifs, ands and buts, however, Table 3 does begin to give us a more accurate picture of what a ton of coal really costs—and hence what consumers should be paying for it. Further and more elaborate research on this question should be a high-priority item on our list of national needs.

STILL COMPETITIVE

Would coal still find a ready market if the selling price were increased by the social costs developed in this article? The answer, according to David Maneval is definitely "yes."

The key, of course, is the price of alternative fuels, particularly those used to produce electricity, since this is the only use of coal that is subject to serious competition from other fuels.[58]

In producing electricity, there are currently four possible fuels that can be used to produce heat to run the generators: coal, oil, gas and nuclear fuel. "The importance of nuclear energy is in the future, but not now," says Maneval. "Natural gas isn't a real alternative because it's in such scarce supply. So that leaves oil and coal. And between these two, coal has an enormous cost advantage. To make a cost comparison that means something in practical terms, the cost per ton of steam coal (the type of coal used to generate electricity) and the cost per barrel of crude oil must both be translated into the cost of producing a given number of BTUS. (A BTU, or British thermal unit, is a measure of heat.) At the May 1974 average market prices of steam coal ($10 per ton) and crude oil ($10 per barrel), it costs much more to produce 1 million BTUS of heat with oil than it does with coal—four to five times as much, in fact."

There is no question about it. Coal can compete with oil, even if the selling price of coal is raised to cover the social costs developed in this article. As a matter of fact, these social costs could increase substantially before coal's competitive advantage would be lost. Pay-as-you-go coal has always been an ethical imperative. It can now be a practical reality.

NOTES

1. Perry, Harry, "Environmental Aspects of Coal Mining," *Power Generation and Environmental Change*, Berkowitz and Squires, eds., MIT Press, 1969, p. 323.

2. Personal conversation with L. W. Westerstrom of the Bureau of Mines.

3. *Surface Mining*, Hearings before the Subcommittee on Mineral, Material and Fuels of the Committee on Interior and Insular Affairs, U.S. Senate Nov.-Dec. 1971, p. 215.

4. Maneval, David, "Coal Mining vs. Environment: A Reconciliation in Pennsylvania," *Appalachia*, Feb.-Mar. 1972, p. 35.

5. Perry, Harry, and Harold Berkson, "Must Fossil Fuels Pollute?," *Technology Review*, Dec. 1971, p. 215.

6. Perry, op. cit., p. 325.

7. Ibid., p. 331.

8. Ibid., p. 332.

9. Ibid., p. 334.

10. Ibid., p. 335.

11. Personal conversation with Dr. David Maneval.

12. Perry, op. cit., p. 330.

13. Ibid., p. 323.

14. Personal conversation with Dr. David Maneval. A description of the four projects is contained in a paper given by him at the annual meeting of the American Institute of Mining Engineers in February 1967. Readers who wish to obtain copies of the paper may write to Dr. Maneval at the Appalachian Regional Commission, 1666 Connecticut Ave., N.W., Washington, D.C. 20235. Cost figures for treatment of acid mine drainage are contained in a study (funded by the Appalachian Regional Commission) by Michael Baker, Jr., Inc. Information about the study is available from Dr. Maneval at the above address.

15. *Surface Mining,* op cit., pp. 592, 594.

16. Personal conversation with James O'Brien of AFL-CIO.

17. "Injury Experience and Worktime in the Solid Mineral Mining Industries, 1970-71," Mineral Industry Survey, U.S. Dept. of the Interior, Bureau of Mines, Aug. 1972, p. 8.

18. Average productivity for 1971 (the latest year for which national data were available when this article was completed) is based on data for bituminous and lignite: shown in "Weekly Coal Report No. 2898," Mineral Industry Surveys, Bureau of Mines, Table 8, p. 14. This was done since the data on anthracite (shown in the *Minerals Yearbook, 1971,* Bureau of Mines, Table 21, p. 398) shows only an average and does not distinguish between deep-mined and surface-mined coal. Productivity in anthracite mining is significantly lower than in bituminous and lignite, but it is also of relatively minor significance, since anthracite accounted for only 14.5 percent of total coal production during 1971 and has declined further in subsequent years (down to 11.2 percent in 1972), as shown in "1972 U.S. Energy Use Continued Upward," a press release from the Bureau of Mines dated March 10, 1973, p. 3, Table 1).

19. *Method of Recording and Measuring Work Injury Experience.* ANSI Z16. 1-1973, American National Standards Institute, Inc., 1973, pp. 8, 12.

20. Ibid., p. 10.

21. Ibid., pp. 7, 9.

22. *Injury Experience in Coal Mining, 1970.* Bureau of Mines Information Circular/1973, U.S. Dept. of the Interior, Table 5, p. 37.

23. "Employment and Earnings," vol. 20, no. 2, U.S. Dept. of Labor, Bureau of Labor Statistics, Aug. 1973, p. 82.

24. *Method of Recording and Measuring Work Injury Experience,* op. cit., p. 9.

25. "Black Lung Victims Fail to Change Jobs," *The Washington Post,* May 29, 1973, p. A18.

26. *Surface Mining,* op. cit., p. 602.

27. Ibid., p. 307.

28. Nephew, E. A., "Healing Wounds." *Environment,* 14:15, Jan./Feb. 1972.

29. *Environmental Quality,* third annual report of the Council on Environmental Quality, Aug. 1972, p. 275.

30. Preate, Ernest D., Jr., "A New Law for an Old Problem," *Appalachia,* Feb./Mar. 1972, p. 48.

31. Croom, Dan F., "Last of a Vanishing Breed," *Soil Conservation,* Jan. 1974, p. 9.

32. *Processing Procedures and Methods to Control Pollution from Mining Activities,* U.S. EPA, Oct. 1973, pp. 113, 121.

33. Ibid., p. 171.

34. Ibid., p. 177.

35. Ibid., p. 165.

36. Ibid., p. 166.

37. Ibid., p. 181.

38. Nephew, op. cit., p. 17.

39. Ibid., p. 18.

40. Ibid., p. 21.

41. Ibid., p. 19.

42. Greensburg, William, "Chewing It Up at 200 Tons a Bite: Strip Mining," *Technology Review,* Feb. 1973, p. 52.

43. "Thickness of Bituminous Coal and Lignite Seams Mined in 1970," unpublished Bureau of Mines Information Circular. Information provided by L. W. Westerstrom of the Bureau of Mines, U.S. Dept. of the Interior.

44. Nephew, E. A., "Surface Mining and Land Reclamation in Germany," Oak Ridge National Laboratory, May 1972, p. 8.

45. Ibid., p. 9.

46. Box, Thadis W., "Land Rehabilitation: Prompt Passage of Federal Reclamation Law Recommended by Ford Foundation Study," *Coal Age,* May 1974, pp. 108-118.

47. *Surface Mining,* op. cit., p. 178.

48. Nephew, "Healing Wounds," op. cit., p. 12.

49. Ibid., p. 18.

50. Maneval, op. cit., p. 15.

51. Maneval, op. cit., p. 19.

52. Spaulding, William M., Jr., and Ronald D. Ogden, *Effects of Surface Mining on Fish and Wildlife Resources of the United States,* Bureau of Sport Fisheries and Wildlife, U.S. Dept. of the Interior.

53. Ibid., p. 41.

54. Ibid., p. 12.

55. Personal conversation with Harry Perry.

56. Personal conversation with Willard Spaulding of the U.S. Fish and Wildlife Service.

57. Personal Conversation with Thomas Flynn of the Bureau of Mines (Division of Environment).

58. "The Economic Impact of Public Policy on the Appalachian Coal Industry and the Regional Economy," Appalachian Regional Commission staff summary of study by Charles River Associates, Inc., April 26, 1973, pp. 2-3.

13
Market Imperfections, Social Costs of Strip Mining, and Policy Alternatives[1]

Robert A. Bohm
James H. Lord
David A. Patterson

Introduction

All but the most hardened defenders of the coal utility interests must admit that the visual and ecological impact of an unreclaimed stripped area is unfavorable. For those who care to look more deeply, there are the thousands of miles of acid or otherwise damaged streams and a myriad of resulting damages and treatment costs. There are in excess of one million acres of land disturbed by strip mining, with only a fraction significantly reclaimed.[2] There are mining counties in Kentucky, West Virginia and elsewhere that simultaneously produce thousands of tons of coal worth millions of dollars while yielding infant mortality and other human welfare plights that we would expect to find in an underdeveloped country—not in the U.S.A.[3] The sum total of these impacts, and what they really mean to the human welfare of those who must live in these areas, is a situation which has led some observers to call for the abolition of strip mining, and the reallocation of resources to more deep mining.

Yet deep mining has its costs too. Most of the acid mine drainage originates in deep mines. Soil subsidence and burning is a deep-mine problem.[4] And worst of all, the human damage of deep mining, as presently carried out, must outweigh any other consideration. In spite of the Mine Safety Act of 1969, figures released by the Bureau of Mines show only a slight decrease in mine accidents.[5] The full extent of black lung among deep miners is just beginning to be realized as state laws and medical examiners officially acknowledge the existence of a disease which has accounted for, or contributed to, crippling disability and early death for thousands of coal miners.[6] Thus deep mining is hardly presentable as an attractive alternative to stripping.

From the *Review of Regional Studies*, vol. 3, no. 2, 1973, pp. 69-81. Reprinted by permission.

Robert A. Bohm and James H. Lord are in the Department of Finance at The University of Tennessee. David A. Patterson is with the Tennessee Valley Authority, on leave from the Department of Economics at The University of Tennessee.

Yet, on the other hand, one must have some grasp for the economic importance of coal, or rather, for its major end product, electricity.[7] At the recent Governor's Conference on Meeting Tomorrow's Energy Needs (Chevy Chase, Maryland, September 23-24, 1971) many of the well known perceptions and facts pertaining to this nation's current and future energy requirements and the concomitant environmental problems were reiterated. Among the many prognostications and available options proffered as solutions to the impending crisis was the suggestion that the market be allowed to find its own balance. By this, we presume, the protagonists of such a strategy were reminding us whether we be economists or chemists, friend of "Big Coal" or foe—that the market mechanism, while not impervious to breakdowns, is capable of handling its own "domestic" problems thank you, and any outside assistance which might be necessary will be solicited—all in good time. It is the considered opinion of the authors of this paper that there is lacking in this approach a realistic appraisal of the effectiveness of the market mechanism in the case of Appalachian strip mining.

We would never quarrel with the overriding concern of the participants at the conference, that concern being for the strained resources situation and the financial and technological difficulties in meeting energy demands. However, the prevailing attitude of the power sector seems to be to put highest priority on meeting the future demands for energy and then, if the public is willing, to meet the considerable additional costs of diminishing environmental spillovers. But this attitude reflects a basic misunderstanding of the economics of the issue, i.e., of the role of price in determining the quantity of energy used. Since society has never been given the chance to vote on the issue in the market place—more cars, air conditioners, etc. vs. cleaner air and water—the power producers and others should not project the current trend in energy demands into future estimates of energy requirements without due recognition of probable changes in quantity demanded, and perhaps changes in the price elasticity of demand, when and if the social costs of producing energy are reflected in market prices. Sometimes it seems that the power people get their projections of energy demand from atop Mt. Sinai. We would hasten to intercept them and point out that the demand for power is a derived demand. If consumers point to a new basket of goods and services—namely, one with fewer hair dryers and cleaner streams in Appalachia—there is no energy crisis implied by that act per se, and industry should respond by producing the indicated bill of goods. Moreover, the change in emphasis could be accomplished with no diminution in our "current per capita consumption of consumables" since what we are going to reallocate from hairdryers to making cleaner streams are the additional resources which will become available to society over and above what will be necessary (given productivity increases) to provide us, each and every one, with a hairdryer. We just won't each have two hairdryers, but rather one hairdryer and clean streams. Let us forever lay to rest the notion that an impending power shortage plays a role in the feasibility of removing environmental spillovers of producing that power. In fact, removing them might alleviate the future power problems.

What is disconcerting in the matter of resource allocation is the failure of the market mechanism to signal producers to supply the socially correct bill of goods in the optimal proportions, and this concern brings up the whole complex problem of externalities i.e., spillover effects, and how to deal with them.[8]

II. Externalities and Market Failures

In his well-known treatment of the subject, Francis Bator categorized externalities according to the rationale of why the market fails under certain conditions.[9] His definition of an externality is a situation where some Paretian costs and benefits remain external to decentralized cost-revenue calculations.[10] Loosely translated, externalities are direct, non-market interactions of a producer-producer, consumer-consumer, or producer-consumer type that affect the physical outputs that producers get from their inputs or the satisfaction that individuals get from consumption. These effects are not reflected in prices, therefore they go unnoticed in a market system which is characterized by the decentralized decision making of business firms and households.

The Ownership Externality

In Bator's classification of externalities there is the ownership externality, characterized by situations where producers are unable to physically exclude users (beneficiaries), or to control the rationing of their product among them, an example being J. E. Meade's "apple blossom and honey production" case.[11] In Meade's classic example the owner of an apple orchard is greatly benefited by his neighbor's activity of keeping bees. The bees crosspollenate the apple blossoms, thus rendering a valuable service to the orchard owner. The latter would realize an increased yield from his orchard if only he could persuade his neighbor to purchase more hives, but the beekeeper is not led to behave in a manner which would maximize their joint welfare because he is unable to physically appropriate the benefit that his activity confers on the orchard. The distinguishing characteristic of this case is that difficulties reside in institutional arrangements. The valuable service is rationable and finely divisible and there are no difficulties with "total conditions," i.e., at the bliss-configuration every activity would pay for itself.[12] If payment for all of the benefits were only enforceable, then the service in question would be produced in the right amount and would be rationed efficiently. The problem is due exclusively to the difficulty of keeping accounts on recipients of the benefits; the exclusion principle of economics is here inoperative and the market fails because of a lack of proper incentive.

The Public Good Externality

Another type of externality described by Bator is the public good externality. As analyzed by Paul Samuelson it exists where each man's consumption of a good is related to the total output of that good (or service) rather than as a part of a summation—as in the case of private goods.[13] Often referred to as collective goods, a public good once supplied to one consumer is supplied to all others at zero marginal cost.[14] Moreover, one individual's consumption of the public good does not decrease the amount of the total available for another person's consumption, and this leads to a special problem in pricing public goods because consumers will not reveal their true preferences. Each individual will understate his preference for a public good, expressed by his reluctance to pay a full-value price, because he knows that if others will reveal their true preferences and pay full-value prices he will enjoy equal benefits and pay less than others will pay. This situation is different from that of the ownership externality where one man's consumption of the external benefits (which indeed are not feasibly appropriable) reduces the quantity of these benefits which can be consumed by others.[15] Certainly national defense is a public good that closely conforms to the above definitions.

Externalities: The Coal Industry and Legal Conditions

Environmental spillovers usually embody both the ownership and public good externalities. For instance, acid mine drainage once eliminated for one person is eliminated for all others at no additional cost, and if it is eliminated for one person it is necessarily eliminated for everyone. Therefore, the strip miner is not led to strip coal in an environmentally compatible way since consumers will not reveal their true preferences for a clean natural environment and pay full-value prices. The exclusion principle is inoperative. Moreover, one person's pleasure from viewing a mountain stream in a pristine state does not diminish the esthetic enjoyment available for others, and any price charged for looking will cause people to underpurchase "viewing" since there is no need to ration it. Thus we can conclude that eliminating acid mine drainage (thus restoring Appalachia's waterways) has public good characteristics, but it also gives rise to an ownership externality. For example, operators will not strip-mine coal in an environmentally compatible way because the benefits of doing so would accrue to others, and there is no market arrangement, nor practical private institutional device for charging beneficiaries. How could mining operators appropriate the benefits to a farmer whose land borders a stream? Clean water is a divisible good and its use can be rationed, but how could users along a waterway be charged for their consumption?

Viewing the environmental spillover in the above manner clearly shows how the market has failed to provide a good which presumably is in great demand, and this failure has led to the current situation, i.e., a situation where laws have been tolerant of pollution and the courts have treated property rights to the environment on a first-come, first-served basis. But the ownership and public good externalities are no less evident when one considers the proposition of reversing the above situation, making the laws on pollution rigorous, with property rights to the natural environment clearly vested in public ownership. If mining companies are required by law to reduce the problem and bear the costs, it is no less important to determine the optimal degree of abatement than in the historical case. Either way it is society's resources which are the stakes; the resources used to prevent acid mine drainage cannot be used to manufacture hairdryers. In this case, however, the public good externality causes each individual to overstate his true preference for clean streams in an effort to maximize his compensation, and this fact may lead a responsive public authority to invoke a much too rigorous set of pollution control laws. The result could be an overinvestment of society's resources in pollution control laws. The ownership externality in this case is manifest in the fact that the pollution-control authority cannot ascertain to what extent each offender is responsible for the total damages. How could the authority keep accounts on damages to each party and how could it establish how much each strip mine was responsible.

III. Alternative Policy Responses to Market Failure

Failure of the market mechanism to achieve the socially optimal allocation of resources brings up the question of how to correct for its deficiency. To develop the answer let us begin by calling X a group of strip mining interests which impose an external diseconomy, and Y the public residing in Appalachia.[16] How much Y suffers depends not only on the scale of X's activity, but also upon the *nature* of the activity and upon Y's *reaction* to it. If the activity in question is strip mining land, Y's loss will depend not only on the number of acres left unrestored but also on the condition of the disturbed land and the economic cost to Y of overcoming the effects of acid mine drainage, siltation, landslides, etc. To determine the proper commitment of society's resources for alleviating the problem requires an investigation into the nature and cost of alternative activities open to X and devices by which the public can reduce the impact of each activity.[17] The optimum scale and nature of restoring the land and the proper adjustment to it by Y is that outcome which maximizes the algebraic sum of Y's gain to Y's loss.

The diagram in Figure 1 shows the relevant variables. The assumption here is that two types of land restoration are available to X—Activity I, no effort and Activity II, thorough reclamation. Even in the event of a complete reclamation effort, however, some damage to Y occurs. For example, reclaimed land is ugly and erosion prone until reforestation takes effect. The scale of each activity and the losses sustained by Y are measured as continuous variables. The area under X's curves reflects the gain to X of leaving strip-mined land in two alternative conditions while the area under Y's curves gives the total loss to Y, in each case, after he has made the best possible adjustment to the damages; thus it measures the direct loss to Y after the adjustment plus the cost to Y of making the adjustment.

If X is unhampered by any restrictions he would choose Activity I (no reclamation) at a scale of ON. From the social point of view OM is better than ON. However, the socially optimal outcome is for X to engage in full restoration, Activity II, at a scale of OT since area OZW is greater than area JKL. With the Appalachian public in a weak position but able to bargain with X, it should subsidize the proper restoration of land by paying up to (OJLN + LGN – OWT) to induce X to Activity II at a scale OT. Certainly X will accept as little as (JKL + OJLN – OZW – OWT) to cooperate. The difference between the maximum bribe offered and the minimum acceptable payment is (LGN – JKL + OZW) which is the maximum social gain to be shared between them.[18] Changing the legal environment so that X is liable to Y for external damages would induce mining companies to move to Activity II at a scale OT (the optimum allocation) pay OWT to Y and retain a net gain of OZW. Thus the result is the same as when there is no liability, although the distribution of the gain is very different; as demonstrated earlier Y would pay up to (OJLN + LGN – OWT) to induce X to move to Activity II at a scale OT. It follows that, if bargaining can take place, questions of property rights to the use of the natural environment in Appalachia are a matter of income and wealth distribution and not a matter of efficient use of society's resources. Realistically, of course, state intervention is necessary because the Appalachian public is too large a group for members to get together for direct bargaining with the coal industry. However, we hasten to point out that government intervention can be a very complex and costly undertaking where the goal is to achieve

FIGURE 1

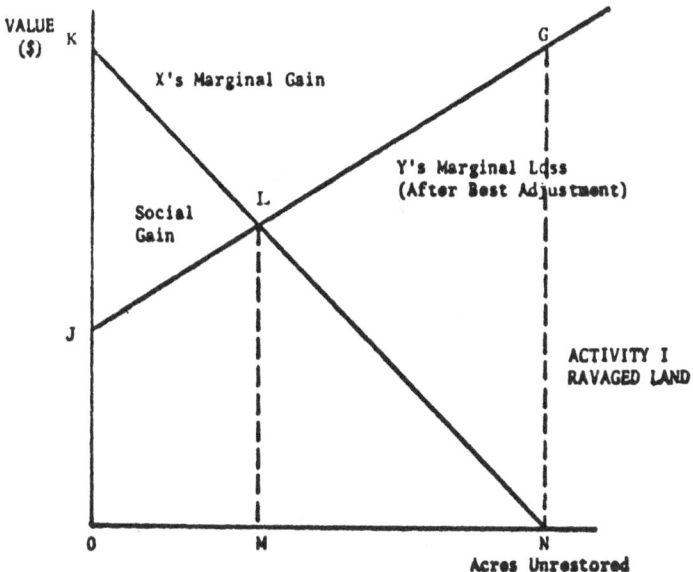

VALUE
($)

K

X's Marginal Gain

Social
Gain

J

L

Y's Marginal Loss
(After Best Adjustment)

G

ACTIVITY I
RAVAGED LAND

O   M   N

Acres Unrestored

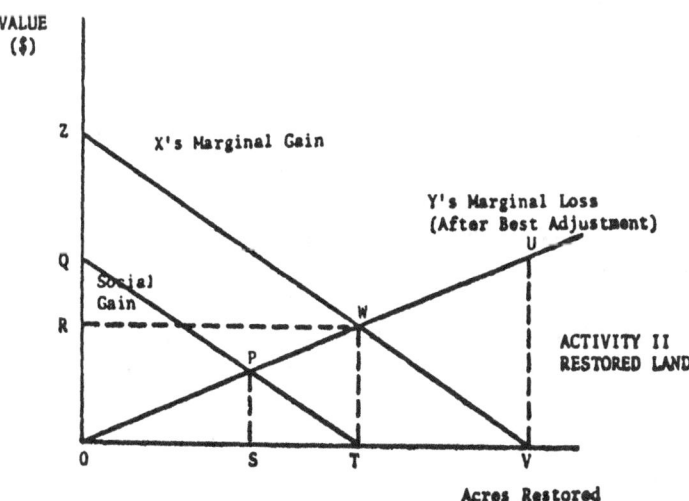

VALUE
($)

Z

X's Marginal Gain

Q

Social
Gain

R

P

W

Y's Marginal Loss
(After Best Adjustment)

U

ACTIVITY II
RESTORED LAND

O   S   T   V

Acres Restored

Adapted from Ralph Turvey, "On Divergences Between Private and Social
Cost," _Economica,_ (1963), p. 311.

the optimal resource allocation, which is Activity II at a scale of OT.

The problems faced by the state authority are (1) to list and evaluate all of the alternative activities available to X and examine their effects on Y; (2) to determine adjustments available to Y for reducing the direct loss suffered and the costs of these adjustments; and (3) to determine what measures are necessary to move the market toward the optimum and the costs of determining and enforcing the change. Moreover, before the state will move to take action, it should (1) be sure that the gains from resource allocation outweigh the costs of effecting and maintaining the change and (2) be sure that any income redistribution can be justified.

In the examples just discussed, the nature of the activity as well as the scale of the activity are variable so it is necessary to control both. No amount of tax per acre of strip-mined land would induce coal interests to move from Activity I to Activity II, because X's net marginal gain after paying the tax would be greater with Activity I than II.[19] It would be necessary to enforce the practice of Activity II (full land reclamation) and then impose a tax of WT per acre of restored land to move X to scale OT.

What could be true is that the costs of achieving this optimum resource allocation outweigh the gains, the latter given by (LGN – JKL + OZW) and this is an empirical question. The costs of restoring the land, of course, are paid for by the strip-mining interests—reflected in Figure 1 by the shift downward of X's marginal gain (from Activity I to Activity II). There are two assumptions here: (1) that none of these costs can be passed on through higher prices for strip-mined coal because deep-mined coal is a nearly perfect substitute, and (2) the number of strip mining firms remains unchanged. Certainly there would be considerable costs of affecting and enforcing Activity II as well as the tax WT. Total tax receipts would be equal to the area ORWT. The collection of this tax would give rise to X's net marginal gain curve which is the line QT. If X then moves to output OT (optimal for society and X) it will pay X to induce him to move to position S (which is sub-optimal) as it is this position which maximizes X and Y's joint net gain.[20] Even in this case Y still sustains a diseconomy equal to OPS. This problem of a sub-optimal outcome as well as the problem of Y sustaining a diseconomy could be taken care of by using the tax revenue ORWT to fully compensate Y for the diseconomy (OWT) sustained in the socially optimal circumstance. This leaves a residual tax revenue equal to ORW to contribute to the administration and enforcement costs of both the reclamation standards and the tax WT.

It may well be that the administration and enforcement costs are greater than LGW – JKL + OZW, the gains from the optimal solution; moreover, it may be that strip-mining cannot operate profitably under the reclamation standards and the tax WT (assuming no increase in coal price).

If either is true in the Appalachian strip-mining situation, then a strong case can be made for disallowing all strip-mining rather than permit ON acres to remain ravaged (since JKL is less than LGN). However, if the administrative costs of enforcing a standard reclamation policy (with no tax, WT) are less than (OZW + LGN - JKL - WUV), another solution would be to continue strip mining but invoke the flat requirement that all acres be fully restored as in Activity II. This solution may be referred to as the "standards" solution and usually takes the form of a severe penalty if land is not restored in a standard manner.[21] Thus, strip-mine operators would choose Activity II at a scale OV rather than endure the penalty and would avoid mine sites that are unrecoverable.

Within the context of the partial equilibrium analysis presented here, the "standards" solution is no more than a second best solution from the point of view of economic efficiency. However, it may well be a more feasible solution. To achieve the socially optimal solution would require the assembly of enormous amounts of information to determine WT as well as to continually readjust the tax anytime X's marginal gain curve shifted (which would happen if X passed on to consumers any of the costs of restoration, any of the tax, or if firms were to drop out of the industry) or Y's marginal loss curve shifted.

It should be noted that for a "tax" measure alone to achieve the optimum without the standards policy, charges would have to be levied in accordance with the quantity of damages inflicted. But this system would be almost impossible to administer since it would require a constant monitoring of damages. As a consequence the usual suggestion is a flat rate charge per ton of strip-mined coal (e.g., a severance tax).[22] This is certainly "a less than second best solution" since minor offenders under a flat-rate system (say partial restoration) would subsidize the major violators (say no restoration) if both were paying the same tax per unit of output. Furthermore, as pointed out earlier, the same output tax levied on major and minor offenders would do nothing to encourage strip miners to engage in better restoration techniques. On the basis of the preceding analysis, it is our intuitive judgment that the "standards" solution is the preferred policy alternative.

IV. The Benefits and Costs of Abatement

Irrespective of the final choice between an optimal solution or "second best" solution, the first decision that must be made is whether a shift from Activity I to Activity II is indeed economically warranted. Such a shift requires that the value of the external costs which are controlled or eliminated (i.e., the net benefits of any environmental protection policy)

equal or exceed the cost of effecting and maintaining the change. Again
we should point out that net benefits are here defined as the benefits
to the public of removing the spillovers over and above the negative
effects inflicted on the mining interests. In the graphical analysis presented above this was assumed to be the case (i.e., area OZW is greater
than area JKL). However, in the real world, it is not entirely clear that
such a case has been firmly established. Rather, the approach of the
environmentalists has been to first make a value judgment that pollution
is "bad" from which follows the logical conclusion that "something must
be done."

Any environmental policy aimed at the elimination or reduction of
the external environmental costs of coal production must enumerate, in
value terms, the benefits and costs to be achieved. If this is not done,
society has no way of knowing if it is better off after the spillovers have
been eliminated than it was before. In other words, it is possible that the
reallocation of society's resources towards cleaner streams is an inferior
position in terms of the overall level of welfare attained.

Benefits Estimation

On the benefits side, a considerable literature is rapidly appearing
which identifies a large number of detrimental environmental effects which
result from coal mining and coal use. The number of unreclaimed stripped
acres (1,024,000 in 1970), the number of tons of mine acid produced per
day due to strip mining (1,500), and the number of acres of ponds and
lakes adversely affected by silt and acid (145,146), etc. are rapidly becoming familiar figures.[23] We hasten to point out, however, that only in a few
instances has the value of these spillovers been estimated. Such a "costing
out" of both benefits and costs of environmental action is perhaps the
most important task at hand today in the area of environmental research.

Some estimates of the value of benefits, of course, do exist. For example,
the Department of the Interior estimates that strip mining destroys 35 million dollars of outdoor recreation resources annually.[24] The enormous
potential benefits to be gained from control of environmental spillovers
of coal, however, are best illustrated by reference to external costs imposed
on humans by current methods of coal usage. A recent estimate based on
the use of coal to produce electricity indicates that accidents in mining
and processing of all coal lead to the death of one man per yer per 1,000
MWE produced.[25] To the individual in question, of course, the value of
his life is infinite. However, a finite estimate of the market value of his
future earning power can be derived. If we assume that the fatality in question has 20 working years left and his average wage during that future
period is between $8,000 and $10,000 a year then a direct economic loss

to society of $160,000 to 200,000 (undiscounted) ensues over the period.

Lung disease, excluding lung cancer, from coal mining and processing may result in up to 15 premature deaths per year per 1,000 MWE.[26] The number of deaths is probably indicative of a much larger number of sufferers of chronic black lung disease and other respiratory ailments. In this case, the earning power of the afflicted is greatly reduced if not zero. In addition, society bears the further cost of medical care and compensation.

The Costs of Abatement

The costs of abating the ill effects of coal related environmental spillovers are seldom discussed. These costs include the direct costs of abatement plus information costs and enforcement costs. All three are important because they involve resources which have alternative uses; they are a measure of the benefits society forgoes when it decides to engage in abatement.

In choosing the actual abatement policy, information and enforcement costs are indeed crucial variables. Recall in part III above, four acceptable policy options were discussed as alternative ways of dealing with environmental spillovers from strip mining. These alternatives are: (1) do nothing; (2) a tax plus standards solution (efficiency optimum); (3) establish standards (second best); or (4) outlaw the activity entirely. The information and enforcement costs of doing nothing are, of course, zero. If this option is chosen, society continues to absorb the external costs of coal production which are presumably less than the costs of doing anything. Likewise, information and enforcement costs of "outlawing" the activity are probably slight. In this case, however, a major mining process, which may well be the industry's most economical technique, would be outlawed on non-economic grounds.

The information and enforcement costs of effecting the standards plus tax solution are probably prohibitive. Not only must a basic benefit-cost study be undertaken but marginal cost and benefit schedules must be derived from the data and constantly updated, since they are necessary for proper enforcement. As mentioned earlier, a uniform "output" tax is feasible but sub-optimal in terms of its effect on pollution and equity. The "standards" solution requires only the basic benefit-cost study perhaps periodically updated. Information costs of the "standards" solution may be high but certainly not as high as those of the standards plus tax solution. Moreover, the enforcement costs of the "standards" solution would be considerably less than those of the standards plus tax solution—since the detection of any violation can be accomplished visually. The costs entailed in complete wage and price administration World War II style, as opposed to the 1971 wage and price freeze, are probably indicative of the difference in magnitude of the information and enforcement costs of the standards plus tax solution versus the standards solution.

V. Income Redistribution Effects

Policies to control pollution and other spillovers of coal production and consumption are likely to have side effects of their own. These effects will largely be transfer effects or what are known in economics as pecuniary external effects. From the point of view of economic efficiency they are irrelevant. From the point of view of the distribution of income and wealth in society, however, pecuniary external effects may be extremely important.

Pecuniary external effects are nothing more than the market reacting to changes in demand or to a change in institutional arrangements. At any point in time a set of these effects is embedded in the economic system. As a result of the existing effects some individuals have gained and some have lost over time. By changing institutional arrangements, for example, a new set of pecuniary external effects is generated which create a new set of gainers and losers. As noted above, society must make a judgment regarding the desirability of any change, presumably through the political process.

Income redistribution effects are probably of greatest concern when the suggested policy is to completely outlaw some activity. For example, consider the possible consequences of outlawing all strip mining of coal. The initial effect would be unemployment for all strip miners and an increase in demand for deep mined coal. Since the Mine Safety Act of 1969 undoubtedly reduced the elasticity of supply of deep-mined coal, its price would rise appreciably resulting in windfall profits for the owners (somewhat counteracting the windfall profits earned by strip mine owners as a result of the Mine Safety Act) since wages typically react slowly in such cases. The high price of deep-mined coal will result in attempts at substitution in production by heavy coal users. To the extent that production functions cannot be altered instantaneously, however, extensive substitution by consumers will occur away from goods with a heavy coal component.

The unemployed strip miners will not completely forsake their current location for at least as long as required for the Supreme Court to decide the constitutionality of outlawing strip mining. In the interim, they will be unemployed and drawing unemployment compensation. In Appalachia alone there were 11,697 strip miners in 1967.[27] If they draw an average of 52.15 dollars per week for 24 weeks (6 months), the cost will be $14,639,974 assuming no reemployment during the period and the collection of full benefits.[28] These payments amount to an income redistribution from the general public to a special group.

Consider further the plight of many Appalachian communities for which strip mining is the major source of basic employment and income. Successful outlawing of strip mining will undoubtedly doom them as viable economic units, poor as they may now be. On the other hand, communities that depend on deep mining will benefit. It is perhaps here that potential contradictions in public policy and the importance of distribution effects in environmental programs may be seen. While irrelevant in terms of national economic efficiency, pecuniary external effects are quite real and may result in severe hardships at specific locations. Public sector concern for the income distribution problem has recently manifested itself in the formation of such agencies as the Appalachian Regional Commission, the Economic Development Administration, etc. These agencies have devoted considerable resources to the economic development of depressed areas. In fact, a popular technique employed to foster economic development has been the use of pecuniary external effects of such public investments as highways in Appalachia. To ignore the potentially contradictory distribution effects of environmental programs certainly seems criminally naive.

VI. Policy Recommendation

As already indicated, there are four basic policy options. The do-nothing alternative seems indefensible, if not already eliminated from consideration by the nation's environmental consciousness. Similarly, the complete prohibition of strip mining is not a valid proposal if techniques of reclamation can be devised which are environmentally compatible.

The tax solution was described as having two forms. One is a flat rate tax per ton or per dollar of sales. This has the obvious disadvantage of being unrelated to the true cost of the polluting activity, except perhaps on an average basis. The optimal tax would be in conjunction with a standard reclamation policy and would overcome this deficiency but require extensive cost analysis as well as post-mining inspection of each site. Of course, a flat severance tax may be proposed and justified simply for its income producing, and in this case income redistributional effects. This would be, in effect, a tax on the users of coal paid to residents of coal producing states. Such a tax would then merely join a long list of taxes with redistributional effects but with no predictable effect on strip mining pollution.

The final alternative is the "standards" solution. This is the option already adopted in most states.[29] However, standards have been so low and enforcement provisions so weak that benefits have been nonexistent. For this reason, standards have earned an undeserved bad name.

Standards can be set to meet whatever goal society decides is economically and environmentally justified. The Tennessee Valley Authority is currently experimenting with new mining techniques that would leave the terrain closer to its original position than generally believed possible in the past at a cost of less than two dollars per ton or probably about equal to the German cost experience. In any case the effectiveness of standards depends essentially on only one variable; the public will for a sufficiently strong law and effective enforcement.

Assuming that effective standards can be made into law and enforced and considering the cost and benefits of implementing the various control policies, it is our judgment that the standards solution is the preferred policy alternative.

Notes

1. An earlier version of this paper was presented at a Symposium on Coal and Public Policy held at the University of Tennessee, Knoxville, October 13-15, 1971.

2. United States Department of the Interior, *Surface Mining and Our Environment*, 1966, pp. 32-49. Also Appalachian Regional Commission, *Acid Mine Drainage in Appalachia*, 1969; Marion Edey, "Strip Mining Legislation," No Man Apart (July 1971), pp. 7-8 and statement of the Honorable Ken Hechler before subcommittee on Mines and Mining, *Congressional Record*, September 20, 1971, p. E9885.

3. See the discussion in Mary Jean Bowman and W. Warren Haynes, *Resources and People in East Kentucky* (Baltimore, Maryland: The Johns Hopkins Press, 1963) esp. pp. 180-242.

4. See the interesting comments of M. Gordon Walman, *Science* (Nov. 26, 1971), p. 909. Also *Nashville Tennessean* (December 3, 1971).

5. United States Department of the Interior, Bureau of Mines, *Mineral Industry Survey*, October 1971.

6. United Mine Workers of America, *Black Lung*, 1969.

7. See for example the data presented in *Bituminous Coal Facts 1968* (Washington, D.C.: National Coal Association, 1968).

8. The assumption here, of course, is that the recent clamor about environmental spillovers reflects a true (effective) demand for their removal.

9. Francis Bator, "The Anatomy of Market Failure," *Quarterly Journal of Economics*, LXXII (1958), pp. 351-379.

10. Ibid.

11. J. E. Meade, "External Economies and Diseconomies in a Competitive situation." *Economic Journal*, LXII (1952), pp. 54-67.

12. In equilibrium of production and exchange there is a set of shadow-prices implied by the marginal-rate-of-substitution equalities which would efficiently ration the bill of goods; moreover, this set of prices will signal the socially correct bill of goods and will lead to their optimal production. This conclusion is predicated on the assumption that there are no indivisibilities and/or increasing returns in production. The consequence of indivisibility or increasing returns (in inputs, outputs, or processes) is to render the set of feasible points in production (input-output space) nonconvex. Nonconvexity disallows duality and is referred to by Bator as a technical externality. For a further discussion see Bator, op. cit., pp. 63-69.

13. P. A. Samuelson, "The Pure Theory of Public Expenditures," *Review of Economics and Statistics*, XXXVI (1954), pp. 387-89 and P. A. Samuelson, "Diagrammatic Exposition of a Theory of Public Expenditures," *Review of Economics and Statistics*, XXXVII (1955), pp. 350-356.

14. Warren Robinson, "Benefits Received Financing in the Federal System," *National Tax Journal*, XXVII (1964), p. 242.

15. The public good externality exists in what is otherwise a neoclassical world, i.e., preference and production functions are convex throughout. If y is the numeraire, the marginal y-cost of x in production, p_x, is the efficient price for x. If x is offered for sale at p_x and consumers adjust their purchases so that their individual marginal rates of substitution are equal to p_x, they will necessarily underconsume x because additional consumption would not interfere with another person's (who is also paying p_x and underconsuming) consumption. There is no set of prices associated with the bliss point which will sustain the configuration; i.e., a set of prices which would induce competitive producers to produce the optimum bill of goods would be inefficient in allocating that bill of goods. There are no nonconvexities involved nor would any accounting device, if one were only feasible, solve the problem. See Bator, op. cit., pp. 69-71, and Samuelson, "Diagrammatic Exposition of a Theory of Public Expenditure," op. cit., p. 350.

16. It is important to point out that X, the strip mine operator, by externalizing some costs to the environment, is able to (1) earn excessive profits and/or (2) sell coal for "environmentally subsidized" prices. If the latter is true then X represents the public interest as well as private strip-mine ownership interests. Viewed in this way the external costs of strip-mining amount to a welfare transfer: a transfer of welfare from the relatively few who endure the external costs to the larger public who benefit from subsidized coal prices.

17. The following analysis is based on Ralph Turvey, "On Divergences Between Social Cost and Private Cost," *Economica* (1963) pp. 309-313.

18. The realism of this theory is easily missed by too narrow an understanding of the term "bribe." If one views the theory in the contexts of the current debates raging in several state legislatures and in the Federal Congress, the analog between "bribe" and the political compromise usually necessary to achieve new legislation is more obvious.

19. A severance tax per ton of coal indirectly amounts to a tax per acre.

20. Clearly a stable solution occurs only at the origin.

21. The most popular proposed penalty is forfeiture of a performance bond if adequate reclamation is not undertaken. Strip-mine bills which include a bonding provision have recently been proposed or enacted in Missouri, Virginia, Alabama, Kentucky, Tennessee, West Virginia and by Senators Baker (R, Tennessee) and Cooper (R, Kentucky) and Senator Aspinall (D, Colorado). See St. Louis Post-Dispatch, April 4, 1971; Bristol Herald Courier, January 7, 1972; Decatur Daily, October 1, 1970; Louisville Courier Journal, April 4, 1971 and December 14, 1971; Memphis Press-Scimitar, October 12, 1971. The amount of the bonds proposed varies considerably and in many cases may not be sufficient to force Activity II.

22. Senator Lee Metcalf (Montana) has proposed a tax of 5% of sales on all extractive minerals (*Congressional Record*, November 29, 1971, S1843). Other proposed taxes range as high as 10%. In Germany restoration of strip-mined land costs $1.25 per ton. Tennessee Valley Authority experience indicates a cost of 15 to 25 cents per ton for reclamation meeting present Tennessee Valley Authority standards (New York Times, December 7, 1971). However, "reclamation" and "restoration" are vastly different, both in terms of visual impact and land-use potential.

23. Statement of the Honorable Ken Hechler before Subcommittee on Mines and Mining, Congressional Record, September 20, 1971, p. E9885.

24. Ibid.

25. Data derived from Chauncy Star, *Nuclear Safety*, V (1964), p. 325. The shift to strip mine coal since 1964 should have reduced this ratio considerably.

26. Ibid.

27. Appalachia is defined here as the coal mining areas of Alabama, Kentucky, Maryland, Pennsylvania, Tennessee, Virginia, and West Virginia. Data derived from *Bituminous Coal Facts 1968* (Washington, D.C.: National Coal Association, 1968), p. 68.

28. This amount is the national average compensation paid between July 1970 and June 1971. Data derived from *Monthly Labor Review*, 94 (September 1971), p. 93.

29. U.S. Department of the Interior, *Study of Strip and Surface Mining in Appalachia*, 1966, Appendix II and footnote 21.

14
Economic Impacts
of Surface Mining Reclamation

Alan Schlottmann
Robert L. Spore

Federal legislation to control surface mining, particularly for coal, is a response to the adverse environmental impacts of coal surface mining which have been discussed in the literature.[1] The explicit concern of most of the sponsors of this legislation is the severe environmental degradation that has occurred as a result of surface mining in Appalachia. Even those in the West fear that a massive coal industry development would "Appalachianize" their region.[2] Yet, for the most part, the current controversy over the proposed legislation concerns its possible adverse secondary economic impacts on coal costs, prices, and output, and on employment and income in the mining regions.

There are three main areas of coal production in the United States. The coal fields of Appalachia stretch from Pennsylvania to Alabama. The majority of total U.S. underground production occurs in this area, and here surface mining occurs generally on steep slopes. Over 70 percent of Appalachian production in 1971 was on slopes over 15° [C.E.Q. 1973, p. 66], and the environmental problems associated with contour mining on such slopes are particularly severe [C.E.Q. 1973, ch. 1]. The major Midwest fields—Illinois, Indiana, western Kentucky—are dominated by surface mining on flat or gently rolling terrain, and this activity is referred to as area mining. The relatively recent development of the western coal fields

From *Land Economics*, vol. 52, no. 3, August 1976, pp. 265-277. Reprinted by permission.

Research sponsored by the National Science Foundation RANN Program under both a grant to the Appalachian Resources Project, University of Tennessee, and Union Carbide Corporation's contract with the U.S. Energy Research and Development Administration. The contributions and comments of John Moore and Sidney Carroll of the University of Tennessee and of Edmund Nephew, Wen S. Chern, and Charles Kerley of the Oak Ridge National Laboratory are gratefully acknowledged.

Alan Schlottmann is with the Department of Economics at The University of Tennessee, Knoxville. Robert L. Spore is with the Energy Division of the Oak Ridge National Laboratory, Oak Ridge, Tennessee.

has been mainly by means of surface mining, used where coal seams are relatively thick and the overlying strata relatively shallow.

The primary purpose of this paper is to analyze the regional economic impact of back-to-contour surface mining regulations on the dominant coal market (the use of "steam-electric" coal by electric utilities) and its coal suppliers. Analysis is accomplished by means of a nonlinear regional programming model of the steam-electric coal market developed to provide quantitative information concerning alternative energy-environmental policy combinations. The effects of alternative policies such as reclamation standards cannot be reduced to a single dimension. There is an impact on total production, but given an emphasis, particularly in recent years, on the importance of federal policy as a source of differential regional effects, the diverse regional output effects are also important.[3] A one-ton change in Appalachia's output is not necessarily equivalent to an identical change in the West in terms of regional development effects. Each consuming and producing region will be affected not only by public policy towards coal in its own area but also by policies applied to mining elsewhere. Thus, regional interaction is explicitly considered. As public policy alters the private cost of mining and use, there will be regional effects on the delivered price of energy derived from coal.

A second purpose of our analysis is to consider the effects of the surface mining regulations on the regional competitive position of underground mining. As L. Hines has written, "The use of strip mining and auger mining in the coal fields has reduced mining costs well below those of underground operations—if the wide-spread degradation and property loss in areas where these practices are employed is disregarded" [1973, p. 53]. The explicit purpose of some sponsors of "original contour" legislation is to stimulate underground production. This increase in underground production is seen as being centered in Appalachia, and such an increased dependence on underground mining has been seriously proposed as a panacea for the production of coal in the United States.[4]

The Model

L. Moses has characterized models of regional interaction and location interdependencies as being of "... two varieties: (1) highly abstract formulations that extend Walras' reasoning to spatial phenomena, (2) models that are more restrictive in their assumptions and less general in their intent but which lend themselves to empirical application" [Moses 1970, p. 15]. One example given of the latter type of model by Moses is Henderson's [1958] linear programming model of the coal industry. In Henderson's study, where coal was differentiated by type of extraction and location of the mine, the regional shipment pattern which minimized the costs of meeting regional demands was found. Our study is an extension of this modeling effort.

The Programming Model

Solutions to the model are characterized as determining the most efficient (minimum cost) network of coal production, distribution and use in steam-electric generating plants. A convex programming model, which results from nonlinearities in the cost estimates, is used to attain this solution.[5] An equilibrium solution across all regions is attained through an iterative procedure which interfaces the set of demand equations by electric utilities with the convex programming model.[6] Such a procedure allows the specification of the equations and the effects of changes in important regional variables to be incorporated into the model in a much more general format than a quadratic program, for instance. Constraints include (a) that coal-fired generating requirements be met in each market (constraint 2); (b) that no coal region can have shipments which exceed capacity (constraint 3); (c) coal use can be constrained by a regulation on the average sulfur emissions per million B.T.U. from coal-fired plants (constraint 4) which limits the average sulfur emissions allowed in any region to the specified level S'_L; and (d) shipments have to be non-negative (constraint 5). The algebraic structure of the model is as follows:[7]

Minimize:

$$\sum_i \sum_j \sum_k \sum_{lm} (C_{ijm} [X_{ijklm}] + t_{ll} X_{ijklm} + \Phi[R_{ij}] X_{ijklm}) \quad [1]$$

where R_{ij} is the incremental production costs per ton resulting from reclamation requirements in the ith supply region. The remaining constraints in the model are:

$$\sum_i \sum_j \sum_k \sum_m b_{ijkm} X_{ijklm} \geq G_L \text{ for all } l.$$

$$G_L = f(P_c, P_o, P_g, E_L) \quad [2]$$

$$\sum_L X_{ijkm} \leq K_{ijkm} \text{ for all } i,j,k,m \quad [3]$$

$$\sum_i \sum_j \sum_k \sum_m S_{ijklm} X_{ijklm} \leq S'_L \text{ for all } l. \quad [4]$$

$$X_{ijklm} \geq 0 \quad [5]$$

where:

X_{ijklm} is the coal extracted by the *j*th
mining method in new or existing
mines (*m*th) and the *i*th district with
the *k*th sulfur level and delivered to
the *L*th demand region, measured in
1,000-ton units.

C_{ijm} is the per-unit extraction cost of
the *j*th mining method in new or existing mines (*m*th) in the *i*th supply district, and *til* is the transportation cost
from the *i*th supply region to the *L*th
demand region.

K_{ijkm} is the physical capacity of coal of
the respective types in each district. In
Appalachia, we also consider these by
the additional classification of slope
angle.

b_{ijkm} is the energy value measured in
B.T.U.s for the respective activities, in
millions of B.T.U.

S_{ijklm} is the average emitted sulfur per
million B.T.U.s in the *L*th demand
region equivalent to $S_k b_{ijkm} / G_L$,
where G_L is the demand by utilities
for steam-electric coal measured in
millions of B.T.U.s in the region *L*. S_k
is the sulfur percent by weight of the
*k*th sulfur class. The five sulfur classes
in the model, in sulfur emissions per
million B.T.U., are 0-.6, .6-1.0,
1.0-2.0, 2.0-3.0, 3.0 or greater.

P_c = price of coal per million B.T.U. on
an "as-burned" basis.
P_o = price of oil per million B.T.U. on an
"as-burned" basis.
P_g = price of natural gas per million
B.T.U. on an "as-burned" basis.
E_L = steam-electric generation in region
L in billions of kilowatt hours.

Until recently, there was no data base sufficiently adequate to estimate
the regional characteristics of production. The basis for our estimates is
the Federal Power Commission (FPC) Form 423 data, which is a disaggregated set of monthly responses by 267 electric utility companies with 744
plants in the continental United States. These data report the quality and
source of their fossil fuel deliveries. The coverage represents 98 percent of

all delivered fuel. For existing mines, the data for 1973 were used to generate information in this study.

As an example of the use of these data, consider the B.T.U. values, denoted as b_{ijkm}. In our study, they are differentiated not only on a regional basis but also by mining method and sulfur content. The average B.T.U. value of underground coal is 8 percent higher than for surface production. This differential is more distinct in the lower sulfur coals, and the lower sulfur coals in the West in particular tend to have lower energy values, regardless of mining method, when compared to higher sulfur coals.

Another difficulty solved by use of FPC data is the identification of the sulfur content of coal from a particular coal supply district. The FPC data allow an aggregation for each district of all shipments by sulfur content for the five sulfur classes used in this study. Similarly, they allow the differentiation of shipments from each district by mining method for each sulfur class.

The transportation of coal is based on actual rail-barge routes between major concentrations of utility power plants and coal supply regions. Thus transport activities are *not* based on linear, "crow flight" distances as is customarily the case.[8] The transportation sector in the model attempts to determine the actual routes that would be followed in coal shipping. Rather than use the shortest route indicated by railroad maps, we have secured information on the shortest feasible routes (i.e., obtained information on factors such as track conditions that would affect routing).[9]

In previous studies, the main data problem with production and reclamation costs has been the values for contour mining in Appalachia. In this study, a linear programming model employing process analysis techniques is used to estimate the short-run supply curve for coal surface mining in Appalachia.[10] In process analysis, the production function is described in a linear programming manner where each of the column vectors describes a process in engineering or physical terms. Thus, the overall mining operation can be represented by the ($n \times 1$) vector x where each element, x_j, represents one alternative production process expressed in tons of processed coal. In addition, let c be an ($n \times 1$) vector whose elements, c_j, are the unit variable costs of each process. The mine operator's decision regarding how to produce a specified level of output consists of selecting among alternative production processes such that total variable costs are minimized and all required tasks are performed; i.e.,

minimize
$\pi = c'x$

Subject to

$A_1 x \leqslant b_1$

$A_2 x \geqslant b_2$

$x \geqslant 0$

where:

k = a constant integer varying para-
metrically from zero to a maximum
possible value encompassing the relevant output range; and

δ = a percentile increment in output for
each observation along the cost curve.

π = total variable cost

A_1 = a $(p \times n)$ matrix of technical coef-
ficients consisting of elements a_{ij} denoting the amount of the ith input
required per unit of process j;

b_1 = a $(p \times 1)$ vector of elements specifying input (labor and equipment)
constraints;

A_2 = a $(q \times n)$ matrix of elements $-1, 1$,
and 0 denoting material (inter-
process) balance constraints; and

b_2 = a $(q_k \times 1)$ vector of all zero elements except b_i which specifies output.

The complete cost-output relationship
can be derived by varying b_i^k, i.e., $b_i^k = b_i \pm k\delta$

where:
b_i = base output level;

The impact of regulation is analyzed by expressly considering its direct
effects. Thus, the imposition of reclamation requirements is reflected by
expanding the set of required tasks to be performed. Such additional tasks
can include topsoil removal and replacement, backfilling, and revegetation.

The solution to equation [1] requires a specification of c, A, and b.
A linear programming algorithm will give both the primal and dual solutions and the objective function, which is equivalent to total variable costs
for each output level. The marginal cost at each output can be found by
the shadow price corresponding to the output constraint or merely by
dividing the change in costs by the change in output from the previous
output level. Under the assumption of perfect competition, the marginal
cost/output relationship defines the short-run supply curve. In this study,
representative firm process analysis models of coal production were con-

structed for each of three subregions (North, Central, and South) of the Appalachian coal surface mining industry.

Table 1 presents a sample of the model's output relative to 1972 actual production levels. The backfilling requirements of approximately 50 percent represent current levels of reclamation, with "back to contour" simulated by 100 percent backfilling. For use in this study, these results were then adjusted to reflect (1) estimated 1974 production costs, and (2) the planned addition of new mines.[11]

Mining costs for surface mining in the remaining districts and for all deep mining are based upon the coal supply methodology of ICF.[12] They have attempted to derive regional cost estimates which explicitly consider the characteristics of reserve base data on a regional basis along with smaller mine size categories rather than the use of standard Bureau of Mines "million tons per year or more" mine techniques. An important characteristic of their work is that, unlike the traditional Bureau of Mines model mine approach, there is no unreasonable dichotomy in a supply district which forces surface mining to be "always" cheaper than an underground mine. Appalachian underground mine production is assumed to be limited to the levels of a normal working year.[13] This may not be an unreasonable assumption given the fairly constant levels of underground production from non-unionized mines during the 1971 strike and the lack of output response during the recent variability in coal prices.

In summary, the highest reclamation costs occur in the eastern fields, with the lowest in the producing areas of the West. The average reclamation costs per ton in Appalachia were $2.25 (1974 dollars). Reclamation of area mining in the Midwest had a value of $1.20 a ton, with the average value in the West at $0.32 a ton. The model's supply estimates, particularly from the process analysis of production in Appalachia, result in nonlinear production costs. Since they are convex, however, as is well known, a piecewise linear approximation can reduce a convex programming model into linear form.[14]

The demand for coal-fired generation, G_L, reflects the price impacts of reclamation.[15] In Appalachia, the price elasticity of steam coal is based on the estimated demand equation in Lin, Spore and Nephew [1975]. For other regions, we use the price elasticity from the study by Bohm and Vaughn [forthcoming] :

Appalachia:

$$\ln G_L = -1.258 - 0.165 \ln Pc + 0.216 \ln Po$$
$$(0.316)\ (0.073)(0.091)$$

$$+\ 0.144 \ln Pg + 0.899 \ln E_L$$
$$(0.081)(0.022)$$

$$R^2 = 0.997$$

TABLE 1
ESTIMATED AVERAGE AND MARGINAL COSTS OF COAL SURFACE
MINING IN APPALACHIA: 1972

Production Level (1,000 Tons)	Zero backfill, $/ton		50% backfill, $/ton		100% backfill, $/ton	
	Average Cost	Marginal Cost	Average Cost	Marginal Cost	Average Cost	Marginal Cost
Northern Appalachia						
56,701 (0.80)[a]	3.91	4.50	4.26	4.89	4.58	5.22
60,245 (0.85)	3.95	4.50	4.29	4.89	4.62	5.22
63,789 (0.90)	3.98	4.50	4.33	4.89	4.66	5.46
67,332 (0.95)	4.01	4.50	4.35	4.89	4.70	6.22
70,876 (1.00)	4.03	4.50	4.39	5.11	4.78	6.22
74,420 (1.05)	4.05	4.50	4.42	5.11	4.88	7.08
77,964 (1.10)	4.08	4.70	4.49	6.59	b	b
81,508 (1.15)	4.11	4.70	4.58	6.59	b	b
85,051 (1.20)	4.18	8.15	b	b	b	b
88,595 (1.25)	b	b	b	b	b	b
Central Appalachia						
35,643 (0.80)	2.93	2.99	3.38	3.45	3.60	3.67
37,879 (0.85)	2.94	2.99	3.38	3.45	3.60	3.67
40.098 (0.90)	2.94	2.99	3.39	3.45	3.62	3.82
42,326 (0.95)	2.95	2.99	3.39	3.59	3.63	3.82
44,553 (1.00)	2.95	2.99	3.40	3.59	3.64	3.82
46,781 (1.05)	2.95	2.99	3.41	3.59	3.65	3.82
49,009 (1.10)	2.96	2.99	3.42	3.59	3.66	3.82
51,236 (1.15)	2.96	3.11	3.43	3.59	3.67	3.82
53,464 (1.20)	2.97	3.11	3.44	3.59	3.68	3.82
55,692 (1.25)	2.98	3.11	3.45	3.59	3.69	3.82
57,919 (1.30)	2.99	3.11	3.46	3.59	3.71	4.95
60,147 (1.35)	2.99	3.11	3.46	3.59	3.76	4.95
62,375 (1.40)	3.00	3.11	3.47	3.59	b	b
64,603 (1.45)	b	b	b	b	b	b
Southern Appalachia						
11,043 (0.80)	3.94	4.84	4.45	5.38	4.82	5.90
11,734 (0.85)	3.99	4.84	4.51	5.53	4.88	5.90
12,424 (0.90)	4.04	4.84	4.57	5.53	4.95	7.01
13,114 (0.95)	4.09	4.98	4.62	5.53	5.06	7.02
13,804 (1.00)	4.13	4.98	4.67	5.53	5.16	7.02
14,494 (1.05)	4.17	4.98	4.72	5.77	b	b
15,185 (1.10)	4.21	4.98	4.79	7.49	b	b
15,875 (1.15)	4.25	5.19	b	b	b	b
16,565 (1.20)	4.29	5.19	b	b	b	b
17,255 (1.25)	4.41	9.13	b	b	b	b
17,945 (1.30)	b	b	b	b	b	b

[a]Figures in the parentheses represent fractions of the 1972 production level.
[b]Infeasible with specified labor and equipment constraints.

Elsewhere:

$$\ln G_L = 1.96 - .349 \ln Pc + 0.161 \ln Po$$
$$(0.465)\ (0.074)\qquad (0.036)$$

$$+\ 1.23 \ln E_L + .001\ (\text{Time Trend})$$
$$(0.041)\qquad (0.003)$$

$$R^2 = 0.993$$

There are 52 demand regions in the model. These include the 48 states in the continental United States plus the District of Columbia. Three urban areas with generating patterns markedly different from the rest of their state are treated separately—Chicago, New York City, and Philadelphia. The supply regions correspond to the 23 coal supply districts delineated by the Bureau of Mines.[16] Given that complete phasing in and enforcement of the federal surface mine regulations will take time, we have considered the effects of the regulations for 1978. We also assume that the requirements of the Clean Air Act of 1970 will not be met, although coal use in each region is constrained in the model to the average regional sulfur emission level of coal use in 1973.

Results

The main results of the model are shown in Table 2. Our analysis does indicate that the reclamation regulations should be a stimulus to underground production to some extent. Overall, underground's share of production increases by 5 percent. In Appalachia, underground production increases by 4 percent, while surface production falls by 7 percent. The main importance of surface mining reclamation requirements for underground production may be for new investment on steep slopes in Appalachia, since reclamation costs are the highest for steep slope surface mines. Thus, new underground mines would narrow their competitive disadvantage with new surface mine operations on a cost basis. Whether or not it is the intent of such legislation to stimulate the competitive position of underground mining, it does in effect appear to occur with "original contour" reclamation.

An initial hypothesis was that the largest effect to be expected from the reclamation requirements would be in Central Appalachia. Not only do the highest reclamation costs in the United States occur in this region, but the steepest supply curve generated by the process analysis submodel is for Central Appalachia. Further, Central Appalachian markets in the Middle and South Atlantic states are subject to competitive pressure from

TABLE 2
COMPARATIVE REGIONAL EFFECTS OF THE SURFACE MINE RECLAMATION
POLICY ON COAL PRODUCTION FOR 1978

(millions of tons)

Region*	With Current State Requirements**	"Original Contour" Requirements	Net Change	Percent Change
Northern Appalachia				
Surface	95.8	89.6	−6.2	−6.5
Underground	69.9	71.9	+2.0	+3.9
Total	165.7	160.5	−4.2	−2.5
Central Appalachia				
Surface	63.0	60.0	−3.0	−4.8
Underground	59.1	63.1	+4.0	+6.8
Total	122.1	123.1	+1.0	+0.8
Southern Appalachia				
Surface	14.6	11.5	−3.1	−21.2
Underground	5.5	0	0	0
Total	20.1	17.0	−3.1	−21.2
Total, Appalachia				
Surface	173.4	161.1	−12.3	−7.1
Underground	134.5	140.5	+6.0	4.5
Total	307.9	300.6	−7.3	−2.4
Midwest				
Surface	126.4	120.4	−6.0	−4.7
Underground	23.6	26.6	+3.0	12.7
Total	150.0	147.0	−3.0	−2.0
West				
Surface	45.6	45.6	0	0
Underground	3.3	3.3	0	0
Total	48.9	48.9	0	0
United States				
Surface	345.4	327.1	−18.3	−5.3
Underground	161.4	170.4	+9.0	5.6
Total	506.8	496.5	−9.3	−1.8

*The regions are identified as: Northern Appalachia = Districts 1, 2, 3, 4, 5, 6; Central Appalachia = Districts 7, 8; Southern Appalachia = District 13; Midwest = Districts 9, 10, 11, 12, 14, 15; West = Districts 16, 17, 18, 19, 20, 21, 22, 23.

**Regional production cost estimates include 1969–1972 levels of reclamation. One factor that has resulted in these levels representing essentially little or no reclamation lies in the inadequate levels of bonding for strip mining. Most state laws require performance bonding for restoration on a per acre basis. It is not unusual for mines in Central Appalachia to have the estimated cost of adequate backfilling for reclamation to run higher than $5,000 an acre, and revegetation adds additional costs. The 1972 Tennessee surface mine reclamation law, for example, requires only $1,000 per acre for performance bonding.

Northern Appalachia producers. Yet the change in Central Appalachia, as seen in Table 2, is relatively minor. This result can be explained by considering the sulfur characteristics of Central Appalachian production. From information obtained from the FPC, Form 423 data, approximately 62 percent of all production in Central Appalachia has a sulfur content of less than 1 percent. Indeed, 97.3 percent of this region's coal production has a sulfur content less than or equal to two percent. The reclamation policy, designed to improve the land use impacts of surface mining, is *not* independent of public policy towards air quality issues. Given the high price and limited availability of alternative low sulfur fuels such as oil and natural gas, the demand for low sulfur production from Central Appalachia in order to meet the model's sulfur emissions constraints at coal-fired power plants remains high. Indeed, Central Appalachia is presently the largest single source of coal which can meet proposed air quality standards. Our results suggest that there is an interaction between air quality proposals and those dealing with land use impacts which *must* be recognized rather than treating these proposals as separate issues.

This is also an important result for the effects of reclamation on the future patterns of surface mining development in Appalachia. Since reclamation costs increase with increasing slope angles of production, one might expect an increased use of lower slope strippable reserves. The sulfur content of Central Appalachian coal reserves, however, is inversely correlated with increasing slope angle [C.E.Q. 1973]. Lower sulfur reserves are mined on steep slopes, most on slopes exceeding 20°. This contour surface mining on steep slopes in Appalachia has been the most controversial environmental issue for the land use impacts of surface mining. In particular, the problem of soil stability after mining and current levels of reclamation is generally considered to be extremely difficult to deal with. Thus there is no reason to expect a decline in the intensity of the use of higher slope surface reserves as long as public policy results in a premium to low sulfur coal.

The results of imposing a reclamation policy on the western coal industry as currently developed leads to the conclusion that the use of western coal is relatively unaffected by additional production costs associated with reclamation.[1] In the West, coal is a least-cost generation source, given natural gas shortages and the price of oil. Its use outside of the West is determined as a consequence of sulfur emission issues, not by a relative cost comparison to alternative production, which is so marginal that slightly raising the lowest production costs in the United States would make its use noncompetitive.

The increases in the equilibrium costs of energy due to reclamation are relatively small. The largest increases of 9 to 10 cents per million B.T.U. occur in the eastern states, where the current average price of coal is 80-90 cents per million B.T.U. Since these markets are supplied by Appalachian producers, with the largest reclamation costs, this is as expected. In the Midwest, cost increases would be around 4 cents per million B.T.U. The maximum in the West would be only 1 or 2 cents per million B.T.U. These cost increases, assuming they are passed on to the electric utilities, would have a minor effect on interfuel substitution. Given the costs of converting plants designed to burn coal only to oil, etc., the price effects are insignificant. Demands for coal-fired generation were decreased by approximately 2 percent in the model, but this was essentially the only impact.

The long-run effects from indirect impacts of unreclaimed, scarred surface mined lands should be minimized by the reclamation requirements. Economists such as William Miernyk have argued that improving the environmental quality of Appalachia would increase the potential for "amenity-oriented economic growth."[18] Since the wilderness appeal of many of the Appalachian areas is recognized as a basis for this growth, large areas of unreclaimed surface mining would obviously decrease this growth possibility. Others have argued that the disruption of wilderness areas by surface mining is not socially optimal from a benefit-cost evaluation, an activity which our results indicate could be expected to decrease.[19]

The reduction in the intensity of surface mining in Appalachia suggested by our analysis could have several benefits difficult to measure quantitatively. The first relates to the social effects of surface mining in rural Central Appalachian communities.[20] The historical "development" of unregulated surface mining in rural mountain areas previously undisturbed by surface mining has placed severe strain on the local mountain culture. Local community institutions, often informal, which were able to coordinate public decisions such as road repair, etc., have broken down. One manifestation of these effects is the sale of land to individuals who are not members of the extended kinship group in the community, a pattern which has generally not occurred for over two hundred years except for the sales of land companies around the turn of the century. An additional benefit, related to the first, would be an expected decrease in migration both within and from Appalachia. Migration to urban centers and between rural counties in Appalachia, and areas outside of the region, can be affected by the intensity of surface mining activity.[21]

Conclusions

We have seen that the incremental production costs from reclamation are not severe, particularly since the adverse land use impacts of surface mining can be controlled by adequate reclamation. The intent of some proponents of this legislation to increase our reliance on underground mining would lead us to suggest that a comprehensive policy analysis of environmental impacts and possible environmental regulations for underground mining be undertaken.

Underground mines have historically been associated with several additional environmental damages—fires both underground and in refuse piles, subsidence, acid mine drainage, erosion and aesthetic problems from mine wastes.[22] Though engineering studies have been made relating abatement costs to most of these factors if regulation were to occur, particularly for acid mine drainage, there has been no extensive public policy awareness of the social costs of these factors.[23] If public policy is oriented toward stimulating underground production, it seems imperative for such factors to be considered.

Notes

1. See a summary of these effects and references in Council on Environmental Quality (C.E.Q.) [1973, ch. 1].

2. See Environmental Policy Center [1974].

3. For a study concerned with such development benefits, see Leven [1969].

4. See Coalition Against Strip Mining [1974].

5. For a discussion of convex-separable programming problems, see Dantzig [1963, ch. 16].

6. For a similar description of the interregional model emphasizing interfuel substitution, see Schlottmann and Abrams [1976].

7. The basic model without the process analysis submodel has 305 row constraints, the objective function, and 3,006 activities (excluding artificial variables) including a slack activity for each row constraint.

8. Bob Young, Traffic Department, Peabody Coal Company, was quite helpful. Transportation rates are calculated by using a combination of data provided by Peabody Coal Company and a transport cost equation using a composite rail-barge coal freight rate. See Schlottmann [1975, Appendix B] for a discussion of the equation. Verification work on the transportation sector has been as follows. Included on the magnetic tape summary of the Form 423 data is a code for the county and state in which the purchasing utility plant is located, and the state and coal-producing district from which the coal shipment originated. Using the Federal Railway Administration 1972 Waybill Statistics data on the county-to-county level of disaggregation, estimates of transportation cost per ton of bituminous coal were calculated by aggregating the statistics for origin counties to the coal-producing district and state level, while maintaining the county level of disaggregation on the destination. The Waybill Statistics data are based on approximately a 1 percent sample, and therefore many of the coal-producing district—state-to-county combinations in the Form 423 data did not appear in the Waybill data or, if they did, were based on such a small tonnage figure that the transportation cost estimate could not be considered reliable. For the figures which were reliable, we used the combination of these two sources as a cross-check on our estimates in the original transportation sector. Our original work checked out to be extremely consistent.

9. Interregional differences in coal tariffs are included in our transportation costs. These have been supplied to us by the Traffic Department of Peabody Coal Company. Thus, for example, western shipments to western utilities would, even if we assumed equal transportation distances, be less than for western shipments into the Midwest.

10. For a complete discussion of this process analysis model of production, see Spore, Nephew and Lin [1975].

11. New mines were added to the model's sectors in each region as follows. First, new mines planned, under construction, and ready to open were identified from (1) *Coal Age* publication of mine additions as the main source; (2) any revisions to the *Coal Age* forecasts we could secure through the *Keystone News Bulletin*, the FEA Coal Advisory Committee, and our correspondence with the National Coal Association; and (3) more detailed information on expansion plans in the Midwest which our cooperative contacts at Peabody Coal Company could provide. Given the company and the mine name and/or location, we attempted to identify the coal seam which would be mined from Keystone's *Coal Directory* and *Producers of Coal from Low Sulfur Seams in the Appalachian States.* In a few cases, multiple seams in the same area occur, and our method has been (1) to locate the mine at the same seam as other mines owned by the same parent company, since this appears to have occurred in the past with little variability; (2) to locate the mine in the heavily mined seam if there are no present mines owned by the same parent company, which again appears to have occurred in the past with little variability. Seam characteristics not covered in the above two publications were based on the tipple samples of the Bureau of Mines.

12. Over the past year the ICF has attempted the development of supply curves having consistent methodology which can be reviewed and critiqued rather than using estimates representing Bureau of Mines best guesses of the development pattern of the coal industry. There are twenty-four curves. Their work has been reviewed by Richard Gordon, Martin Zimmerman, and the senior author of this paper. Professor Zimmerman's work on coal supply at the M.I.T. Energy Laboratory has been particularly valuable. For an example of their methodology as applied to one curve in Illinois-Indiana, a twenty-five-page outline is available from A. Schlottmann upon request. For complete details see the study by ICF [1976].

13. Although standard Bureau of Mines practice is to define this capacity as 280 working days per year, few underground mines run at this level. Our work suggests that a more reasonable short-run limit is in the range of 240 working days per year. A similar conclusion is reached from the technological-coefficient matrix in Tabb's study based on individual mines [1968].

14. This problem is extensively analyzed in Dantzig [1963, ch. 16].

15. The demand forecasts have been derived as follows. The starting point is the disaggregated data available from the FPC on regional "Projected Generating Unit Retirements and Additions" which are used as the basis for future regional demand projections. Historically, regional differences in electric power consumption growth have been substantial. Based on FPC data, current interregional power pool sharing arrangements are also taken into account. We have tried to recognize alternative scenarios with regard to the share of nuclear power, as for example in Reiber and Halcrow [1974]. There is, however, a problem in deciding the "proper" load factors to apply to these plants. This point is particularly critical for any region that will add substantial amounts of nuclear capacity. Given the low running costs of such plants, they may capture substantial parts of any increased market. In this study we have used a "low" assumption of 60 percent with nuclear plants used first in meeting baseload requirements. It was generally assumed that 75 percent of total generation is met by baseload units, 20 percent by intermediate load units, and 5 percent by peaking units. Assumed electrical demand growth rates from the present were 5 percent per year. These are based on the preliminary forecasts of the FEA Electric Utility Task Force and are more consistent with last year's electricity growth than assuming the traditional 7 percent (i.e., 2 percent in the first six months of 1975). For an analysis similar to ours see I.C.F. [1975].

16. A delineation of these districts is given in Bureau of Mines, *Coal— Bituminous and Lignite*, annual issues.

17. We are assuming for western reclamation (where coal seams can run to 100 feet in thickness) that a general rolling topography be restored able to be crossed by cattle or farm machinery depending on whether the future use of such land is for livestock or agriculture. Obviously, a strict interpretation of "original contour" would require a large hole to be dug elsewhere in order to fill the relatively deep western surface mines. The current bills reject such an absurdity.

18. See the paper by Miernyk [1971].

19. As discussed in Spore and Nephew [1974, pp. 209-224].

20. The following is based on the social anthropological study by Bryant [forthcoming].

21. See Washburn [1975].

22. The human capital costs associated with the high accident and fatal injury rates and black lung disease have been significantly decreased due to enforcement of the 1969 Federal Coal Mine Health and Safety Act. In 1973, for the third consecutive year, fatalities were the lowest in the statistical history of the industry. In addition, 94 percent of the sampled

sections met the required standard for respirable dust, with the number of sections sampled doubled from 1971 inspection levels. See Department of the Interior [1974].

23. This is a conclusion reached also in the unpublished Department of the Interior report, "Environmental Effects of Underground Mining and Mineral Processing," where it is concluded that no comprehensive study of the environmental effects of underground mining has ever been done. It has been suggested that this conclusion stopped the report from being published.

References

Bohm, Robert, and Vaughn, Garrett. Forthcoming. "Regional Demands for Steam-Electric Coal." In *The Future of Coal in Appalachia*, ed. J. Moore. Knoxville: University of Tennessee Press.

Bryant, Carlene. Forthcoming. "Community Impacts of Surface Mining." In *The Future of Coal in Appalachia*, ed. J. Moore. Knoxville: University of Tennessee Press.

Bureau of Mines, *Coal—Bituminous and Lignite* (annual issues).

Coalition Against Strip Mining. 1974. "Policy Letters" (July-Sept.), Washington, D.C.

Council on Environmental Quality. 1973. *Surface Mining and Reclamation.* Washington, D.C.

Dantzig, George. 1963. *Linear Programming and Extensions.* Princeton, N.J.: Princeton University Press.

Department of the Interior. 1971. "Environmental Effects of Underground Mining and Mineral Processing." (Unpublished report.)

_____. 1974. *The Administration of the Federal Coal Mine Health and Safety Act.* Washington, D.C.: Government Printing Office.

Environmental Policy Center. 1974. "Newsletter" (Aug. 10).

Federal Energy Administration. 1975. "Documentation of the *Project Independence* Assessment Model." Washington, D.C.: Government Printing Office.

Henderson, James. 1958. *The Efficiency of the Coal Industry: An Application of Linear Programming.* Cambridge, Mass.: Harvard University Press.

Hines, Laurence. 1973. *Environmental Issues: Population, Pollution, and Economics.* New York: W. W. Norton and Company, Inc.

I.C.F. 1976. *Coal Supply Curve Methodology.* Washington, D.C.

I.C.F. 1975. *Demand for Coal for Electricity Generation, 1975-1984.* Washington, D.C., prepared for Office of Coal, F.E.A.

Leven, Charles. 1969. *Development Benefits of Water Resources Invest-ments,* IWR Report 69-1 (Nov.). Arlington, Virginia.

Lin, W. W.; Spore, R. L.; and Nephew, E. A. 1975. "Land Reclamation and Strip Mined Coal Production in Appalachia." Paper presented at the American Agricultural Economics Association Annual Meeting, Columbus, Ohio, Aug. 10-13.

Miernyk, William. 1971. "Environmental Management and Regional Economic Development." Paper presented at Southern Economic Association Meetings. Miami Beach, Florida, Nov.

Moses, L. 1970. "The General Equilibrium Approach." In *Spatial Economic Theory,* ed. R. Dean, et al. New York: The Free Press.

Reiber, Michael, and Halcrow, Ronald. 1974. *Nuclear Power to 1985: Possible Versus Optimistic Estimates.* Urbana: Center for Advanced Computations, University of Illinois.

Schlottmann, Alan. 1975. "Environmental Regulation and the Allocation of Coal: A Regional Analysis." Ph.D. dissertation, Washington University.

____ and Abrams, L. 1976. "Sulfur Emission Taxes and Coal Resources." *Review of Economics and Statistics* 58 (Nov.), forthcoming.

Spore, R. L., and Nephew, E. A. 1974. "Opportunity Costs of Land Use: The Case of Coal Surface Mining." In *Energy: Demand, Conservation and Institututional Problems,* ed. M. Macrakis. Cambridge, Mass.: M.I.T. Press.

____, ____, and Lin, W. W. 1975. "The Costs of Coal Surface Mining and Reclamation in Appalachia: A Process Analysis Approach." Paper presented at the Western Economic Association's 50th Annual Conference, San Diego, California, June 25.

Tabb, William. 1968. "A Recursive Programming Model of Resource Allocation and Technical Change in the U.S. Bituminous Coal Industry." Ph.D. dissertation, University of Wisconsin.

Washburn, Bruce. 1975. "The Effects of Strip Mining on Migration in the East Tennessee Coalfield." Mimeo available from B. Washburn, Department of Economics, University of Tennessee.

Part 4
Legal Impacts

15

Strip Mining of Coal: A Federal Response to State Legislation

Richard E. Fox

There are three methods of surface or strip mining,[1] each developed for a particular type of landscape, area stripping, contour stripping and auger stripping.[2] During the past thirty years, the amount of coal produced by strip mining has quadrupled.[3] Along with this increased production,[4] however, strip mining has brought a corresponding increase in the disruption of the surrounding environment.

Recently there has been much discussion and debate concerning the need for a federal agency to oversee the regulation of strip mining operations.[5] State regulation has, for various reasons, proved ineffective in reducing the environmental and health dangers created by surface mining. This note will set forth the dangers created, analyze two state attempts to cope with the problem, discuss proposed federal legislation, and suggest a program which could effectively respond to the problems inherent in the strip mining of coal.

Environmental and Health Problems of Strip Mining

Strip mining may give rise to some air pollution problems.[6] However, the water pollution caused by strip mining operations creates a more serious threat in the form of sedimentation and chemical pollution of rivers and streams. Sedimentation results from the washing of dirt, silt and other materials into streams and rivers. This erosion is caused by heavy rains falling on the barren slopes of ridges and spoil banks[7] and on substandard access and haulage roads.[8] Sedimentation destroys fish habitats,[9] erodes bridges and roadways,[10] clogs culverts and seriously hampers flood control and water storage projects.[11] More serious than sedimentation, however, is chemical pollution. As surface water drains off the slopes and

From the *Indiana Law Journal*, vol. 47, summer 1972, pp. 771-787. Reprinted by permission.

ridges left by the mining operation and seeps from auger holes into streams and lakes, acids are formed.[12] These chemicals corrode and poison the banks and water located downstream from their point of origin.[13]

In addition to air and water pollution, other adverse effects result from strip mining. Abandoned or unguarded strip mines create a danger to livestock and children. In addition, dynamite blasting in strip mining operations often results in considerable damage to the environment by affecting the water table and available water supply.[14] If such water tables, once deep in underlying rock strata, are exposed, surface drainage containing silt or chemical pollutants may intermingle with the underground flow of the water supply.[15] As a result, homeowners live in fear of losing their well water, having it rendered unfit to drink or losing their homes in land slides.

Perhaps the most visible evidence of the adverse effects of strip mining is the despoliation of vast acreages. According to the United States Geological Survey, by 1965 an area of land the size of the state of Delaware had been disturbed by the strip mining of coal. Today an area nearly the size of the states of Delaware and Rhode Island has been so disturbed.[16]

However, the nation is now faced with temporary shortages of coal, natural gas and residual oil.[17] Such shortages are the result, not of a lack of fossil fuel resources, but of a diminishing supply of recovered fuel. This supply began to diminish in 1965 when the government decided to favor nuclear energy as the primary source of electric power and to open the East Coast to unrestricted imports of cheap residual oil.[18] These decisions eliminated incentive for the coal industry to develop and maintain reserve production capacity. In 1969, however, delay in the development of nuclear power plants increased the demand for coal[19] and aggravated the already existing supply problem.[20] As a result, strip mining expanded since it was simply more productive and less expensive than other methods.[21] Furthermore, the Federal Coal Mine Health and Safety Act of 1969[22] requires such huge monetary investments that only large-scale underground operations are economically feasible.[23]

The Shortcomings of Present Regulations

To be effective, the regulation of strip mining must both reduce detrimental side-effects and provide for restoration of the mined area to some productive use.[24] State legislation, because of its failure to deal adequately with the problems inherent in surface coal mining, has failed to do this. In an effort to attract new industry, some states have oriented their statutes to favor industrial development over environmental protection. This bias is often reflected in the improper vesting of responsibility for regulatory enforcement in an industry-oriented agency. The conflicts of interest thus created may inhibit effective regulation. Additionally, state regulations have not been stringent enough to prevent such adverse effects of strip mining as acid mine drainage, mudslides and sedimentation. Some states have adequate legislation; however, their governing agencies often do not have sufficient authority to implement the regulations.

Although reclamation provisions are an integral part of any strip mining statute, some states have failed to enact adequate provisions. As a consequence, many acres of strip-mined land have not been restored to productive use. Further, insufficient bond requirements have failed to induce compliance with reclamation provisions.

In considering these problems, it must be recognized that regulation may create other problems. Strict control of surface mining can result in mine shutdowns and unemployment. Additionally, strict standards might further intensify the energy crisis. All of these problems are illustrated by a discussion of the Alabama and Kentucky regulations and sample federal proposals.

Two State Strip Mining Statutes—An Analytical Overview

Probably the best example of an industry-oriented strip-mining statute is the 1969 Alabama Surface Mining Act (ASMA).[25] The Alabama Mining Institute, in an effort to forestall proposed federal legislation, drafted and successfully lobbied for this law.[26] Emphasis throughout the Act is on the economic welfare and industrial development of the state,[27] as opposed to environmental concerns.

Within the administrative provisions of the Alabama Surface Mining Act, there are several fundamental deficiencies. First, the administrative responsibilities are inappropriately vested in the Department of Industrial Relations. The Alabama Department of Conservation or the State Department of Agriculture would have been a more logical choice to handle the administrative duties, since these two agencies are better suited to administer technical provisions relating to surface mining methods, water pollution, soil stabilization and other conservation measures.[28] A second deficiency is the powerlessness of the Department of Industrial Relations. ASMA authorizes the Department Director to "adopt and promulgate reasonable rules and regulations"[29] respecting the administration of the Act. The inclusion of the word "reasonable" indicates the use of restraint and caution. The Act's lack of specificity makes it difficult for the Department or the courts to identify any specific violation. Also, any rules and regulations promulgated by the Department must remain "consistent with . . . maximum employment and the economic and industrial well-being of the state."[30] While the Director has certain legal remedies for violations,[31] he is not empowered to terminate a violator's mining operation.[32] Indeed, the Director of the Department of Industrial Relations lacks the authority to deny an operator a license even though that operator has a past history of violations.[33] Nor can the Director institute criminal proceedings against delinquent operators.[34]

A further deficiency in ASMA is its failure to require the operator to submit a sufficiently detailed plan of reclamation prior to the issuing of a license. Only a statement by the applicant describing the manner in which he intends to reclaim the affected land need be provided. When compared with the comprehensive advance plans required in other states,[35] the permissiveness of the Alabama Act becomes all the more apparent.

In Alabama, every application for strip mining must be accompanied by a reclamation bond of 150 dollars per acre to be mined. Determining the sufficiency of a bond is difficult because reclamation costs depend on the particular land mined. The Alabama bond, however, is almost twenty dollars per acre below the lowest feasible cost.[36] Because of the low bond requirements[37] and the lack of criminal sanctions, there is an incentive for many operators to forfeit their bonds rather than reclaim the land.

ASMA's reclamation requirements are also inadequate. The foremost deficiency is the inability of the Director to reject a proposed reclamation plan.[38] While operators must submit maps and aerial photos of the mined land after their permit expires, no standards exist by which to approve or reject them. The maps and photos should give the Director an accurate accounting of what land was mined, to what extent it was disturbed and what progress is being made toward its reclamation. If the maps or photos fail to depict the area in sufficient detail, they will be meaningless.[39]

An important part of reclamation is revegetation requirements. While the revegetation provisions of ASMA are quite extensive, they are ineffectual because of the broad exemptions provided. An operator is not required to revegetate land used for the disposal of refuse, land within depressed haulage roads or final cuts or other areas where pools or lakes may be formed by rainfall. Further exceptions include land which is toxic, deficient in plant nutrients or which is composed of sand, gravel, shale or stone to such an extent that plant growth would be hindered.

While ASMA is a state statute which creates substantial advantages for the surface mining industry, the Kentucky statute[40] is directed toward improving environmental quality. Kentucky's statute has been hailed by conservationists as "probably the strongest" surface mining act in the country.[41] The statute has been unsuccessful, however, in solving problems in the eastern part of the state.

Kentucky is the only state which has two distinct coal mining regions, the western plains area and the mountainous eastern region.[42] Contour strip mining methods are used in the hills and mountains of eastern Kentucky, while in western Kentucky area strip mining is employed. Due to the different methods of strip mining, the statute has had mixed success.

In western Kentucky, adherence to the Act has improved environmental conditions and cured the deficiencies which existed under former statutes. While the earlier law called for grading "where practicable," the present statute requires restoration of all affected land to its original contour. The Act further provides that backfilling and grading must have been completed by the operator and approved by the regulatory agency before the bond can be released. The Director retains the required bond until planting and revegetation are completed and approved. There are also extensive planting and revegetation requirements which call for a planting report to be filed by the operator and approved by the regulatory agency.

An important feature of the Act permits any citizen of the state, having knowledge that any provision of the Act is willfully or deliberately not being enforced, to bring an action of mandamus against the appropriate officer to require him to enforce such provision. This concept may prevent bureaucratic delay and favoritism. The Act also provides for fines of up to 1,000 dollars for each day a violation continues. The Attorney General is vested with the power to bring both civil and criminal actions for recovery of these penalties.

In the mountainous Appalachian region of eastern Kentucky, the Act has been less successful. Indeed, the problem of strip mining in mountainous areas may, outside of the complete banning of the operation, be beyond effective regulation.

> Erosion and sediment damage to streams, even under the best grading conditions, cannot be prevented. Freshly disturbed earth placed on a downhill slope, even where stable, invites erosion. Particularly during the first six months to one year after the soil is disturbed, heavy rain and snowfall, along with freezes and thaws, will cause large deposits of sediment to choke and fill the nearby creeks and streams.[43]

Thus, from a conservationist's viewpoint, even if the law and subsequent regulations are successful in preventing landslides, serious damage to the environment will still occur daily. If strip mining is allowed to continue in eastern Kentucky, it appears that the cost to the environment will be staggering.[44]

Federal Intervention—Three Approaches

Unregulated strip mining operations not only adversely affect the environment but also create competitive advantages for persons operating in a given market area. Therefore, experience has demonstrated that reliance upon the states alone for effective regulation is not practical. Only a federally sponsored program with strict standards and a strong enforcement policy will end competition that leads to environmental decay. In this manner the full restoration of the land can become a reality.

Various bills have been introduced in Congress which attempt to solve the problems created by strip mining. These bills vary in scope and application from the cautious proposal of the Nixon Administration[45] to the radical suggestion by Congressman Ken Hechler (D.-W. Va.) that strip mining be completely abolished.[46] There are three basic approaches embodied in the proposed legislation. Each will be independently analyzed.

The Administration's Proposal

The administration's proposal, entitled the Mined Area Protection Act of 1971, would regulate both surface and underground mining of coal and other minerals.[47] The bill provides for a two-year period during which the states must formulate environmental regulations for mining operations on all lands within the state. Fifteen criteria are listed in the bill,[48] and the Secretary of the Interior would be required to furnish additional guidelines within thirty days of enactment. Also, an advisory committee with representation from the Departments of Agriculture and Commerce, the Environmental Protection Agency, the Tennessee Valley Authority and the Appalachian Regional Commission would be established to advise the Secretary in developing the guidelines.

If, in reviewing state regulations, the Secretary determines that: (1) the state has failed to enforce the regulations adequately; (2) the state's regulations require revision as a result of experience or the guidelines issued by the Secretary or (3) the state has otherwise failed to comply with the purposes of the Act, he will notify the state suggesting appropriate action. If the state fails to take such action, the Secretary may withdraw approval of the state regulations and issue federal regulations. In the event a state does not submit satisfactory regulations within the initial two-year period, federal regulations will control. The administration's proposal provides for inspections and investigations of the mining operation by the Secretary, injunctive relief to prevent violations, and both criminal and civil penalties for failure to comply after notice of violation.

The administration's proposal, however, has several weaknesses which render it inadequate. The two-year waiting period created by the bill is not practical. According to Russell Train, Chairman of the Council on Environmental Quality, "[e]ach day that effective regulation is delayed, mining sears an additional 750 acres of land—adding to the Nation's backlog of unreclaimed land."[49] It is, therefore, imperative that regulations be implemented as quickly as possible.

A lack of sufficient standards and remedial guidelines is also a weakness of the administration's bill. The proposal provides that the Secretary "suggest appropriate action" if the state regulations are not suitable and authorizes the state a "reasonable time" to take appropriate action before the federal government interferes.[50] However, this "reasonable time" period is in addition to the initial two-year waiting period provided by the proposal. This slow-moving implementation procedure and the inadequate remedial provisions would cause the bill to fail in its basic purpose—"to encourage a nation-wide effort to regulate mining operations to prevent or substantially reduce their adverse environmental effects."[51]

An additional problem with the administration's bill is that it is too broad in scope. The proposal attempts to deal with the environmental challenge from both surface and underground mining. The regulation of underground mining should not be included in surface mining legislation since conflicting rules would tend to disrupt the effective administration of both schemes.[52]

Further, there is no provision in the administration's bill for reclamation of areas already mined and abandoned. A total of 1,024,000 acres of strip-mined land was unreclaimed as of the end of 1970; and this acreage is rapidly increasing.[53] Finally, tbe administration's bill improperly vests the governing responsibility in the Department of Interior, a body notorious for its conflict of interest problems.[54]

Congressman Hechler's Proposal

The proposals outlined in Congressman Hechler's bill[55] are the most drastic that have been introduced in Congress. This bill provides for a program, administered by the Environmental Protection Agency, which would close all surface coal mines within six months and require their reclamation. The states would be required to adopt, within six months after the establishment of the federal standards, a state implementation plan. The bill also authorizes the Administrator of EPA to enter into agreements to reclaim abandoned and inactive surface and underground coal-mined lands currently owned by a state or local jurisdiction. Both federal enforcement of implementation plans and civil suits against violators are authorized. Civil and criminal penalties are provided, along with an informer's fee of one half of the fine for those providing information resulting in a conviction.

There are several problems with this bill, however, which make it an inadequate solution to the surface mining problem. By calling for the total abolition of surface mining for coal, the Hechler proposal ignores the fact that technology exists for effective reclamation in some areas. Indeed, reclamation can restore the land to such useful purposes as farming, cattle grazing,[56] and recreation.[57] Also, the prohibition of surface coal mining induces a stronger reliance on other forms of mining. However, other methods are not necessarily less detrimental to human values. Surface mining has a far better safety record than does underground mining and is less expensive.[58] "In light of the cost advantages of surface mining, it may prove cheaper in human and economic terms to require surface miners to be environmentally responsible than to rely solely on underground mining."[59]

The most serious problem with the Hechler bill, however, is its disregard of the fact that nearly 44 percent of the coal produced last year came from surface mining.[60] If this source of fuel were eliminated, it would be extremely difficult to locate a substitute. Atomic power has not developed as expected, and there is a shortage of domestic oil and gas reserves.[61] Nor is it realistic to believe that the slack could be taken up by increasing coal production from underground mines. While there are ample underground reserves, it would require 132 additional underground mines of two million tons annual capacity each and a capital investment of 3.2 to 3.7 billion dollars, to compensate for the loss of strip mine production. Also, an additional 78,000 trained underground miners would be required to produce the surface coal mined last year. It would take three to five years before full production could be anticipated.[62] Further, termination of surface coal mining would adversely affect coal prices:

> The effect on the cost of coal would be tremendous—the coal industry would be required to virtually duplicate its present underground mine capacity, calling for an enormous capital investment, and at the same time be required to write off as a loss its existing investment in surface mining equipment and reserves.[63]

Compromise Proposals

There are several proposals more moderate than Congressman Hechler's which avoid most of the weaknesses found in the administration's program. One such bill was introduced by Senator Frank Moss (D.-Utah).[64] The Moss proposal provides that within ninety days following the date of its enactment, the Secretary of the Interior, in consultation with the Administrator of EPA and the Secretary of Agriculture, will formulate mandatory standards covering mining operations and reclamation requirements. While there are no provisions prohibiting surface mining,[65] the Secretary is authorized to establish special standards governing the method of surface mining used on steep slopes. However, one weakness of this measure is its provision for tripartite administration by the Secretaries of Agriculture and the Interior, and the Administrator of the Environmental Protection Agency. Administration by three different agencies can only result in jurisdictional conflicts and cumbersome administration.

The basic problem with this bill, however, lies not in its individual regulatory standards and guidelines, but rather in its application to a given situation. The Moss proposal, as well as the other compromise bills,[66] fails to recognize the problems peculiar to each form of strip mining. Most of the proposed federal regulations are overly broad, weak and slow in implementation.[67] This might very well postpone a real resolution of a serious environmental problem.

A Possible Solution

It is evident that a new federal regulatory scheme must be quickly developed. Such legislation must be characterized by maximum simplicity, as well as by uniform environmental quality standards consistent with regional differences in topography and climate. A program essentially based on state control, with the federal government assuming supervisory responsibility, will provide an adequate regulatory mechanism. Recognizing the vulnerability of such a program to delays, it is necessary that the federal program give the state legislatures six months within which to draft suitable laws, regulations and implementation procedures. If a state should fail in this requirement, it would automatically become subject to federal regulations until such time as it developed a suitable program of its own.[68] The Conservation Foundation has formulated a program[69] which has the "merit of being flexible in terms of particular regional needs and differences while being consistent with traditional federalist approaches."[70]

Under the Foundation proposal, the appropriate federal agency to enforce strip mining regulations and their implementation is the Environmental Protection Agency. Conflicts of interest within the Department of the Interior should disqualify it from assuming such responsibilities. EPA, which is responsible for enforcing most of the federal environmental protection laws, is the logical choice.

Within six months after enactment, there should be complete abolition of all contour and other surface mining where, in the judgment of the Administrator, reclamation is not feasible or where such mining would violate existing environmental standards.[71] Ideally, this type of mining should be abolished immediately because of the irreparable damage it inflicts. However, employment hardships and the country's energy needs present competing values which call for the six-month grace period.

Further, new surface mining on lands with an average slope of less than thirteen degrees would be subject to a six-month moratorium to evaluate both the feasibility and enforceability of reclamation.[72] EPA should begin extensive studies of specific technical problems associated with strip mining, and should classify all lands: "by acidity of the seams, by feasibility of stripping, by the effect of the climate on reclamation, and components contributing to air pollution."[73]

Reclamation should be accomplished on an acre-by-acre basis and should be performed concurrently with the mining activity. Each day the land lies unreclaimed the possibility of acid mine drainage, erosion and aesthetic blight increases. A performance bond should be required and the amount should be large enough to be an effective incentive for carrying out the reclamation program. The term of liability under this bond should also be sufficiently long to insure proper reclamation of the affected area. Prior to the opening of a surface mining operation, a reclamation plan should be required which will assure that the land will be restored to a condition allowing its original use and potential to be fulfilled.

Presently there are more than 1,000,000 acres of land which have yet to be reclaimed.[74] These lands must eventually be made productive. Perhaps a joint state-federal program to fund the reclamation of this land could be implemented. Such a reclamation program would not only turn ugly, dangerous acreage into useful lands, but would also help provide employment for the 28,000 workers who might lose their jobs because of the prohibition of contour stripping.[75]

Any surface mining legislation should also provide for citizen participation in the implementation process. "Environmental regulation of coal stripping would be improved by citizen participation in EPA decision making, and by giving citizens standing to bring suit against private parties as well as against State and Federal governments."[76] In addition, there must be effective civil and criminal sanctions for violation of the regulation and implementation programs. An informer's fee might also be authorized for information leading to a conviction.

Recognizing that contour strip mining accounts for approximately twenty per cent of domestic coal production, new sources of fuel must be developed if contour mining is to be prohibited. However, it must be remembered that the abolition of contour mining, which is only one method of surface mining, will result in a loss in annual domestic coal production of only 53 million tons. The Foundation believes that new sources can be developed to offset the banning of contour mining and has offered three suggestions as to how this might be accomplished.

First, many electric utilities could be converted from coal to oil or natural gas. Although there is a shortage of domestic oil and gas reserves, the Foundation believes, based on data from the National Coal Association, that conversion to other fossil fuels is feasible and at least temporarily would not cause severe hardship to the utilities industry.[77] Second, the production of coal from deep mines could be increased. One study indicates that with a three-shift operation and a six-day production schedule, underground mines could produce an additional 150 million tons of coal annually—far more than would be required.[78] It is unrealistic to believe that all of the 264 million tons of strip-mined coal produced last year could be supplanted by increasing underground production. The increased production that would be required due to the prohibition of contour mining, however, is possible from mines presently in operation. An increase in underground mining would also provide extra jobs for those who were working at the contour mining sites.[79] Third, an expanded market for residual oil might be created by prohibiting contour strip mining. In the past, oil companies have not been able to compete with the coal stripping industry. By abolishing contour mining, the price of stripped coal will increase, making residual sales more profitable, thus encouraging increased production of residual oil from domestic sources.[80]

Conclusion

Effective regulation of surface mining for coal is long overdue. With 77 percent of the surface coal reserves in this country located in thirteen Western states, and with leases for coal stripping already having been obtained on 3,500 square miles of public and acquired Indian lands,[81] the situation has become urgent. It is imperative that prompt and effective action be taken by the Congress to cope with this imposing threat.

Notes

1. Strip mining is a method of mining by which large power shovels "strip" off the soil and rock overlying coal beds, dump it to one side and then load the underlying coal onto trucks. Brooks, *Strip Mine Reclamation and Economic Analysis*, 6 NATURAL RESOURCES J. 13 [hereinafter cited as Brooks].

2. Area strip mining, which is used in relatively flat terrain, involves making a boxcut through the overburden to expose the coal seam, which is then removed. As each succeeding parallel cut is made, the spoil (overburden) is deposited in the cut just previously excavated. The final cut may be a mile or more from the starting point of the operation and several miles in length. KENTUCKY ENGINEER, Nov. 1967, at 11.

Contour strip mining is used in steep or mountainous country. It consists of removing the overburden above the coal seam by starting at the outcrop and proceeding along the hillside. After the exposed coal is removed in the original cut, additional cuts are made until the ratio of overburden to coal produced makes additional cuts impractical. Contour mining creates a shelf on the hillside. The inside of the shelf is bordered by a "highwall," which may range from a few feet to more than 100 feet in height. The opposite side is a precipitous slope that has been covered by spoil material cast down the hillside. Id. at 11-12.

Augur mining entails boring into a seam from the surface and leaving it perforated with a series of holes from which the coal has been removed. Id.

3. Fifty years ago strip-mined coal represented only 1.2 percent of all coal produced. This percentage had risen to 10.7 in 1941, to 22.7 in 1951 and to 43.8 in 1970. 117 CONG. REC. E9885 (daily ed. Sept. 22, 1971) (remarks of Cong. Ken Hechler (D.-W. Va.)) [hereinafter cited as Hechler].

4. Nine years ago the average productivity of bituminous coal and lignite strip mines was almost 27 tons per man per day. The average for underground mines was twelve tons. The absolute difference between these two rates has been increasing. 2 BUREAU OF MINES, MINERALS YEARBOOK, FUELS 71, 86 (1962).

5. See Hechler, note 3 supra.

6. The danger of air pollution is limited to densely populated areas. Dust may be stirred up by the digging, hauling and blasting involved in strip mining. In addition, some smoke may be caused by burning coal seams. But since strip mining is localized, air pollution does not seem to be a major problem. Note, *The Regulation of Strip Mining in Alabama: An Analysis of the 1969 Alabama Surface Mining Act*, 23 ALA. L. REV. 420 (1971) [hereinafter cited as Note, *Regulation*]. See also Basselman, *The Control of Surface Mining: An Exercise in Creative Federalism*, 9 NATURAL RESOURCES J. 137, 140 (1969); Brooks, supra note 1, at 23; DEP'T OF THE INTERIOR, SURFACE MINING AND OUR ENVIRONMENT 33, 56 (1967) [hereinafter cited as OUR ENVIRONMENT].

7. Spoil banks are a mixture of soil, and bits of coal and rock piled next to an exposed coal seam. This mix is an unnatural one because the material with potential for producing vegetation becomes buried at the bottom, while the material with the greatest potential for acid and mineral pollution is brought to the top. This mixture lacks coherence; as a result, rain will cause silt to be carried into streams and rivers where it destroys the aquatic invertebrates upon which fish feed. Testimony presented by Cong. J. Edward Roush (D.-Ind.) on H.R. 4556, before the Subcomm. on Mines & Mining of the House Interior Comm., Sept. 20, 1971, at 3 (on file at Indiana University Law Library, Bloomington) [hereinafter cited as Roush].

8. Loosely piled spoil banks and crude dirt roads leading to the mining operation create the danger that silt and mud will be carried by surface water into streams and rivers. This type of physical pollution is most serious in hilly regions where high-intensity storms frequently occur. OUR ENVIRONMENT, supra note 6, at 63.

9. Roush, supra note 7, at 3.

10. This type of sedimentation often results in massive landslides which block streams and highways. G. Siehl, Mined Land Reclamation Requirements Pro and Con 7 (Legislative Reference Serv., Library of Congress, Apr. 18, 1968) [hereinafter cited as Siehl].

11. Brooks, supra note 1, at 24.

12. See Note, *Regulation,* supra note 6, at 423.

13. Brooks, supra note 1, at 23.

14. Dynamite blasting could cause a decrease in underground water flow which, in turn, could deprive people of their supply of well water. On the other hand, the underground flow could increase to such an extent that underlying rock and soil are carried away, thereby creating the possibility of landslides. Hechler, supra note 3.

15. Roush, supra note 7, at 3.

16. Hechler, supra note 3. Although they emphasize the loss of arable lands to strip pits, agronomists will concede that poor farming practices can and do result in a far greater economic loss. Brooks, supra note 1, at 17. See also H. Moore & R. Headington, Agricultural Land Use as Affected by Strip Mining of Coal in Eastern Ohio 34 (Bull. No. 135, Ohio State Univ. Agricultural Experiment Station, 1940).

17. *Hearings on S. 4092 before the Subcomm. on Minerals, Materials & Fuels of the Senate Comm. on Interior & Insular Affairs,* 91st Cong., 2d Sess., at 54 (1970) [hereinafter cited as *Fuels and Energy Hearings*].

18. Id. at 54.

19. Id.

20. There is also a natural gas shortage which has been attributed to the decision of the Federal Power Commission to regulate rates of interstate sales of natural gas on a regional basis. This action, it is claimed, reduced the incentive for exploration of natural gas for interstate sale. Id. at 55. Another factor contributing to the energy crisis was the rapid increase in East Coast imports of foreign residual oil. This increase has forced U.S. companies producing residual oil out of the market because they could not compete with the foreign imports. Id.

21. Strip mining results in the recovery of ninety percent or more of the coal in the mine, whereas underground mining seldom leads to recovery of more than fifty percent. Brooks, supra note 1, at 17. See also Siehl, supra note 10, at 6.

22. 15 U.S.C. Secs. 633, 636, 30 U.S.C. Secs. 801-04, 811-21, 841-46, 861-78, 901, 902, 921-24, 931-36, 951-60 (1970).

23. See generally *Fuels and Energy Hearings* supra note 17, at 55.

24. Various means are available to achieve reclamation. One approach is the limitation of strip mining through the zoning process:

> That a governmental unit through zoning could prohibit the use of property without compensation, and without justifying it as being a common law nuisance or creating a risk of imminent injuries was recognized for the first time by the United States Supreme Court in the 1926 decision of *Village of Euclid v. Ambler Realty Co.* [272 U.S. 365 (1926)].

Schneider, *Strip Mining in Kentucky*, 59 KY. L. J. 652, 667-78 (1970) [hereinafter cited as Schneider]. There has been increasing judicial acceptance of this type of zoning legislation since *Euclid*, and there is now a definite trend in support of the idea that aesthetic considerations alone may justify this exercise of the zoning power.

Another method of achieving reclamation is to restrict strip mining severely on slopes of a certain steepness. This would reduce the risks of landslides and other damage brought about by contour strip mining.

Another alternative is the institution of watershed regulation and extensive preplanning. Such a procedure might consist of, first, recovering any marketable timber from the area to be mined; second, requiring operators to build earthen dams with concrete stand pipes and adequate spillways in designated locations within the watershed for the purpose of containing sediment damage to the areas immediately adjacent to the mining operation; finally, implementing additional measures to control sediment and water run-off. Id. at 669-70.

25. ALA. CODE tit. 26, Secs. 166 (116)-(117), (119), (121)-(124), (127) (Supp. 1969).

26. At the time, pending before the Senate was S. 3132, a federal regulatory scheme for surface mining proposed by the Johnson Administration. The mining industry was able to establish strong opposition to the bill, and it never got out of committee. Nevertheless, the strip mining interests in Alabama felt that action at the state level was necessary to forestall further agitation for federal legislation. Note, *Regulation*, supra note 6, at 429.

> 27. The objective of this article is to provide for the safe and reasonable reclamation of lands upon which surface disturbances will be created by certain types of surface mining so as to protect the taxable value of property and preserve natural resources within the state and to protect and promote the health and safety of the people of this state, consistent with the protection of physical property and with maximum employment and the economic and industrial well-being of the state.

ALA. CODE tit. 26, Sec. 166 (116) (Supp. 1969).

28. Note, *Regulation*, supra note 6, at 432. Most other state statutes

vest this responsibility in an environmentally oriented agency. See, e.g., W. VA. CODE ANN. Sec. 20-6-6 (Supp. 1971):

> There is hereby created and established in the department of natural resources a reclamation commission which shall be composed of the director of natural resources, serving as chairman, the chief of the division of reclamation, the chief of the water resources division and the director of the department of mines.

See also PA. STAT. ANN. tit. 52, Secs. 1396.3, .4b (Supp. 1971); TENN. ANN. CODE Sec. 58-1523(i) (1968).

29. ALA. CODE tit. 26, Sec. 166 (127) (Supp. 1969).

30. ALA. CODE tit. 26, Sec. 166 (116) (Supp. 1969).

31. The Act does provide for hearings to be instituted by the Director after receiving complaints. If the hearing confirms existence of a violation of the Act, the Director can issue an order to the offender to take remedial action. This order is appealable to the circuit court. If the order is upheld on appeal the operator can still ignore the order and continue to mine; if he does, the Director can institute civil proceedings for injunctive relief or for forfeiture of the operator's bond. ALA. CODE tit. 26, Sec. 166 (123) (D) (Supp. 1969).

32. The lack of such a provision is rare in this type of legislation. Pennsylvania provides that a mine conservation inspector has authority to order an immediate halt in operations "where safety regulations are being violated or where the public welfare or safety calls for the immediate halt of the operations." The order is effective "until corrective steps have been started by the operator to the satisfaction of the conservation inspector." PA. STAT. ANN. tit. 52, Sec. 1396.1c (1968). See also MD. CODE ANN. art. 66c, Sec. 664(c) (1969); W. VA. CODE ANN. Sec. 20-6-14a (Supp. 1971).

> 33. Upon receipt by the department of such application, bond, or security, and fee due from the operator, the department shall issue a permit to the applicant which shall entitle the applicant to immediately engage in surface mining on the land described in the application for a period of one year from the date of issuance of said permit.

ALA. CODE tit. 26, Sec. 166 (119) (Supp. 1969).

Directors in other states are given discretion when confronted with a license applicant who has past history of violations. See, e.g., GA. CODE ANN. Sec. 43-1408 (Supp. 1970); MD. ANN. CODE art. 66c, Sec. 661(c) (Supp. 1969); PA. STAT. ANN. Sec. 1396.3a(b) (1966).

34. *Contra,* ILL. ANN. STAT. ch. 93, Sec. 180.13 (Supp. 1971); KY. REV. STAT. Sec. 350.990(3) (Supp. 1968); PA. STAT. ANN. tit. 52, Sec. 1396.16 (1968); W. VA. CODE ANN. Sec. 20-6-30(a) (Supp. 1971).

35. In Maryland, every applicant for a permit to surface mine coal must pay the Director a "special reclamation fee" of thirty dollars for each

acre of land affected. This fee, along with an equal amount contributed by the state, is deposited in the "Bituminous Coal Open-Pit Mining Reclamation Fund." The reclamation fee is paid in addition to the normal licensing fee and performance bond. MD. ANN. CODE art. 66c, Sec. 662(c) (Supp. 1969). Pennsylvania requires each applicant for a permit to submit "a detailed proposal showing the manner, time and distance for backfilling." The operator must also offer a plan to prevent surface water from draining into the pit. PA. STAT. ANN. tit. 52, Sec. 1396.4 (1968). Tennessee requires that an operator must submit with his permit application a reclamation plan including, but not limited to, proposals for covering the face of a coal seam, exposed auger holes, and all toxic materials which the commissioner feels are acid-producing or create a fire hazard. In addition, there must be acceptable drainage, water control, and grading plans and provisions for the removal of all refuse resulting from the operation and for revegetation. TENN. CODE ANN. Sec. 58-1529(a) (1968).

36. [A] survey by the Bureau of Mines of reclamation work conducted in 1964 by the major surface mining industries showed that, in the principal coal-producing areas, average costs of completely reclaiming coal lands ranged from $169 per acre in the South Atlantic States to $362 in the Middle Atlantic area. OUR ENVIRONMENT, supra note 6, at 90.

37. Maryland requires a bond of 200 dollars per acre and a minimum bond of 2,000 dollars. MD. CODE ANN. art. 66c, Sec. 663(a) (1969). Pennsylvania's bond requirements range between 500 and 1,000 dollars per acre and call for a minimum of 5,000 dollars. PA. STAT. ANN. tit. 52, Sec. 1396.4(g) (1968). West Virginia demands a bond of not less than 600 nor more than 1,000 dollars for every acre disturbed and a minimum bond of 10,000 dollars. W. VA. CODE ANN. Sec. 20-6-16 (Supp. 1971). Illinois requires at least 1,000 dollars and 200 dollars for every acre under five, or at least 3,000 dollars and 600 dollars for every acre over five, depending on the land to be mined and the relative ease of reclamation. ILL. ANN. STAT. ch. 93 Sec. 180.8 (Supp. 1971). Kentucky leaves the bond requirement to the limited discretion of the Director, who can demand between 100 and 500 dollars per acre. KY. REV. STAT. Sec. 350.060 (7) (Supp. 1968). Tennessee requires a bond of between 100 and 200 dollars for every acre disturbed. TENN. ANN. CODE Sec. 58-1528 (1968).

38. West Virginia law provides: "The [reclamation] plan shall be submitted to the director and the director shall notify the applicant . . . if it is or is not acceptable. . . . [The director] may . . . reject the entire plan." W. VA. CODE ANN. Sec. 20-6-9 (Supp. 1971). See also MD. CODE ANN. art. 66c, Sec. 662(b-1) (1969); PA. STAT. ANN. tit. 52, Sec. 1396.4(b) (1968); TENN. ANN. CODE Sec. 58-1529(a) (1968).

39. In contrast with ASMA, the West Virginia statute establishes pre-

cise standards for determining whether to accept or reject the map. The Act requires that three copies of a progress map, prepared according to rigid specifications, be furnished to the Department of Natural Resources. W. VA. CODE ANN. Sec. 20-6-9 (Supp. 1971). See also PA. STAT. ANN. tit. 52, Sec. 1396.4(a) (1968); MD. CODE ANN. art. 66c, Sec. 662(b) (1969); TENN. ANN. CODE Sec. 58-1529(b)(1) (1968).

40. KY. REV. STAT. Secs. 350.093, .095, .113, .117, .250, .990 (1966).

41. *Three Murdered Old Mountains*, LIFE, Jan. 12, 1968, at 66.

42. The eastern field is part of the Appalachian region and covers approximately 10,200 square miles in 31 countries. The western field covers about 6,400 square miles in fourteen counties. Schneider, supra note 24, at 653.

43. Schneider, supra note 24, at 663-64.

44. According to the Department of the Interior, surfaced-mined land annually destroys outdoor recreation resources valued at 35 million dollars, including 22.5 million dollars worth of fish and wildlife benefits. Hechler, supra note 3. A sizable percentage of this loss would be suffered by the eastern Kentucky region.

45. S. 993, H.R. 4704, 92d Cong., 1st Sess. (1971), are companion bills which lay out the Administration's program.

46. H.R. 4556, 92d Cong., 1st Sess. (1971), proposes the Environmental Protection and Enhancement Act of 1971.

47. Federal and Indian trust lands are exempted. S. 993, 92d Cong., 1st Sess., tit. II, Sec. 201(a) (1971).

48. Some of the more important criteria provide:

(1) the regulations require that each operator of a mining operation obtain a permit from a State agency established to administer the regulations, and file a mine reclamation plan . . .; (2) the regulations contain requirements designed to insure that the mining operation (i) will not result in a violation of applicable water or air quality standards, (ii) will control or prevent erosion or flooding . . .; (3) the regulations require reclamation of mined areas by revegetation, . . .; (4) the regulations require posting of performance bonds in amounts at all times sufficient to insure the reclamation of mined areas . . .; (9) the state agency or interstate organization responsible for the administration and enforcement of the regulations has vested in it the regulatory and other authorities necessary to carry out the purposes of this Act . . .; (13) the regulations are authorized by law and will become effective no later than sixty days after approval by the Secretary.

S. 993, 92d Cong., 1st Sess., Sec. 201(a) (1971).

49. *Hearings on S. 77, S. 630, S. 993, S. 1160, S. 1240, S. 1198, S. 2455 & S. 2777 before the Subcomm. on Minerals, Materials & Fuels of the Senate Comm. on Interior & Insular Affairs*, 92d Cong., 1st Sess., ser.

92-13, pt. 1, at 114 (1971) [hereinafter cited as *Surface Mining Hearings*] .

50. S. 993, 92d Cong., 1st Sess., tit. II, Sec. 201(e) (1971).

51. S. 993, 92d Cong., 1st Sess., tit. I, Sec. 102(e) (1971).

52. To include underground mining in federal legislation which intends to rely on the state's surface mining regulatory structure is inconsistent with the predicate underlying the federal-state approach to this problem. It would require the state surface mine land reclamation inspectors to acquire a complicated new expertise in a completely unrelated field.

Surface Mining Hearings, supra note 49, at 386 (statement of Carl E. Bagge, President of the National Coal Association).

53. Hechler, supra note 3.

54. Interior Department performance in the area of mining on federal and Indian lands has been characterized by vague and conflicting lines of authority and serious understaffing in critical positions. In 1969, the Department formulated regulations placing great responsibility for environmental protection in the Bureau of Land Management and the U.S. Geological Survey. These regulations were completely inadequate. They did not apply retroactively: as a result there is no regulation of the vast acreage of coal land leased prior to 1969 (some 2.4 million acres). These regulatory deficiencies suggest serious administrative inadequacy on the part of the Department. Moreover, there has been no evidence of correction:

For example, the Department has been reluctant to prepare environmental impact statements under the National Environmental Policy Act before issuing strip mine leases or mining permits, despite the major environmental implications of strip mining.

Surface Mining Hearings, supra note 49, at 517-18 (testimony by the Conservation Foundation).

55. H.R. 4556, 92d Cong., 1st Sess. (1971).

56. During the past thirty years, the Hanna Coal Co. has graded approximately 27,000 acres of surface-mined land. Of this total, 12,000 acres have been seeded with native grasses and legumes and another 15,000 have gone to crownvetch. A deep-rooted legume, crownvetch helps to prevent soil erosion and is beneficial to both the animals which feed on it and the ground in which it is planted. *Surface Mining Hearings*, supra note 49, at 320-21 (statement by R. W. Hatch, President of Hanna Coal Co.).

57. Id.

58. See Brooks, supra note 1, at 17.

59. *Surface Mining Hearings*, supra note 49, at 143 (statement of Russell Train, chairman of the Council on Environmental Quality).

60. Coal is the primary fuel for electric generating plants, and surface-mined coal constitutes almost sixty percent of the coal burned by the electric utilities industry. In 1970 the electric utilities generated a total

of 1.5 trillion kilowatt hours, including the amount produced by the great hydroelectric dams, and 28.2 percent of this electricity was produced from surface-mined coal. Id. at 316, 331.

61. Id. at 331.

62. Id. at 332.

63. Id.

64. S. 2455, 92d Cong., 1st Sess. (1971).

65. There is some legislation before Congress which provides for the partial prohibition of strip mining if necessary to protect the environment. A bill introduced by Senator Gaylord Nelson (D.-Wis.) states: "If warranted, the Secretaries [of Agriculture and the Interior] may prohibit strip and surface mining in areas where reclamation is considered unfeasible because of physical considerations, such as ground-surface slope, but not limited thereto." S. 77, 92d Cong., 1st Sess. Sec. 101(b)(7) (1971).

66. S. 77, S. 630, H.R. 6482, 92d Cong., 1st Sess. (1971).

67. *Surface Mining Hearings*, supra note 49, at 507.

68. In view of past experience and the general condition of the state laws in this area, it is quite likely that this six-month requirement would result in direct federal control of many strip mines.

69. *Surface Mining Hearings*, supra note 49, at 506-40.

70. Id. at 507.

71. For purposes of this proposal, mining on land with an average slope of thirteen degrees or more will be considered contour mining. When slopes are greater than thirteen degrees, reclamation becomes infeasible, and mining in such areas should not be allowed to continue. This is recognized in some state regulations. See, e.g., KY. REV. STAT. Sec. 350.093 (1966).

72. Coal production from area strip mines could, in fact, be allowed to increase after six months, when contour stripping would cease, so long as reclamation were both feasible and strongly enforced. The result of this scheme would be that area strip mining would be governed by standards proclaimed (or approved) by EPA, whether the land is Federal, State, Indian or private, six months from the date of enactment of the Act.

Surface Mining Hearings, supra note 49, at 527.

73. Id.

74. Id. at 529.

75. Id.

76. Id. at 528.

77. Data from the National Coal Association reveal that most plants buying coal can convert with relative ease to oil or gas. In the regions likely to use contour-mined coal, 14,161,000 tons of coal could be replaced by oil or gas (43,000 in the Middle Atlantic region, 48,000 in the South, 14 million in the Midwest, 33,000 in New England and

37,000 in the Border States). Id. at 527, 532.

    78. Id. at 532.

    79. Id.

    80. Id. at 531.

    81. Id. at 152.

16
The Control of Surface Mining:
An Exercise in Creative Federalism

Fred P. Bosselman

Economic and technological changes have impelled the mining industry toward increasing reliance on surface rather than underground mining as the most economical method of extracting minerals from the earth.[1] As surface mining has increased in complexity and scope, an inevitable consequence has been a massive rearrangement of the surface features of the land that is being mined.

It is indicative of the scorn for public relations shown by the mining industry of the pre-Ladybird era that the refuse dumped upon the earth in the process of surface mining has historically been known as "spoil," a term more commonly used to define plunder taken from an enemy in war. But today most mining executives, who cringe at both the engineering and public relations techniques of their predecessors, are now spending large sums to improve both their conservation practices and their euphemisms—yesterday's "spoil piles" have become today's "new land."[2]

To the public, however, the huge scars left on the landscape by past surface mining operations are more impressive than promises of future improvement.[3] The result has been a loud and effective popular outcry for governmental regulation of the surface mining industry and its effect on the landscape.

The response to this outcry has been felt at all levels of government. The resulting contest, in which federal, state, and local authorities each vie for position while the conservation groups and affected industries push and shove from the sidelines, is an interesting test of the ability of federalism to produce regulatory systems at three levels of government which neither duplicate each other nor leave gaping loopholes.

Reprinted by permission from 9 *Natural Resources J.* (1969), published by the University of New Mexico School of Law, Albuquerque.

Fred P. Bosselman is a member of the Illinois Bar. Mr. Bosselman is associated with the firm of Ross, Hardies, O'Keefe, Babcock, McDugald & Parsons, 122 South Michigan Ave., Chiago, Ill. 60603.

I. The Effects of the Industry on the Environment

Regulation of the surface mining industry can only be understood in the economic context of the surface mining industry as a whole, the benefits that it produces, and the harm that it causes.

The term "surface mining" refers to processes by which minerals are uncovered from the surface of the earth and then extracted. It is distinguished from "deep mining," in which a shaft is constructed to the mineral vein and the mineral is extracted through the shaft.[4] Typically, the surface mining process begins with the use of heavy equipment to remove the overburden and allow access to the minerals. The equipment may range from a bulldozer used in a small sand and gravel operation to a mammoth shovel over fifteen stories high used to move 200 cubic yards of overburden at a bite in a large coal mine. When the overburden is removed, the minerals are extracted, usually by smaller mechanical equipment.[5] In some cases blasting may be necessary.[6] When the minerals have been extracted, the land consists of excavated areas plus piles or ridges of deposited overburden (spoil). If the mine is deep in relation to its surface area, the process may be called open pit mining or quarrying.[7] In hilly areas surface mining is often conducted by the "contour method"; the overburden above the outcrop is removed and dumped down the hill below the outcrop. The minerals are then removed, leaving a steep "highwall" and a flat "bench."[8]

Although surface mining is used to remove many types of minerals, its most common uses, and those which have created the greatest problems, are the mining of coal, sand, gravel and stone,[9] and it is these industries that will receive primary consideration here.[10] The extraction of other minerals by surface mining is usually on a localized basis, as in the case of phosphates,[11] iron[12] and copper.[13] While these industries are subject to some of the same problems as the rest of the surface mining industry, this article will not deal with their special characteristics which may require individualized consideration.

All types of surface mining have a direct impact on the immediately surrounding area and also, in more indirect fashion, on larger areas. Surface mining's immediate effects may be classified into five categories: air pollution, water pollution, safety and health hazards, noise and vibration, and aesthetics.

The air pollution problems resulting from surface mining are caused by dust and other fine particulate matter generated by the mining process.[14] These problems can result in an extremely disagreeable nuisance for residences and businesses in the immediate area;[15] the pollution is localized, however, and amounts to no substantial proportion of the regional air pollution problem.[16]

In many parts of the country the mining industry is the primary cause of water pollution problems,[17] and water pollution is often said to be the most serious of the problems caused by surface mining.[18] Acid-water pollution typically results from the uncovering of materials which, when exposed to air or surface water, form acidic solutions of intensity sufficient to destroy fish life and create problems of odor and water purification.[19] The control of acid-mine pollution is a technical problem that remains to be solved.[20]

The extent of acid-water pollution caused by mining is closely related to the type of minerals found in the soil. Many areas of the country are thus not affected by this problem because the overburden contains no significant amounts of acid-forming minerals.[21] But even where such minerals are present, careful surface mining operations can often insure that they are buried in such a manner that they do not come into contact with air or water.[22]

Much of the acid-water pollution problem is caused by abandoned mines in which acid-forming materials have been allowed to remain on the surface.[23] Some of these are abandoned surface mines, but a substantial portion of the acid pollution problem is also caused by abandoned shaft mines.[24]

Sedimentation is another type of water pollution caused by certain types of surface mining.[25] Sedimentation resulting from surface mining is frequently the cause of flooding in downstream areas.[26] The methods of controlling sedimentation, frequently through water impoundments to allow settling, are usually easier than the control of acid-water pollution.[27]

Safety and health hazards of various types frequently result from surface mining operations. Such operations are characterized by steep slopes and, in many cases, by deep pools of standing water, which constitute an attraction and a hazard for young children.[28] In hilly areas landslides are an everpresent danger for those who live in the valleys.[29]

In agricultural areas farmers complain that grazing cattle often fall on the precipitous and unstable terrain of strip mining areas,[30] that the areas harbor foxes and other predators which prey on agricultural livestock,[31] and that mosquitoes breed on the stagnant pools.[32]

In addition, surface mining operations involve heavy equipment transporting large quantities of materials from the mine to market, spreading dust and debris, and increasing traffic hazards on roads and highways in the area.[33]

The process of surface mining is inherently noisy; there is no quiet way of moving tons of earth and rock. The operation of surface mining equipment can typically be heard with annoying clarity at a distance of several hundred yards from the site,[34] and it is difficult to decrease the noise level except by increasing the distance between the listener and the source of the noise.[35]

The blasting associated with quarrying operations may cause annoying vibrations for a substantial distance, although the extent of actual harm resulting from blasting is a matter of frequent contention.[36]

Perhaps the most serious but least tangible effect of surface mining is in the amorphous area of aesthetics. It is generally agreed that an 80-foot high pile of gray rubble rising from the flat countryside like the tunnel of some monstrous mole is aesthetically undesirable.[37] Nor can an aesthetic case be made for the denuding of mountains and the destruction of scenic valleys by the gashes of contour mines.[38] Disagreement arises, however, when an attempt is made to determine the extent to which the aesthetic disadvantages of surface mining should be weighed as an important factor in determining where its location should be permitted.[39] The current public mood indicates, however, that aesthetics will be a strong factor in regulatory decisions, perhaps to a greater degree than would have been anticipated by unemotional cost-benefit analysis.

Surface mining also affects the economy of the nation in a more indirect fashion. To the extent that strip mining leaves land in an unusable condition, it reduces the total available supply of land for productive uses. With the booming population growth of the nation, the existing supply of land in the urbanized regions of the country is proving inadequate to handle all of the uses which society deems desirable.[40] To the extent that surface mining is carried on in such areas, it reduces the supply of land available for other uses.

In agricultural areas, also, the amount of high quality arable land is a finite quantity; to the extent that such land is withdrawn for surface mining and not reclaimed for agricultural use it brings closer the day when such land will be in short supply.[41]

In addition, extensive surface mining can have a catastrophic impact on the tax base of individual counties by creating derelict, non-taxpaying land that makes it difficult for the county to meet its obligations.[42]

II. Pressure Points

Successful regulation depends upon the application of just the right amount of pressure at the right time and place to alleviate the industry's impact on the environment. There are three crucial points at which regulatory pressure can be applied to the surface mining industry: (1) at the time of selecting the mining sites, (2) at the time of choosing the method of operating the mine, and (3) at the time of picking the method of reclaiming the land.

A. Selecting the Site

The selection of a site is the initial step in the surface mining process and a crucial one. The indispensable prerequisite, of course, is the presence of minerals of a quality and quantity to justify their extraction, and under feasible conditions of depth, geology, drainage, etc.[43] The suitability of the land for other purposes also affects the selection of sites for surface mining. If, because of its location, land is valuable for urban uses, it becomes less feasible to use the land for surface mining.[44] A third factor of great importance is the proximity of markets and methods of transportation, because transportation costs are very significant elements in the total delivered price of the extracted minerals.[45] The nuisance aspects of surface mining must also be considered in determining mining locations; the more densely populated the area around the mine the more people will be affected by noise and dust.[46]

Ideally, site selection should be regulated to insure the optimal location of mining sites on the basis of a comprehensive plan taking all these factors into consideration.[47] In practice, however, to the extent there is any public control over site selection it is usually motivated by everybody's desire to keep the mine out of their area. Increasingly, however, regional planners are becoming aware of the need for regulation both to preserve valuable mineral resources as well as to control the nuisance aspects of surface mining, and some preliminary attempts are being made to regulate land use in such a way as to maintain the accessibility of mineral resources.[48]

B. Choosing the Mode of Operation

Many of the harmful effects of surface mining can be controlled by regulating the way in which the mine is operated, particularly if the regulation can be imposed before equipment is purchased and operations begun.[49] For example, the noise of heavy equipment can be somewhat alleviated through the choice of proper equipment and through the planning of excavation timing to create sound barriers using the spoil material;[50] properly planned burial of acid-forming materials can prevent water pollution;[51] proper location and fencing of pits can alleviate safety hazards,[52] and even better forms of regulation might be devised if ingenuity were applied at an early stage.

At this time, however, regulation of operating methods is practiced only to a very limited extent, and in relatively primitive fashion. In fact, the operations of most mines are unregulated except for state safety regulations designed for the protection of mine labor.[53]

C. Determining the Reuse of the Land

Reclamation is a term rather loosely used to cover a wide variety of methods of improving the condition of the residual material that remains after minerals have been extracted by the surface mining process. In its strictest sense reclamation can mean the careful grading of the land and replacement of the topsoil in such a manner that one would not be aware that any mining had taken place.[54] Unfortunately, however, the term's coinage has been devalued by operators who throw a few grains of annual rye at a pile of gray slag and then depart alleging that they have "reclaimed" the land. As Senator Nelson described it, some reclamation "is really a kind of a green lie. When you look at it closely it is crabgrass and quack grass and brush. . . ."[55]

Reclamation varies greatly depending on the future reuse that is planned for the reclaimed land. In most cases, the reclamation process involves two operations. The first is grading: ordinarily, surface mining of level land leaves the land in a series of steeply sloped ridges plus one or more long, narrow bodies of impounded water.[56] Even the most rudimentary reclamation usually requires some degree of leveling of the sharp tops of the ridges in order to obtain access to the area by vehicles or animals.[57] In many cases any growth of vegetation and retention of soil is impossible without intensive grading and terracing to reduce soil erosion.[58] Of course if crops are to be grown, land must be leveled sufficiently to allow agricultural equipment to be operated.[59]

Where hilly or mountainous land is mined using the contour technique, the result is long sinuous slashes winding around the sides of hills and valleys.[60] The highwall rises at a steep gradient above the ruined area, which may contain impounded water, while below lies a landslide-prone slope of tree stumps and mud.[61] In such areas reclamation grading requires the restoration of something approaching the original slope, a much more difficult operation than on flat land.[62]

The second part of the reclamation process is the establishment of vegetation.[63] In the typical surface mining operation the topsoil, being removed first, is dumped on the bottom of a pile, which is subsequently covered by subsoil and then by large quantities of rock and minerals.[64] This bare rubble will not ordinarily support plant life until many years of weathering produces a new and rudimentary form of soil,[65] and if acid-forming minerals are contained in the rock it may never be possible to establish vegetation.[66]

The extent to which it is physically possible to establish vegetation depends on a number of factors. If during the original excavation the topsoil and subsoil are retained separately and spread back over the rock when the mining is completed, the growth of vegetation becomes much more feasible.[67] Other artificial methods, such as hydraulic seeding, may also be used to promote the growth of vegetation.[68] Where the grades are steep, it is difficult to establish vegetation before it erodes away.[69] Perhaps of primary importance is the condition of the soil in the area to be mined— stoniness and acidity are deterrents to plant growth.[70]

Regardless of external conditions, however, it is uniformly true that the earlier in the mining process that reclamation plans are made, the more feasible reclamation becomes, while on land that has already been mined by primitive methods reclamation is much more difficult.[71]

There are a number of beneficial uses which may be made of reclaimed land. Perhaps the most valuable is to make the land available for the growing of crops. While there have been occasional experiments in which mined land has been made to produce reasonable quantities of corn and other grains,[72] the economics of this type of reclamation is probably advantageous only in special situations.[73] More generally feasible is the reclamation of land for the growth of less demanding crops such as alfalfa;[74] reclaimed land has also been used for orchards, nurseries, and tree farms.[75] Still more common is the use of reclaimed land for the grazing of cattle, which requires less grading than is necessary for the growing of crops.[76]

In many cases the most economical and useful method of reclamation is to convert the land to recreational use.[77] The man-made lakes can be used to support wildlife or for boating and swimming, while the rolling topography can be a natural asset for many purposes.[78] If the land is in proximity to urban areas, a wide variety of other uses may be considered.[79]

III. The Effect of Regulation on Industry

The impact of these various forms of regulation—site selection, mode of operation, and reclamation—is felt in a variety of ways. While the "surface mining industry" may be thought of as a single industry in terms of its effect on the environment, it cannot be thought of as a single industry when considering the impact of regulation. The two major mined commodities, aggregates (sand, gravel, and stone) and coal, are affected by regulation in much different ways.

A. Aggregates

Sand, gravel, and crushed stone are the elements that form the primary ingredients of concrete, the major structural element of streets and highways and of most large buildings.[80] These minerals are sufficiently common to be relatively inexpensive, but because of their great weight the cost of transporting them constitutes a large segment of the total cost of concrete construction. The average cost of sand and gravel runs only about $1.16 per ton,[81] but the average cost of moving it to construction sites is about 20 cents per ton-mile.[82] Thus, if it had been necessary to move each of the 680,000,000 tons of sand and gravel used by the construction industry in 1966 just one more mile, the cost of construction through the nation would have increased by $136,000,000.[83]

It is apparent, therefore, that when sources of sand and gravel are removed greater distances from construction sites the increased costs are paid for by the public in the form of increased costs of public roads and buildings. The economic benefits of having supplies of sand and gravel close to urban areas constitute a substantial dollar amount which must be weighed in the balance against whatever injuries to the environment are caused by the sand and gravel industry.[84]

Because of the key importance of site location to the sand and gravel producer, it is frequently economic for him to spend substantial sums to alleviate the nuisance characteristics of his operation rather than move to a more distant site.[85] Furthermore, if his site is located in an urbanizing area it often has a high potential value for other uses once the minerals are exhausted.[86] For these reasons the crucial regulatory decision for the sand, gravel, and stone producer is the grant or denial of permission to use a particular site. In exchange for such permission he may be willing to undertake substantial modifications in his operations and a large degree of land reclamation.[87]

B. Coal

The other major product of surface mining, coal, occupies a much different economic position. Formerly the nation's most commonly used fuel for heating and transportation purposes, coal has gradually been replaced by other fuels in a technological revolution that shows no sign of abating.[88] At present the major use of coal is as a low-cost, low-quality fuel in the generation of electric power and for other heavy industrial uses.[89] The past decade has seen a substantial reduction in the delivered cost of coal for power generation as the coal and railroad industries have fought to maintain their position in this highly competitive fuel market.[90] Even this price reduction, however, may not be sufficient as increasingly efficient nuclear power plants enhance their economic advantages at a faster rate than anyone had predicted. During the first half of 1967, plans were announced for 23 new electric generating plants to be powered by nuclear fuel.[91] While coal expects to rebound, there is little question that the coal industry is in a dangerously competitive position in which it must cut its costs to the bone in order to remain alive.

The nation's coal reserves are extensive and widely distributed; a company seeking to open a new mine usually can find a number of alternative sites located in under-populated areas where permission to mine could not reasonably be refused.[92] Thus control over site selection is not as crucial to the coal producer as to the producer of aggregates. To the extent, however, that government regulation increases the cost of mining through control over operations and reclamation it seriously weakens coal's ability to compete, and threatens the possibility of substantial economic dislocation in coal mining areas.[93] For this reason, the regulatory decision which is most crucial to the coal industry is the extent to which governmentally imposed requirements increase the cost of operations.

C. Cost Impact

It is thus apparent that the differing economic positions of aggregates and coal also result in a completely different pattern of absorption of the cost of regulation. The demand for aggregates being relatively inflexible and unaffected by competing products, any increase in cost is borne by society in the form of increased costs of new construction.[94] Because the demand for coal is highly flexible, however, increases in the cost of coal are likely to result in switches to alternate fuel sources causing substantial decreases in the amount of coal mined.[95] Thus, an increase in the cost of surface mining of coal may have less direct impact on the consumer than an increase in the costs of surface mining of aggregates, but may cause much greater economic dislocation. All of these factors must be weighed in determining the overall impact of any regulation of surface mining.[96]

Although the various mining industries are highly susceptible to injury from unwise regulation, they have not always paid adequate attention to their public image. Some leaders in the coal[97] and sand/gravel industries[98] have proclaimed the importance of "cleaning up" mining operations to prevent overreaction in the regulatory sphere, but it is questionable whether the industry as a whole has really responded.[99]

The inability of industry leaders to control the many operators has created substantial public pressure upon governmental agencies to exert greater control over the industry. This pressure has been felt at state, local, and federal levels, and there have been responses at each level.

IV. The Division of Regulatory Power

A. State Regulation

State action in enforcing reclamation did not begin until 1939 when West Virginia passed a statute imposing nominal reclamation requirements.[100] During the late 40s and early 50s a number of other states passed surface mining legislation.[101] Such state legislation was quite "mild" in nature, containing numerous exemptions,[102] and was typically motivated by the industry's desire to convince local governments that something was being done about the surface mining problem, thus deterring them from the adoption of strict regulatory ordinances.[103] Overall, state reclamation legislation may have reduced rather than increased the total amount of reclamation undertaken because it may have successfully deterred many local governments from imposing their own reclamation requirements.[104]

More recently, however, efforts have been made by conservation groups to strengthen the state legislation in almost every state in which surface mining is a substantial industry.[105] In some cases these efforts have been successful in substantially toughening the reclamation requirements.[106]

In addition to reclamation, the field of water pollution control is another one in which state agencies are the primary regulatory authorities. The federal Water Quality Act of 1965 delegated to the states the power to set and enforce standards of water quality with the Interior Department authorized to prod lagging states and mediate interstate disputes.[107] Mine-caused pollution is typically handled by a state agency with overall responsibility for water pollution control rather than being treated as a separate problem.[108]

The legal power of the states to impose reasonable reclamation requirements has been upheld by the courts and now seems clearly established.[109] In the area of water pollution control the constitutional impediments to state control of interstate waters seem to have been overcome by the 1965 Act.[110]

B. Local Regulation

The selection of appropriate sites for surface mining activities and the control of mining operations are largely free from state or federal control, being traditionally handled on the local level.[111]

A wide variety of forms of regulation are available to local governments. Well planned local regulation of the surface mining industry will have two goals. The primary goal is the protection of surrounding property from undue adverse effects from the surface mining process through control of site selection, and by regulation of mining methods. The secondary goal is to insure that sites for extraction of sand, gravel and other necessary products are located within reasonable distance of construction sites in order to minimize construction costs.

Local regulation of surface mining usually raises constitutional questions relating to whether the regulation is so severe as to constitute a taking of private property without just compensation or in violation of due process of law. In the early cases the owner of land containing undeveloped minerals typically argued that these minerals were property owned by him, and that if the government prohibited their extraction it constituted a taking of his property without compensation.[112] As early as half a century ago, however, the United States Supreme Court adopted what was generally interpreted as a very permissive view toward regulation of surface mining,[113] and as recently as 1964, in one of its extremely rare decisions in the field of land use control, the Court continued its benevolent attitude toward the regulation of surface mining by local governments.[114] In general, it may be said that the regulation of surface mining is not unconstitutional *per se*. An ordinance which simply prohibits surface mining, however, while it may be upheld in a densely developed area, is unlikely to survive judicial scrutiny where the jurisdiction encompasses any substantial amount of rural or undeveloped area.[115]

A technique more sophisticated than simple prohibition is a "mining ordinance" specifically regulating the detailed methods of operation of the surface mining industry.[116] Such ordinances raise the initial question of whether local governments have authority to regulate surface mining, either by statute or under a general grant of police power. It appears that the answer to this question will vary with the "home rule" philosophy of the particular jurisdiction.[117]

Another legal question raised by a mining ordinance is whether it conflicts with state regulation of such subjects as reclamation of mined land, mine safety practices which affect employees, and water pollution control. When local governments enact "similar" regulations a question is raised as to whether the state intended to preempt the field and prevent such regulation by local governments.[118] The difficult factual question, of course, is to decide when local regulations are "similar" to state regulations.[119] If the subject matter of the regulations is found to be "similar," but the local government is attempting to require compliance with a stricter standard than is required by the state legislation, the courts must construe the state statute to determine whether the state intended that local governments be permitted to exercise such power.[120]

It has been suggested that local governments employ exactly the same standards for reclamation as the state government, merely adding their own enforcement powers to assist those of the state agency.[121] Whether this is permissible depends on the language of the statute and the law of the particular jurisdiction.[122]

The other common method of local control of surface mining is as part of a comprehensive zoning regulation. The advantage of using the zoning technique, as opposed to the mining ordinance, is that it eliminates any question as to the authority of the municipality to regulate surface mining, and reduces the possibility that the ordinance might be held to conflict with state regulations.[123] It also allows the municipality to use zoning's existing administrative procedures and precedents as a means of exercising control, and to take into consideration other aspects of overall planning which might be improper subjects of consideration in a mining ordinance.[124]

Zoning has a number of disadvantages, however. It is severely restricted in its ability to regulate existing uses. While a municipality has been permitted to put an existing quarry out of business through a mining ordinance,[125] the courts have generally refused to allow zoning ordinances even to prevent the expansion of existing mining operations.[126]

In most cases, however, the crucial legal question, whether a mining ordinance or a zoning ordinance is involved, is the reasonableness of the ordinance as applied to the particular facts of the case. The courts have found some ordinances reasonable[127] and others not.[128] While there is some indication that a court will look more closely at the reasonableness of a specific application of a zoning regulation than of a regulation designed to promote public safety,[129] the cases do not indicate that this is typically the controlling factor.

To summarize, if (a) control of existing uses is especially important, (b) the question of reasonableness is likely to be acute, and (c) the local government has statutory or "home rule" authority to enact an ordinance created specifically to control the mining industry, the local government may be on safer ground with such a separate mining ordinance. But where preemption may be a problem and comprehensive planning factors may be highly important, then zoning may be the preferred method.[130]

When zoning is used to regulate surface mining a number of different techniques may be employed. The two most common methods of controlling site selection are through the use of a special district for surface mining activities[131] and through the use of the special exception or special use permit for surface mines.[132]

The nuisance problems of mining operations can be regulated through the use of performance standards designed to control all industrial processes including surface mines;[133] additional specific limitations may be applied to surface mines regulating, e.g., the hours that machinery may be operated, the fencing of dangerous areas, etc.[134]

The secondary goal of surface mining regulation—the use of zoning to preserve land for surface mining use—is a relatively new phenomenon for which there is little direct precedent. The most direct way of accomplishing such a result would be "exclusive zoning" of particular areas in which surface mining and no other intensive use would be permitted.[135]

As a practical matter the preservation of land for surface mining uses is often achieved by zoning land exclusively for industrial uses in an area in excess of that immediately demanded by industry.[136] If economically productive mineral deposits are located in such an area and if industrial development is slow, these deposits will remain available for a considerable period.

Another method of reserving mining sites is to permit residential development in mineral resource areas but to reduce allowable densities to rural levels.[137] This would amount to a program of development timing in which low density residential uses would be permitted until such time as development of the sand and gravel deposits becomes economically feasible.[138]

As this sampling indicates, modern zoning is a highly flexible regulatory method offering a considerable variety of zoning techniques for control over the selection of mining sites and the methods of mining operations.

C. Federal Regulation

Federal regulation of strip mining has historically been minimal. As the owner of extensive lands the federal government has certain rights to control mining operations on those lands, but has only recently begun to exercise them.[139] Only about 3 percent, however, of the land being surface mined is under federal ownership.[140] The federal government has also affected site selection on federal lands to a degree.[141] The Tennessee Valley Authority, as the nation's largest consumer of coal, has recently instituted the practice of obtaining contractual requirements from its suppliers that certain reclamation practices be instituted,[142] and some land has been reclaimed under the "Appalachia" program.[143] In the main, however, the federal government has affected only the fringes of the surface mining industry.

Pursuant to Congressional direction in 1965 the Department of the Interior conducted an extensive investigation of strip mining, first in the Appalachian states[144] and then in the nation at large.[145] As a result, the Department of the Interior proposed new federal legislation which would increase the federal funds available for the reclamation of derelict mined land and polluted waters, but which would impose only nominal regulatory requirements on future mining except in those states which have not adopted their own mining regulations.[146] Senator Lausche of Ohio, who was for years an adamant foe of irresponsible surface mining, had also sponsored a bill which would enact the substance of the Interior Department's recommendations into law,[147] and hearings were held in the spring of 1968.[148] Whether or not the federal legislation passes, it appears likely that no substantial change will be instituted in the current practices. The primary responsibility for regulation of surface mining will remain with the state and local governments.

V. What Has Federalism Created?

We are currently witness to "creative federalism" in the process of creation. In response to the stimulus of increased public pressure for control over the surface mining industry, the federalistic governmental process is creating regulatory power and dividing it among the various units of government through some Darwinian process that, magically, seems likely to result in a reasonably rational pattern of governmental control over surface mining. While, in general, the prognosis is favorable, there are at least two gaps that appear in the current pattern of regulation:

1. *More thorough consideration should be given to the future use of mined land.* The present state reclamation statutes typically provide that the mine operator files with the state conservation department a plan of reclamation showing a proposed reuse of the land after the mining is completed.[149] The future use of the land is then negotiated between the operator[150] and the state conservation agency without any consultation with the local government or with any other state agency. The mine operator is interested in saving money. All too often the state conservationist is primarily interested in getting some sort of quick cover on those embarrassing piles of rubble. The result has been a tendency toward immediate planting of cheap grass or seedling trees on the bare rock so that the operator and the state agency could transfer those acres to the column titled "undergoing reclamation." The long range results, however, are shown by the recent survey by the Department of the Interior—only 15 percent of the sampled sites were adequately covered by vegetation to provide protection for the sites.[151]

Reclamation should involve a longer range and much more comprehensive look at the land. Such a careful evaluation will often improve the final result. For example, in some instances a delay permitting land to settle and weather before planting might improve the land's prospect of ultimate rehabilitation.[152] In other cases consideration should be given to using a tract of land for crops or orchards rather than, as state conservation agencies are likely to favor, game preserves and recreation areas.[153] Advice might be obtained regarding other uses which, with a slight increase in time and investment, might greatly increase the value of the land to the operator.[154] Those sites which can be seen from areas used by the public—60 percent of the Interior Department's sample[155]—might be accorded different treatment than isolated areas.[156] These and many other factors could be considered if a team of conservationists, land planners, agronomists, resource economists, landscape architects and other professionals were given adequate time and budget to plan the reuse of the land on a long range basis.[157]

The local governments should also be consulted. At present the interests of the local governments are being neglected in the decision-making regarding the reuse of mined land. In many cases whole counties have been impoverished by the creation of vast derelict acreage.[158] If the land is not returned to productive, tax-paying use the remaining landowners in the jurisdiction must bear a highly inequitable burden of taxation to maintain local services.

The control of land use has traditionally been the province of local government. While state participation is necessary to insure that each local government doesn't insist that each spoil pile be turned into an industrial park, there is no excuse for denying local governments any voice whatsoever in the future use of large tracts of private land within their jurisdiction.

In summary, the long-range interests of the state are best served if the state conservation agency does not assume the sole burden of regulating the future use of mined land. While the professional expertise of conservationists is essential to the supervision of the reclamation process, the conservationist is not the only expert in the reuse of land. This power might best be given to a committee appointed by the state planning agency, the conservation agency, and the local governments, assisted by a competent staff representing a number of professions. More thorough study by such a staff would greatly increase the likelihood that the long-range rehabilitation of the land would be productive.

2. *The State should exercise more control over the selection of mining sites.* Whereas local government has too little to say about the reuse of mined land, it has too much to say about whether land should be mined in the first place. This is particularly true in metropolitan areas. As urban development spreads farther and farther out into the countryside surrounding metropolitan areas the deposits of sand, gravel, and other low value minerals become covered by urban uses, and the sources of such minerals are pushed farther and farther away from urban centers, thus increasing costs for the construction of roads, buildings, and public improvements.[159] But each local government tends to look to the others to solve the problem. Each wants the cheap concrete but not the mine. So each local government "zones out" the available deposits of sand and gravel and encourages their permanent interment under residential subdivisions and shopping centers.

Responsible regulation of surface mining should not permit its extinction. Responsible regulation should not only control the harmful effects of surface mining but should conserve the availability of mineral resources.[160] This function must be assumed by the state because the parochial nature of local governments decreases the likelihood of any one of them voluntarily accommodating an unpopular strip mine needed to serve the larger area.

To accomplish effective state conservation of mineral resources requires, first, a comprehensive survey of the state's mineral resources to determine their location and amount.[161] Second, the state must enact legislation authorizing the creation of conservation zones similar to those now employed in Hawaii.[162] Within such zones the land could not be devoted to intensive use until the underlying mineral resources had been extracted.[163]

Hawaii, with its high density of population and shortage of land is thus far the only state that has sufficiently appreciated the need for conserving its land resources to adopt such legislation. But as population continues to increase it will become more and more apparent, at least in our more populous states, that some form of state legislation for the conservation of privately owned land resources is essential. The sooner this realization becomes translated into law the sooner we will be able to halt the indiscriminate burial of valuable mineral resources, and the spiralling costs that ensue.

Notes

1. In 1965 surface mining produced about four-fifths of all the solid fuels and ores in the nation. U.S. Dep't of the Interior, Surface Mining and Our Environment 42 (1967).

2. Seastrom, *New Land Orchards,* in Proceedings, Pennsylvania State University Coal Mine Spoil Reclamation Symposium 129, 130 (1965) [hereafter referred to as Pennsylvania State Symposium].

3. Reporter Ben A. Franklin recently described the industry's reclamation efforts as "pocket handkerchief parks on a broad badland of unreclaimed razorback ridges of strip mining spoilbanks. . . ." N.Y. Times, July 16, 1967, at F13, col. 1.

4. It is recognized, of course, that the distinction between surface and deep mining is one of degree. See Emery, *What Surface is Mineral and What Mineral is Surface,* 12 Okla. L. Rev. 499 (1959); Wiltsee, *A Proposed Interstate Mining Compact,* in Proceedings of the Ky. Dep't of Natural Resources Symposium on Strip Mine Reclamation 31, 35 (1965) [hereafter referred to as Kentucky Symposium]. Technological advances increasingly blur the distinction. For example, Kennecott Copper Company has proposed the use of an underground nuclear blast to mine low-grade copper ore. The blast would be set off 1,200 feet underground and is expected to create a pile of ore 440 feet high and 220 feet wide. Wall St. Journal (Midwest ed.), Oct. 12, 1967, at 5, col. 3.

5. U.S. Dep't of the Interior, supra note 2, at 37.

6. See Phelps, *Current Practices of Strip Mining Coal,* in Proceedings of Univ. of Ariz., College of Mines Symposium Surface Mining Practices 1,

5 (Krumlauf ed., 1960); Kochanowsky, *The Coal Stripping Operation,* Pennsylvania State Symposium 72, 76 (1965).

7. U.S. Dep't of the Interior, supra note 2, at 33. This is obviously an overgeneralized description of a highly complex process that has many variations. For a more detailed description of the various surface mining methods see U.S. Dep't of the Interior, supra note 2, at 30-37.

8. The process is vividly described in Caudill, *An Offense Against America,* 68 Audubon 356 (Sept.-Oct. 1966).

9. U.S. Dep't of the Interior, supra note 1, at 110.

In the future the use of surface mining in the production of oil may be a substantial problem. Initial development of the Athabasca Sands in Alberta through surface mining is already underway. See Oil & Gas J., Oct. 23, 1967, at 69. See also Oil & Gas J., June 17, 1968, at 52; *Hearings on S.3132, S.3126 and S.217 Before the Senate Committee on Interior and Insular Affairs,* 90th Cong., 2d Sess., at 56.

10. In some instances surface mining is also accompanied by certain industrial processes. Coal, sand and gravel are customarily washed and transferred to rail cars or trucks for shipment. In addition, the presence of these minerals may attract manufacturing processes dependent upon them such as brickmaking, cement manufacture and mine-mouth electrical generating plants. See A. Bauer, *Simultaneous Excavation and Rehabilitation of Sand and Gravel Sites* 16-24 (N.S.G.A., 1965). This article does not attempt to cover the additional problems created by manufacturing processes. Because it is generally held that the right to operate a surface mine does not necessarily carry with it the right to operate a processing plant, such uses may be regulated separately. See Paramount Rock Co. v. County of San Diego, 180 Cal. App. 2d 217, 4 Cal. Rptr. 317 (Dist. Ct. App., 1960); Appeal of Mignatti, 403 Pa. 144, 168 A.2d 567 (1961). Cf. Hadacheck v. Sebastian, 239 U.S. 394 (1915). But see Silliman v. Falls City Stone Co., 305 S.W.2d 322 (Ky. Ct. App., 1957).

11. Of the 84,436 tons of phosphate mined in 1965 over 72,419 tons came from Florida. U.S. Dep't of the Interior, supra note 2, at 115. Phosphate spoils are rich in minerals and make good soil. Miller, *The Impact of Surface Mining on Land and Land Resources,* in Surface Mining—Extent and Economic Importance, Council of State Governments 20, 22 (1964). *Hearings on S.3132, S.3126 and S.217,* supra note 9, at 189-202, 222-63.

12. Iron mining is a particular feature of northern Minnesota, where in 1965 over half the iron ore was mined. U.S. Dep't of the Interior, supra note 2, at 116. See Zube, *A New Technology for Taconite Badlands,* Landscape Architecture 26 (Jan. 1966). See *Hearings on S.3132, S.3126 and S.217,* supra note 9, at 163-75, 263-75.

13. Over seventy percent of the copper ore is mined in Arizona and

Utah and almost ninety percent from the five leading states. U.S. Dep't of the Interior, supra note 1, at 115.

14. See generally Industrial Hygiene Foundation of America, Inc., *Evaluation of Dust and Noise Conditions at Typical Sand and Gravel Plants* (N.S.G.A., undated).

15. Consolidated Rock Products Co. v. City of Los Angeles, 57 Cal. 2d 515, 370 P.2d 342, 20 Cal. Rptr. 638, *appeal dismissed*, 371 U.S. 36 (1962). See Industrial Hygiene Foundation of America, Inc., supra note 13, at 19; City of San Diego Planning Dep't, Planning for Sand and Gravel in San Diego 40 (1966).

16. Surface mining does not ordinarily cause the type of chemical air pollution that affects property at substantial distances from the site or that constitutes severe hazards to health. Brooks, *Strip Mine Reclamation and Economic Analysis*, 6 Natural Resources J. 13, 23 (1966); U.S. Dep't of the Interior, supra note 1, at 56.

17. See, e.g., Bramble, *Reclamation of the Landscape*, in Beauty for America, Proceedings of the White House Conference on Natural Beauty 320 (1965); *Hearings Before a Special Subcomm. on Air and Water Pollution of the Senate Comm. on Public Works*, 89th Cong., 1st Sess., pt. 1, at 34 (May 1965) [hereafter referred to as Hearings on Air and Water Pollution].

18. "Most problems associated with strip mining are either directly or indirectly related to water." Johnson, *Research in Strip-Mine Reclamation*, Kentucky Symposium 11, 16. An Interim Report by the Secretary of the Interior to the Appalachian Regional Commission, Study of Strip and Surface Mining in Appalachia 26 (June 30, 1966).

19. ". . . I was 12 years old before I learned that creeks flowed in some colors other than black or orange." Address by J. M. Quigley, Proceedings of the National Symposium on the Control of Coal Mine Drainage, Div. of Sanitary Eng'r, Bureau of Environmental Health, Pennsylvania Dep't of Health, Harrisburg, Pa. 8 (1962) [hereafter referred to as Symposium on the Control of Coal Mine Drainage]. See also Sanitary Water Board, Pa. Dep't of Health, Control of Acid Drainage from Coal Mines 11 (1958); Klingesmith, *Technical Aspects of Control of Drainage from Active Mines*, Symposium on the Control of Coal Mine Drainage 35-38. Note, however, the following comments: "It is sometimes suggested, uncharitably, that it would be better for the State's economy to supply at public expense a few pounds of fish for each sportsman than to shut down entirely coal stripping projects along a stream which was never of much account 'fishwise' in any event." Dixon, *Report of the Mineral Law Section*, 34 Pa. Bar Ass'n Q. 456, 458 (1963).

20. Environmental Pollution, a Challenge to Science and Technology,

Report of the Subcomm. on Science, Research and Development to the Comm. on Science and Astronautics 29, U.S. House of Reps., 89th Cong., 2d Sess. (1966): "Millions of dollars have already been spent in abortive attempts to deal with the problem. . . . Mine drainage, whatever its effect on environmental quality, should not have funds spent on action programs until more palatable and sensible solutions can be devised." See Klashman, *Improvement of Acid Polluted Streams by Water Resources Development,* Symposium on the Control of Coal Mine Drainage 75; Sanitary Water Board, supra note 19; U.S. Bureau of Mines, *Mine Water Research: Neutralization* (Report of Investigations 6987, 1967); R. Meiners, *Strip Mining Legislation,* 3 Natural Resources J. 442, 462 (1964).

21. An Interim Report by the Secretary of the Interior to the Appalachian Regional Commission, supra note 18, at 23. See U.S. Fish and Wildlife Service, Extent of Acid Mine Pollution in the United States Affecting Fish and Wildlife (Fish and Wildlife Circ. 191, 1964). In some areas the pollution may be caused by caustic solutions rather than acidic. Klingesmith, supra note 19, at 37. Uranium mining wastes may create a danger of radioactive water pollution. See Federal Water Pollution Control Administration, Disposition and Control of Uranium Mill Tailings in the Colorado River Basin (March 1966); *Hearings on S.2947 Before the Comm. on Public Works,* Federal Water Pollution Control Act Amendments and Clean River Restoration Act of 1966, 89th Cong., 2d Sess., pt. 1, at 15-18 (July 11, 1966).

22. Sanitary Water Board, supra note 19, at 14; Klingesmith, supra note 19, at 38; Smith, *Strip Mine Reclamation and Water Pollution,* Kentucky Symposium 1-2; Meiners, supra note 20, at 462.

23. Kentucky Dep't of Natural Resources, Strip Mining in Kentucky 47 (1965).

24. U.S. Dep't of the Interior supra note 1, at 63. See also Hearings on Air and Water Pollution, supra note 17, at 34, 35. In some cases strip mining operations succeed in sealing off abandoned shaft mines that had been causing acid pollution. Jones, *Coal for Today, Timber for Tomorrow,* The Northern Logger 1-2 (June 1964).

25. An Interim Report by the Secretary of the Interior to the Appalachian Regional Commission, supra note 18, at 23-24. U.S. Dep't of the Interior, supra note 1, at 55.

26. H. M. Caudill, Night Comes to the Cumberlands 322-23 (1962).

27. See generally Kentucky Dep't of Natural Resources, supra note 23, at 44-47. See also Jones, supra note 24, at 3; G. D. Beal, *Common Fallacies About Acid Mine Water,* Sanitary Water Board, Pa. Dep't of Health 5-9 (1953); Federal Interdepartmental Task Force on the Potomac, Potomac Interim Report to the President 10 (Jan. 1966); 6 Air/Water Pollution Report 298 (1968).

In addition to polluting flowing water, surface mining often upsets the hydrologic environment by lowering the water table in surrounding areas. See H. D. Graham, The Economics of Strip Coal Mining 50 (Bureau of Economic and Business Research, Univ. of Ill., Bull. No. 66, 1948). But sometimes surface mining has beneficial hydrologic effects. U.S. Dep't of the Interior, supra note 1, at 64. *Hearings on S.3132, S.3126 and S.217,* supra note 9, at 315.

28. See Note, *Reclamation of Strip Mine Spoils,* 50 Ky. L. J. 524-536 (1962).

29. In Appalachia massive landslides have occurred on an estimated 1,400 miles of stripping contour. An Interim Report by the Secretary of the Interior to the Appalachian Regional Commission, supra note 18, at 22. U.S. Dep't of the Interior, supra note 1, at 24. Cf. Feiss, *Coal Mine Spoil Reclamation Scientific Planning for Regional Beauty and Prosperity,* Pennsylvania State Symposium 12, 21.

30. *Hearings on S.3132, S.3126 and S.217,* supra note 9, at 77.

31. See, e.g., Graham, supra note 27, at 50-51.

32. See City of San Diego Planning Dep't, supra note 15, at 41. But see Graham, supra note 27, at 70.

33. K. L. Schellie & D. A. Rogier, *Site Utilization and Rehabilitation Practices for Sand and Gravel Operations* 9 (N.S.G.A., 1963).

The safety of mine laborers themselves has historically been treated in a separate manner through a system of existing regulation which seems to have been unaffected by, and to cause no effects on, the other types of regulation herein discussed. In general, strip mining entails less risk for mine employees than underground mining. Graham, supra note 27, at 20; Feiss, *Surface Mining—Minerals, Metals and Divots,* in Surface Mining— Extent and Economic Importance, Council of State Governments 2, 6 (1964); Testimony of E. R. Phelps, *Hearings on S.3132, S.3126 and S.217,* supra note 9, at 124.

34. City of San Diego Planning Dep't, supra note 15, at 39.

35. See Industrial Hygiene Foundation, supra note 14, at 22. But see City of San Diego Planning Dep't, supra note 15, at 39.

36. Phelps, supra note 6. City of San Diego Planning Dep't, supra note 15, at 40.

37. "Vegetation of any kind is an improvement over no vegetation." McQuilkin, *Reclamation for Aesthetics,* Pennsylvania State Symposium 97, 100.

"Personal interviews brought from several people the statement that strip mining is 'downright wicked'." Graham, supra note 27, at 43.

See also Faltermayer, *How to Wage War on Ugliness,* Fortune, May, 1966, at 130, 132.

38. The Pennsylvania Power & Light Company has expended its own funds to screen spoil piles from highways because it believed prospective industries were repulsed by the general appearance of the area. Davis, *Mine-Spoil Revegetation Research in Pennsylvania—A Story of Coopera- tion in Action,* Symposium on the Control of Coal Mine Drainage 58, 65-66.

". . . strip mining on steep, mountainous terrain is wholly inconsistent with the preservation of natural beauty and the natural balance of life." Caudill, supra note 8, at 359.

39. See Brooks, supra note 16, at 24-25. Compare Hon. R. Ottinger, *Reclamation of the Landscape,* in Beauty for America, Proceedings of the White House Conference on Natural Beauty 332 (May 24-25, 1965) with S. M. Colby, *Reclamation of the Landscape,* id. at 343.

40. "[U]rban land is scarce, costly and getting scarcer and more costly every day." D. O'Harrow, *The Urban Future* 5 (N.S.G.A., 1962).

41. See E. Higbee, The Squeeze: Cities Without Space 168 (1960); J. Gottmann, Megalopolis 261-63, 321-22 (1961); Brown, *The World Outlook For Conventional Agriculture,* 158 Science 604 (Nov. 3, 1967). G. F. Deasy & P. R. Griess, *An Approach to the Problem of Coal Strip Mine Reclamation,* 33 Mineral Industries 1, 6 (Oct. 1963); Brooks, supra note 16, at 19; Graham, supra note 27, at 43; Ehrlich, *The Coming Famine,* 77 Natural History 6 (May 1968).

42. See Note, *Reclamation of Strip Mine Spoils,* 50 Ky. L. J. 524, 532 (1962); Caudill, *Paradise Is Stripped,* N.Y. Times, March 13, 1966 (magazine) at 26, 83; Graham, supra note 27, at 57-61. *Hearings on S.3132, S.3126 and S.217,* supra note 9, at 284-86; 113 Cong. Rec. A5976 (daily ed. Dec. 5, 1967).

43. O'Harrow, supra note 40, at 8.

44. In some cases deep mining methods can be used to extract minerals while preserving the surface of the land for other uses, but in many other cases valuable minerals remain in the ground because the value of the surface of the land makes it uneconomic to extract them. See, e.g., Colo- rado Sand and Gravel Producers Ass'n, First Complete Aerial Photo Map of Denver Metropolitan Area With an Outline of our Diminishing Gravel Resources (1957).

45. See Bauer, supra note 10, 14-15.

46. Ironically, however, current zoning practices are designed to put more people in the vicinity of nuisances by encouraging apartments in those areas which are not deemed suitable for single-family housing. Muhly v. County Council, 218 Md. 543, 147 A.2d 735 (1959). See R. Babcock & F. Bosselman, *Suburban Zoning and the Apartment Boom,* 111 U. Pa. L. Rev. 1040, 1060-61 (1963).

47. O'Harrow, supra note 40, at 10-11.

48. See e.g., W. Harrison, A Special Report on the Geology of Marion County, Indiana (1958); K. Rainey, *A Regional Approach to Planning and Development Is Necessary to the Solution of the Problems of the Coal Regions*, Pa. State Symposium 7; Address by G. Allen, Annual Convention of the National Sand and Gravel Ass'n, Feb. 7, 1962.

49. See generally, C. Johnson, Practical Operating Procedures for Progressive Rehabilitation of Sand and Gravel Sites (1966).

50. Industrial Hygiene Foundation, supra note 14, at 22.

51. Sanitary Water Board, supra note 19, at 14; Klingesmith, supra note 19, at 38; R. Smith, *Strip Mine Reclamation and Water Pollution*, Ky. Symposium 1-2.

52. See Colo. Rev. Stat. Ann. Sec. 92-10-1 (1960).

53. See Va. Code Ann. Secs. 45.1-34 to -101.

54. "[T]he ultimate objective should be to restore them [the mined lands] to a use similar to that of the lands contiguous to them." An Interim Report by the Secretary of the Interior to the Appalachian Regional Commission, supra note 18, at 46. Compare testimony of Secretary Udall at *Hearings on S.3132, S.3126 and S.217*, supra note 9, at 49.

See also J. Cunningham, *The Pennsylvania Strip Mine Reclamation Program*, Ky. Symposium 21, 22. In fact the displaced overburden swells and can never be leveled quite to the original grade. Phelps, supra note 6, at 9.

55. *Hearings on S.3132, S.3126 and S.217*, supra note 9, at 43. U.S. Dep't of the Interior, supra note 1, at 42, 74.

See also An Interim Report by the Secretary of the Interior to the Appalachian Regional Commission, supra note 18, at 26-29, 32.

56. U.S. Dep't of the Interior, supra note 1, at 34-35.

57. Deasy & Greiss, supra note 41, at 4. In the national survey conducted by the Department of the Interior only 24 percent of the sampled sites had received any grading at all. U.S. Dep't of the Interior, supra note 1, at 76.

58. P. Struthers, *Rapid Spoil Weathering and Soil Genesis*, Pa. State Symposium 86, 88.

59. The Department of the Interior survey found only 5 percent of the sampled sites had been graded to the degree necessary for agricultural equipment. U.S. Dep't of the Interior, supra note 1, at 76.

If the land is not to be used for row crops, complete leveling may be undesirable because it causes greater soil compaction. See Graham, supra note 27, at 68-70; L. Sawyer, *The Strippers*, Landscape Architecture 21, 22 (Jan. 1966); L. Sawyer, *Restoration of Areas Affected By Coal Mining*, Symposium on the Control of Coal Mine Drainage 52, 56; D. Hall, *Strip*

Mine Reclamation Under the 1964 Act (Western Ky.), Ky. Symposium 9, 10.

60. See note 8, supra. About one-half of the surface-mined coal is extracted by the contour methods. U.S. Dept. of the Interior, supra note 1, at 54.

61. Miller, supra note 11, at 21; An Interim Report by the Secretary of the Interior to the Appalachian Regional Commission, supra note 18, at 22.

62. H. Pyles, *Reclamation of the Landscape,* in Beauty for America, Proceedings of the White House Conference on Natural Beauty 317 (May 24-25, 1965); Bauer, supra note 10, at 33; Caudill, supra note 42, at 84; Feiss, supra note 29, at 18.

63. See generally, Schellie & Rogier, supra note 33, at 41-74.

64. Graham, supra note 27, at 19 G. Limstrom & G. Deitschman, Reclaiming Illinois Strip Coal Lands by Forest Planting 208, Univ. of Ill. Agricultural Experiment Station Bull. 547 (1951); G. Limstrom, Forestation of Strip-Mined Land in the Central States 3, Agriculture Handbook No. 166 (U.S.D.A., 1960).

65. L. Sawyer, *The Strippers,* Landscape Architecture 21, 22 (Jan. 1966).

66. In the national survey conducted by the Department of the Interior 48 percent of the examined spoil banks were too acid to grow good vegetation. U.S. Dep't of the Interior, supra note 1, at 56. See Davis supra note 38, at 61; An Interim Report by the Secretary of the Interior to the Appalachian Regional Commission, supra note 18, at 28.

67. See C. Crompton, *The Restoration of Waste Slate Heaps,* 33 Town and Country Planning 344, 346 (Sept. 1965); City of San Diego Planning Dep't, supra note 15, at 49; E. Stearn, *Surface Mining's Conservation Program Pays Off,* Coal Mining and Processing (April 1964).

68. J. Oxenham, Reclaiming Derelict Land, Faber and Faber Ltd., 24 Russell Square, London 115-16, 124, 137-40 (1966); Urban Land Institute, *New Engineering Concepts in Community Development,* 12 Technical Bull. 59 (1967).

69. G. Sullivan, Presentation Before the Illinois Mining Institute 9 (Springfield, Ill., Oct. 1963); An Interim Report by the Secretary of the Interior to the Appalachian Regional Commission, supra note 18 at 26-28. See also supra note 58.

70. Jones, supra note 24, at 1; M. Heddleson, E. Farrand & R. Ruble, Strip Mine Spoil Reclamation 11, 14 Pa. State Univ. College of Agriculture (undated); R. Krause, *Spoil Bank Goes From Waste to Fodder,* Coal Mining and Processing (May 1964); Limstrom, supra note 64, at 3-12. Eastern Region, U.S. Forest Service, A Digest: Strip Mine Reclamation 2-5 (1962).

71. F. Griffith, M. Magnuson & R. Kimball, *Demonstration and Evalua-*

tion of Five Methods of Secondary Backfilling of Strip-Mine Areas 16, Bureau of Mines Report of Investigations No. 6772 (1966); Miller, supra note 11, at 23; Johnson supra note 49, at 5.

See Pyles, supra note 62, at 318-19; Bauer, supra note 10, at 33; U.S. Dep't of the Interior, supra note 1, at 37.

72. See A. Grandt, *Reclamation for Pasture and Agricultural Crops*, Pa. State Symposium 124.

73. Id. at 128; R. Donley, *Some Observations on the Law of the Strip-Mining of Coal*, 11 Rocky Mt. Mineral Law Institute 123, 124, 126 (1966).

74. A. Grandt & A. Lang, *Reclaiming Illinois Strip Coal Land with Legumes and Grasses*, Univ. of Ill. Agricultural Experiment Station Bull. 628 (1958). Grandt, supra note 72, at 124.

75. See W. Sturgill, *Strip Mine Reclamation Under the 1964 Act (Eastern Ky.)*, Ky. Symposium 4, 7. Grandt, supra note 72, at 130-31. Note, *Reclamation of Strip Mine Spoils*, 50 Ky. L. J. 524, 540 (1962).

76. See Phelps, supra note 6, at 8.

77. U.S. Dep't of the Interior, supra note 1, at 64.

78. See generally, National Sand and Gravel Ass'n, Case Histories: Rehabilitation of Worked-out Sand and Gravel Deposits (1961).

79. Schellie & Rogier, supra note 33, at 15-24. See also D. Laird, *The Potential Industrial Use of Abandoned Strip Mines in Allegheny County, Pa.* (unpublished thesis), Univ. of Ill., Dep't of City Planning (1961). The use of strip mine excavations to dispose of solid wastes offers an intriguing possibility of joint planning to solve two different problems. See *Strip Mining Heals Its Own Scars*, Business Week, Nov. 13, 1965, at 140, 146. Cf. U.S. Dep't of the Interior, supra note 1, at 52.

80. See State of Ill. Bd. of Economic Development, *Atlas of Illinois Resources*, Sec. 2, Mineral Resources 24, 26 (1959); E. Davison, *Are the Sand and Gravel Reserves of the United States Adequate for Future Needs?* 1 N.S.G.A. (1965).

81. National Sand and Gravel Ass'n, Production of Sand and Gravel in 1966, at 5 (1967). The average price of crushed stone is $1.42 per ton. *Hearings on S.3132, S.3126 and S.217*, supra note 9, at 149.

82. Letter from Vincent P. Ahearn, Jr., Assistant Managing Director, National Sand and Gravel Ass'n, to author, Aug. 28, 1967.

83. National Sand and Gravel Ass'n, Production of Sand and Gravel in 1966, at 4 (1967). See also City of San Diego Planning Dep't, supra note 15, at 18-21.

84. See Colorado Sand & Gravel Producers Ass'n, First Complete Aerial Photo Map of Denver Metropolitan Area With an Outline of Our Diminishing Gravel Resources (1957).

85. D. R. Jenson, Selecting Land Use for Sand and Gravel Sites (1967). See also Cong. Rec. 214-15.

86. See Schellie & Rogier, supra note 33, at 15-24.

87. See State v. Local Control of Land Use, address by V. P. Ahearn, Jr., 49th Annual Convention of the National Sand and Gravel Ass'n, Feb. 25, 1964.

88. In 1920 coal produced 80.4 percent on a Btu basis of the total energy produced by mineral energy fuels and electricity from water and nuclear power. By 1965 coal's share had dropped to 27.8 percent. National Coal Ass'n, Bituminous Coal Data, 1966, at 91 (1967).

89. U.S. Bureau of Mines, 1-2 Minerals Yearbook, 1966, at 690 (1967).

90. Federal Power Comm'n, National Power Survey, pt. 1, at 59 (1964).

91. Press release issued by Chicago Operations Office, U.S. Atomic Energy Commission, July 13, 1967. See also W. Davis & F. Karlson, Nuclear Energy in the United States and Western Europe (October 1967); Hogerton, *The Arrival of Nuclear Power*, Scientific American, Feb. 1968, at 21.

92. See U.S. Bureau of Mines, 2 Minerals Yearbook, 1965, at 44 (1967); Illinois Dep't of Mines and Minerals, 1966 Annual Coal, Oil and Gas Report 118; Illinois State Geological Survey, Strippable Coal Reserves of Illinois (1963). But see Federal Power Comm'n, National Power Survey, pt. 1, at 54 (1964).

93. To discourage strip mining also has the effect of encouraging deep mining, a much more hazardous occupation. Of the 250 fatal accidents to mine labor in 1965 only 19 occurred at surface mines. U.S. Bureau of Mines, 2 Minerals Yearbook, 1965, at 35 (1967). See Weiss, supra note 33, at 6. Cf. L. Mumford, The Myth of the Machine: Technics and Human Development 239-40 (1967).

94. See Land Management and the Extractive Industries, Address by J. A. Carver, Convention of the National Sand and Gravel Ass'n, Feb. 8, 1966.

95. But as David Brower put it, if the alternative to strip coal mining is Grand Canyon dams, the strip mines might be preferable. Brower, *Reclamation of the Landscape*, in Beauty for America, Proceedings of the White House Conference on Natural Beauty 350 (May 24-25, 1965). Cf. Sturgill, supra note 75, at 4.

96. See generally, Brooks, supra note 16. See also Brower, supra note 95.

97. In many instances reclamation of coal lands has been found to be profitable purely on an economic basis, i.e., the reclaimed land may be sold for a price exceeding the cost of reclamation. Irrespective of immediate dollars and cents, however, the many coal industry leaders have recognized the dangers inherent in an aroused public and are promoting reclamation as an important long range factor of their overall public relations program. See Mahoney, *The Industry and Regulatory Laws—Current and*

Future, Pennsylvania State Symposium 44; Sullivan, supra note 29, at 8-9; A.B.A. Rep. of the Comm. on Coal, Section of Mineral and Natural Resources Law 4-7 (1963); *Strip Mining Heals Its Own Scars*, Business Week, Nov. 13, 1965, at 140.

98. See E. K. Davison, *Reclamation of the Landscape*, in Beauty for America, Proceedings of the White House Conference on Natural Beauty 330 (May 24-25, 1965). Because sand and gravel are usually excavated at shallower depths and often at locations in close proximity to urban areas, the reclamation of sand and gravel sites is often easier than the reclamation of coal mining land, and such property can frequently be reclaimed at a profit for a wide variety of uses. See Bauer, supra note 10, at 2-3; Jensen, supra note 85.

99. See the news report by R. A. Wright of the poorly attended session on public relations at the 1967 meeting of the American Mining Congress. N.Y. Times, Sept. 24, 1967, Sec. F, at 17, col. 3. See also address by Assistant Secretary of Interior J. Cordell Moore, Kentucky Strip Mining Symposium, Owensboro, Kentucky, July 13, 1967: "Not enough men who guide the destiny of the mining industry have been conservationists, and the conservation efforts of those mining industry leaders who do understand the problem have been insufficient. . . ."

100. W. Va. Acts 1939, ch. 84.

101. Law of May 31, 1945, Pa. Laws. 1198; Ohio Laws 1947, p. 730; Ky. Act. 1954, ch. 8. (For citations to statutes currently in effect see note 106, infra.)

Indiana passed a very limited act in 1941. Ind. Acts 1941, ch. 68. Illinois passed a law in 1943 which was subsequently held invalid. Ill. Laws 1943, Vol. 1, p. 912. Northern Ill. Coal Corp. v. Medill, 397 Ill. 98, 72 N.E.2d 844 (1947). Cf. Maryland Coal & Realty Co. v. Bureau of Mines, 193 Md. 627, 69 A.2d 471 (1949). Both Indiana and Illinois have now adopted new statutes. See note 106, infra. See generally Donley, supra note 73, at 153-68.

102. See Meiners, supra note 20, at 450, 455; *Report of the Mineral Law Section*, 34 Pa. Bar Ass'n Q. 456, 459 (1963); Deasy & Greiss, supra note 41, at 5; Caudill, supra note 26, at 316.

103. Harris-Walsh, Inc. v. Borough of Dickson City, 420 Pa. 259, 216 A.2d 329 (1966). See Ahearn, supra note 87, at 2; Address by S. J. Schulman, Empire State Sand Gravel and Ready Mix Ass'n, Messena, New York, Sept. 21, 1962.

104. Cf. Caudill, supra note 42, at 84; *Hearings on S.3132, S.3126 and S.217*, supra note 9, at 204.

105. See Breathitt, *Strip Mining in Kentucky*, Kentucky Symposium 45. A.B.A. Rep. of the Comm. on Coal, Section of Mineral and Natural

Resources Law 7 (1966). N.Y. Times, July 16, 1967, at 46, col. 3. Ade, *Strip Mining in Pennsylvania*, National Parks Magazine, March 1967, at 15-17.

106. See especially W. Va. Code Ann. Secs. 20-6-1 to -32 (Supp. 1968). The other currently effective strip mining regulatory statutes are: Ill. Rev. Stat. art. 93. Secs. 180.1-.15 (Supp. 1969); Ind. Stat. Ann. Secs. 46-1501 to 1528 (Repl. 1965, Supp. 1968); Iowa Code Ann. Secs. 83A.1-.29 (Supp. 1969); Ky. Rev. Stat. Secs. 350.010-.250 (1963, Supp. 1968); Md. Code Ann. art. 66C Secs. 657-74 (Supp. 1968); Ohio Rev. Code Secs. 1513.01-.27 (1966); Okla. Stat. Ann. tit. 45 Secs. 701-33 (Supp. 1968); Pa. Stat. Ann. tit. 52 Secs. 681.1-.22, 1396.1-.21 (1966, Supp. 1969); Tenn. Code Ann. Secs. 58-1522 to 1539 (Repl. 1968); Va. Code Ann. Secs. 45.1-162 to 179 (Repl. 1967, Supp.·1968).

The Council of State Governments is encouraging the adoption of an Interstate Mining Compact to promote the pooling of experience of state regulators. See Wiltsee, supra note 4, at 31; State Government News, June 1966, at 1. The Compact has been enacted in Kentucky and Pennsylvania. Ky. Rev. Stat. Secs. 350.300-.990 (Supp. 1968); Pa. Stat. Ann. tit. 52 Secs. 3252-3257 (Supp. 1969).

107. 33 U.S.C. Sec. 466g(c) (Supp. II, 1965-66).

108. U.S. Public Health Service, Suggested State Water Pollution Control Act (rev. ed., 1965), forms the basis for most state statutes. But see Pa. Stat. Ann. tit. 52 Secs. 682-688 (1966, Supp. 1969). See generally, Carmichael, *Forty Years of Water Pollution Control in Wisconsin: A Case Study*, 1967 Wis. L. Rev. 350.

109. Dufour v. Maize, 358 Pa. 309, 56 A.2d 675 (1948), noted in 96 U. Pa. L. Rev. 703 (1948). See Note, *Reclamation of Strip Mine Spoils*, 50 Ky. L. J. 524, 545 (1962).

110. 33 U.S.C. Sec. 466g(c) (Supp. II, 1965-66); see also 42 U.S.C. Sec. 1962-1962d-11 (Supp. II, 1965-66).

111. See U.S. Dep't of the Interior, supra note 1, at 101-02.

112. Village of Terrace Park v. Errett, 12 F.2d 240 (6th Cir. 1926), *cert. denied*, 273 U.S. 710; *Ex parte* Kelso, 147 Cal. 609, 82 Pac. 241 (1905); This argument has been most effective when the mineral rights are held by a separate owner. See Michaelman, *Property, Utility, and Fairness: Comments On the Ethical Foundations of "Just Compensation" Law*, 80 Harv. L. Rev. 1165, 1193 (1967). Compare United States v. Twin City Power Co., 350 U.S. 222 (1956), *reh. denied*, 350 U.S. 1009, *cert. denied*, 356 U.S. 918 (1958), with United States v. Virginia Electric & Power Co., 365 U.S. 624 (1961).

113. Hadacheck v. Sebastian, 239 U.S. 394 (1915) (The regulation prohibited the operation of a brickmaking plant but permitted the mining of

clay on the property.) Cf. Commonwealth v. Tewksbury, 11 Met. (52 Mass.) 55 (1846); Hodges v. Perine, 24 Hun. 516 (N.Y. Sup. Ct. 1881).

114. Goldblatt v. Town of Hempstead, 369 U.S. 590 (1962). See also Consolidated Rock Products Co. v. City of Los Angeles, 57 Cal.2d 515, 370 P.2d 342, 20 Cal. Rptr. 638 *appeal dismissed,* 361 U.S. 36 (1962).

115. The prohibition of surface mining in developed residential districts is commonly upheld. Goldblatt v. Town of Hempstead, 369 U.S. 590 (1962); Madis v. Higginson, 434 P.2d 705 (Colo. 1967); Village of Spillertown v. Prewitt, 21 Ill.2d 228, 171 N.E.2d 582 (1961); Smith v. Juillerat, 161 Ohio St. 424, 119 N.E.2d 611 (1954). But where counties or municipalities have attempted blanket prohibitions of surface mining regardless of the character of the area the courts have usually held the attempt invalid. Exton Quarries, Inc. v. Zoning Bd. of Adjustment of West Whiteland Twp., 425 Pa. 43, 228 A.2d 169 (1967); City of Warwick v. Del Bonis Sand & Gravel Co., 99 R.I. 537, 209 A.2d 227 (1965); East Fairfield Coal Co. v. Booth, 166 Ohio St. 379, 143 N.E.2d 309 (1957); Midland Elec. Coal Corp. v. Knox County, 1 Ill.2d 200, 115 N.E.2d 275 (1953).

116. See, e.g., Torrance, Calif., Ordinance 1581, Mar. 2, 1965; County of San Mateo, Calif., Ordinance 1416, Sept. 6, 1960; Montgomery County, Md., Ordinance 4-114, Oct. 10, 1961.

117. See generally, 1 Antieau, Municipal Corporation Law ch. III (1965).

118. Harris-Walsh, Inc. v. Borough of Dickson City, 420 Pa. 259, 216 A.2d 329 (1966). See 6 McQuillin, Municipal Corporations 394-98 (3d ed., 1949); *Conflicts Between State Statutes and Municipal Ordinances,* 72 Harv. L. Rev. 737, 744-47 (1959).

119. Cranberry Lake Quarry Co. v. Johnson, 95 N.J. Super. 495, 231 A.2d 837 (1967) *cert. denied,* 50 N.J. 300, 234 A.2d 407. See Antieau, supra note 117, at 279-31; Cf. *In re* Lane, 58 Cal. 2d 99, 372 P.2d 897, 22 Cal. Rptr. 857 (1962), noted in 50 Calif. L. Rev. 740 (1962); Note, *Pre-emption By State Over Penal Ordinances,* 38 N.D.L. Rev. 509 (1962). Regardless of abstract rules of statutory construction, much depends on the form in which the local government casts its legislation. To the extent that local government phrases its legislation in a way that attempts to differentiate its subject matter from that of state legislation, its chances of court approval are increased.

120. Compare Dep't of Licenses v. Weber, 394 Pa. 466, 147 A.2d 326 (1959) with Harris-Walsh, Inc. v. Borough of Dickson City, 420 Pa. 259, 216 A.2d 329 (1966). See Borough of Verona v. Shalit, 96 N.J. Super. 20, 232 A.2d 431 (1967). Antieau, supra note 117, at 290; Note, *Conflicts Between State Statutes and Municipal Ordinances,* 72 Harv. L. Rev. 737, 748-49 (1959); Comment, *The State v. The City: A Study in Pre-emption,* 36 S. Cal. L. Rev. 430 (1963).

121. See Note, *Reclamation of Strip Mine Spoils,* 50 Ky. L. J. 524, 561 (1962).

122. See McQuillin, supra note 118, at 400-11; Antieau, supra note 117, at 286-87; Note, *Pre-emption By State Over Local Penal Ordinances,* 38 N.D.L. Rev. 509, 511-13 (1962).

123. Harris-Walsh, Inc. v. Borough of Dickson City, 420 Pa. 259, —, 216 A.2d 329, 336 (1966) (concurring opinion). See generally, Long v. City of Fort Worth, 333 S.W.2d 644 (Tex. Civ. App. 1960); Brady v. Board of Appeals of Westport, 348 Mass. 515, 204 N.E.2d 513 (1965).

In some states the zoning power may be restricted by special state legislation. See E. Solberg, *Suggestions for Planning and Zoning in Appalachia,* Economic Research Service, U.S.D.A. 9-11 (1967).

124. See Bologno v. O'Connell, 7 N.Y.2d 155, 164 N.E.2d 389, 196 N.Y.S.2d 90 (1959).

125. Goldblatt v. Town of Hempstead, 369 U.S. 590 (1962). Cf. Plymouth Coal Co. v. Pennsylvania, 232 U.S. 531 (1914); Miller v. Schoen, 276 U.S. 272 (1928).

126. Gibbons & Reed Co. v. North Salt Lake City, 19 Utah 2d 329, 431 P.2d 559 (1967); Fredal v. Forster, 9 Mich. App. 215, 156 N.W.2d 606 (1967); County of DuPage v. Gary-Wheaton Bank, 42 Ill. App. 2d 299, 192 N.E.2d 311 (1963); County of DuPage v. Elmhurst-Chicago Stone Co., 18 Ill.2d 479, 165 N.E.2d 310 (1960); Town of Wayland v. Lee, 331 Mass. 550, 120 N.E.2d 641 (1954); Hawkins v. Talbot, 243 Minn. 549, 80 N.W.2d 863 (1957); Moore v. Bridgewater Twp., 69 N.J. Super. 1, 173 A.2d 430 (1961); Town of Somers v. Camarco, 308 N.Y. 537, 127 N.E.2d 327 (1955).

Contra, Town of Billerica v. Quinn, 320 Mass. 687, 71 N.E.2d 235 (1947) (presumably overruled).

But see Bither v. Baker Rock Crushing Co., 438 P.2d 988 (Ore., 1968); Town of Waterford v. Grabner, 155 Conn. 431, 232 A.2d 481 (1967); Teuscher v. Zoning Bd. of Appeals, 154 Conn. 650, 228 A.2d 518 (1967); Davis v. Miller, 163 Ohio St. 91, 126 N.E.2d 49 (1955) (may not be extended to lot across highway); Dolomite Products Co. v. Kipers, 23 App. Div. 2d 339, 260 N.Y.S.2d 918 (1965), *aff'd,* 19 N.Y.2d 739, 225 N.E.2d 894, 279 N.Y.S.2d 19 (may not be extended to tract across railway).

See generally 2 Anderson, American Law of Zoning 413-16 (1968); Annot., 10 A.L.R.3d 1226, 1272-80 (1966).

127. Goldblatt v. Town of Hempstead, 369 U.S. 590 (1962); Consolidated Rock Products Co. v. City of Los Angeles, 57 Cal.2d 515, 370 P.2d 342, 20 Cal. Rptr. 638, *appeal dismissed,* 371 U.S. 36 (1962); Calve Brothers Co. v. City of Norwalk, 143 Conn. 609, 124 A.2d 881 (1956); Abramson v. Zoning Bd. of Appeals, 143 Conn. 211, 120 A.2d 827 (1956); LaSalle Nat'l Bank v. County of Cook, 60 Ill. App. 2d 39, 208

N.E.2d 430 (1965); Village of Spillertown v. Prewitt, 21 Ill.2d 228, 171 N.E.2d 582 (1961); Tankersley v. County Bd. of Appeals, 230 Md. 379, 187 A.2d 302 (1963); Town of Lexington v. Simeone, 334 Mass. 127, 134 N.E.2d 123 (1956); Raimondo v. Board of Appeals, 331 Mass. 228, 118 N.E.2d 67 (1954); Butler v. Town of East Bridgewater, 330 Mass. 33, 110 N.E.2d 922 (1953); Town of Seekonk v. John J. McHale & Sons, Inc., 325 Mass. 271, 90 N.E.2d 325 (1950); Town of Burlington v. Dunn, 318 Mass. 216, 61 N.E.2d 243, *cert. denied*, 326 U.S. 739 (1945); Township of Bloomfield v. Beardslee, 349 Mich. 296, 84 N.W.2d 537 (1957); Wolster v. Borough of Upper Saddle River, 41 N.J. Super. 199, 124 A.2d 323 (1956); L. P. Marron & Co. v. Township of Mahwah, 39 N.J. 74, 187 A.2d 593 (1963); Fred v. Mayor & Council of Borough of Old Tappan, 10 N.J. 515, 92 A.2d 473 (1952); New York Trap Rock Corp. v. Town of Clarkstown, 1 App. Div. 2d 890, 149 N.Y.S.2d 290 (1956), *aff'd*, 3 N.Y.2d 844, 144 N.E.2d 725, 166 N.Y.S.2d 82; *rehearing denied*, 3 N.Y.2d 938, 146 N.E.2d 188, 168 N.Y.S.2d 6 (1957); *appeal dismissed*, 356 U.S. 582 (1958); Incorporated Village of Upper Brookville v. Faraco, 282 App. Div. 943, 125 N.Y.S.2d (1953); Bernhard v. Caso, 19 N.Y.2d 192, 225 N.E.2d 521, 278 N.Y.S.2d 818 (1967); Leichter v. Barrett, 208 Misc. 577, 144 N.Y.S.2d 309 (1955); Village of Willoughby Hills v. Medred, 27 Ohio Op. 2d 154, 189 N.E.2d 164 (1961), *appeal dismissed*, 173 Ohio St. 378, 182 N.E.2d 317 (1962); Miesz v. Village of Mayfield Heights, 29 Ohio App. 471, 111 N.E.2d 20 (1952); Smith v. Juillerat, 161 Ohio St. 424, 119 N.E.2d 611 (1954).

128. Gibbons & Reed Co. v. North Salt Lake City, 19 Utah 2d 329, 431 P.2d 559 (1967); Town of Stow v. Marinelli, 352 Mass. 738, 227 N.E.2d 708 (1967); Herman v. Village of Hillside, 15 Ill.2d 396, 155 N.E.2d 47 (1958); Midland Electric Coal Corp. v. Knox County, 1 Ill.2d 200, 115 N.E.2d 275 (1953); Certain-Teed Products Corp. v. Paris Twp., 351 Mich. 434, 88 N.W.2d 705 (1958); Buckley v. City of Bloomfield Hills, 343 Mich. 83, 72 N.W.2d 210 (1955); Cleveland Builders Supply Co. v. City of Garfield Heights, 102 Ohio App. 69, 136 N.E.2d 105 (1956); East Fairfield Coal Co. v. Booth, 166 Ohio St. 379, 143 N.E.2d 309 (1957); Exton Quarries, Inc. v. Zoning Bd. of Adjustment 425 Pa. 43, 228 A.2d 169 (1967); City of Warwick v. Del Bonis Sand & Gravel Co., 99 R.I. 537, 209 A.2d 227 (1965); Town of Caledonia v. Racine Limestone Co., 266 Wis. 275, 63 N.W.2d 697 (1954).

129. See Bulk Petroleum Corp. v. City of Chicago, 18 Ill. 2d 383, 164 N.E.2d 42 (1960).

The classic experience is that of the Village of Hempstead, N.Y. Its attempt to use a zoning ordinance to control the operations of a quarry was held invalid by the New York Supreme Court. Town of Hempstead v.

Goldblatt, 19 Misc. 2d 176, 189 N.Y.S.2d 577 (1959). But it was subsequently successful in forcing the complete cessation of operations at the quarry through a separate mining ordinance. Goldblatt v. Town of Hempstead, 369 U.S. 590 (1962). See 1 Anderson, American Law of Zoning 470-71 (1968).

130. Zoning can also be used in combination with a mining ordinance. Cf., e.g., City of Hillsdale v. Hillsdale Iron & Metal Co., 358 Mich. 377, 100 N.W.2d 467 (1960).

131. Kozesnik v. Township of Montgomery, 29 N.J. 584, 151 A.2d 537 (1959); earlier opinion: 24 N.J. 154, 131 A.2d 1 (1957); related case: DePew v. Township of Hillsborough, 31 N.J. 157, 155 A.2d 766 (1959); Address by Schulman, supra note 103, at 9.

For examples of zoning ordinances regulating surface mining by special districts, see City of San Diego Planning Dep't supra note 15, at 63-71. Marion County, Indiana, Gravel-Sand-Borrow District Ordinance, Feb. 8, 1966. City of Los Angeles, California, Zoning Code Sec. 13.03 (1966).

132. For examples of zoning ordinances regulating surface mining by special permit see Town of Orangetown, N.Y., Zoning Ordinance Sec. 4.32(c) (1966); County of Henrico, Va., Zoning Ordinance Secs. 7.24, 11.21, 12.21, 13.21, 14.23, 17.8 (1966); E. Miller, *Penn Township—An Example of Local Governmental Control of Strip Mining in Pennsylvania*, Economic Geography 256 (July 1952). See generally J. McCarty & G. Duggar, *Local Regulation of Excavations, Grading and Quarrying in California* 6-8, Bureau of Public Administration, Univ. of Calif. (1956).

133. See generally R. Garrabaut, *Performance Standards for Industrial Zoning: An Appraisal*, 15 Urban Land No. 6, at 1 (June 1956); National Research Council of Canada, *Annotated Bibliography: Performance Standards for Space and Site Planning for Residential Development* (1961).

134. See Los Angeles County, California, Zoning Ordinance Sec. 275 (1966); County of Santa Clara, California, Planning Commission Resolution No. 6178 (1961); B. Sanders, *Zoning for Urban Pits and Quarries*, 24 Tennessee Planner 115, 121 (1965); Bauer, supra note 10, at 28; Ahearn, *Land Use Planning and the Sand and Gravel Producer* 20-24 (N.S.G.A., 1964).

135. See Ahearn, supra note 87, at 4; City of San Diego Planning Dep't, supra note 15, at 63-71.

Consideration might also be given to the formation of a Soil Conservation District to adopt regulations to protect mineral resources. See Note, *Reclamation of Strip Mine Spoils*, 50 Ky. L. J. 524, 564 (1962); J. Beuscher, *Land Use Controls—Cases and Materials* 399 (3d. ed., 1964); U.S. Dep't of the Interior, supra note 2, at 78; Brooks, supra note 16, at 34.

136. See Gruber v. Mayor & Twp. Comm., 39 N.J. 1, 186 A.2d 489 (1962); Camboni's Inc. v. County of DuPage, 26 Ill.2d 427, 187 N.E.2d 212 (1962); People *ex rel.* Skokie Town House Builders v. Village of Morton Grove, 16 Ill.2d 183, 157 N.E.2d 33 (1959); State *ex rel.* Berndt v. Iten, 259 Minn. 77, 106 N.W.2d 366 (1960); Lamb v. City of Monroe, 358 Mich. 136, 99 N.W.2d 56 (1959).

137. See Urban Land Institute, *The Effects of Large Lot Size on Residential Development* (Technical Bull. 32, 1958).

138. See generally H. Fagin, *Regulating the Timing of Urban Development,* 20 Law & Contemp. Prob. 298 (1955); J. Reps, *The Zoning of Undeveloped Areas,* 3 Syracuse L. Rev. 292 (1952).

139. 31 Fed. Reg. 6834 (1966). For a history of federal regulation of mining on federal lands see J. Howerton, *1967—A Critical Year for Mined Land Reclamation Regulation,* 1 Natural Resources Lawyer, No. 4, at 70 (Oct. 1968). See also F. Barry, *Federal and State Regulation and the Legislative Picture,* Pennsylvania State Symposium 35, 36. Address by Hon. Wayne Aspinall, 1967 Metal Mining and Industrial Minerals Convention, American Mining Congress, Denver, Colo., Sept. 10, 1967, at 6-10 (mimeo. ed.). See generally Meiners, supra note 20, at 459-61.

140. U.S. Dep't of the Interior, supra note 1, at 78.

141. See generally Mineral Leasing Act, 30 U.S.C. Sec. 181 (1927). Cf. Comment, *North Cascades National Park: Copper Mining v. Conservation,* 157 Science 1021 (Sept. 1, 1967).

142. See A. Wagner, *The Tennessee Valley Authority,* Kentucky Symposium 23. Tennessee Valley Authority, *An Appraisal of Coal Strip Mining* (1963). Compare Caudill, supra note 26, at 318-21.

143. *Hearings on H.R. 14921 Before a Subcomm. of the Senate Comm. on Appropriations,* 89th Cong., 2d Sess. 191 (1966).

144. An Interim Report by the Secretary of the Interior to the Appalachian Regional Commission, supra note 18.

145. U.S. Dep't of the Interior, supra note 1.

146. The bill is S.524 in the 91st Congress, which is identical to S.3132 in the 90th Congress. See 115 Cong. Rec. 654 (daily ed. Jan. 22, 1969).

See also The President's Council on Recreation and Natural Beauty, From Sea to Shining Sea 138-41 (1968); U.S. Dep't of the Interior, supra note 1, at 90. The Interior Department proposes to require the state legislation to comply with certain federal standards, but would not give federal authorities control of actual enforcement or administration. See Address by Assistant Secretary J. Cordell Moore, Kentucky Strip Mining Symposium, Owensboro, Ky., July 13, 1967, at 7 (mimeo. ed.).

Action by the federal government has been slowed by the typical squabbling between various federal agencies seeking jurisdiction over any surface mining program. See *Hearings on S.3132, S.3126 and S.217*, supra note 9, at 43-58, 94-95.

147. S. 217, 90th Cong., 1st Sess. (1967).

148. *Hearings on S.3121, S.3126 and S.217*, supra note 9.

149. See, e.g., Ky. Rev. Stat. Sec. 350.090 (1966); Ohio Rev. Code Sec. 1513.16(F) (1968). The ability to pre-plan depends on the extent to which the location of the minerals is pinpointed prior to excavation. This varies with the extent of test-bore drilling and with geological conditions. See *Hearings on S.3121, S.3126 and S.217*, supra note 9, at 258.

150. In many cases the problem of the future reuse of the land is complicated by the fact that the operator is not the owner of the land. See U.S. Dep't of the Interior, supra note 1, at 103.

151. U.S. Dep't of the Interior, supra note 1, at 56.

152. See Sawyer, supra note 59, at 22. But see Davis, supra note 38, at 61.

153. See, e.g., J. Roseberry & W. Klimstra, *Recreational Activities on Illinois Strip-Mined Lands*, 19 J. of Soil & Water Conservation 107 (1964).

154. See generally The Kentucky Dep't of Natural Resources, supra note 23, at 55; Schellie & Rogier, supra note 33, at 15-24; Laird, supra note 79.

155. U.S. Dep't of the Interior, supra note 1, at 52.

156. See Faltermayer, supra note 37, at 132; McQuilkin, supra note 37, at 104.

157. Deasy & Greiss, supra note 41, at 6.

158. See note 42, supra.

159. See notes 81-84, supra.

160. Sanders, supra note 134, at 116; see U.S. Dep't of the Interior, supra note 1, at 102.

161. See, e.g., State of Ill., Bd. of Economic Development, supra note 80.

162. See Hawaii Rev. Laws Sec. 98.11-2 (1965).

163. Cf. Sanders, supra note 134, at 119; Ahearn, supra note 87, at 4; J. Dunn & J. Broughton, *A Mineral Conservation Ethic for New York State*, State Government, Summer, 1965, p. 191.

17

Bonding Requirements
for the Surface Mining of Coal

Jerrold Becker
Susan Collins

Introduction

This report is concerned with state bonding requirements for strip mine operators. It is limited in scope to coal mining. Also included are two proposed bills which are currently under consideration by the Congress.

The materials used for this study are primarily the state statutes of eight Appalachian states concerning the strip or surface mining of coal; and, to some extent, the regulations set up by the state agencies concerned with the enforcement of those statutes. In addition, a letter was sent to each of the appropriate agencies on August 20, 1973. The questions in that letter were as follows: (1) From your experience, what is the average time consumed from initiation of a complaint or notification by the Director of a violation of reclamation requirements to a decision by a hearing examiner? (2) Again from your experience, what is the time necessary for the exhaustion of administrative appeal remedies? (3) In legal terminology, what is the presumption accorded the agency decision? In other words, when the case is appealed to the court, what degree of deference does the court pay to the agency decision? (4) How large is the mine inspection bureau and what is its annual budget?

Pennsylvania was the only state which did not respond to this inquiry. Some of the other states had had no experience with some of the questions, and this is reflected by the information presented here.

Finally, in November 1973, a phone call to each of the states produced the information included here concerning the number of acres permitted in a given year, the total amount of the bonds held by the states for the same period, and the total amount of bonds forfeited for a given period. Most of this information was received during the phone call. In the case of West Virginia and Ohio, a follow-up letter was received.

From Appalachian Resources Project publication no. 25, May 1974, pp. 1-13. Reprinted by permission.

This report is not concerned with the states' mining standards in any specific sense. It is not an evaluation of their restrictions on the type of mining operations carried on in each state. In other words, the report is not concerned with the allowable height of the highwall or the allowable bench width or other specific limitations included in the statute, except as they reflect possible violations and therefore basis for the forfeiture of a bond. For this reason, these standards as such are not dealt with specifically in this document.

Summary of State Statutes and Enforcement Procedures

The discussion which follows is concerned with each individual state's statutes and enforcement mechanism.

Alabama: The Alabama statute is administered by the Director of the Department of Industrial Relations. The bond is $150 per acre with no minimum total. The operator is allowed three years from the expiration of his permit to reclaim the land. The state does provide some aid for the reclamation. There is no appeals board set up in Alabama. After the departmental hearing on any violation of the law, the appeal is to the Circuit Court sitting in equity. Alabama holds a total of about 300 bonds valued at around $1,200,000. There are 8,000 acres under permit. There have only been three bond forfeitures in the state. (Alabama Surface Mining Act, effective October 1, 1970)

Kentucky: The Kentucky statute is administered by the Director of the Division of Reclamation of the Department of Environmental Protection. The administration involves 70 employees with a budget of 1.4 million dollars. The bond is set up by the statute to range from $200-$1,000 per acre with a minimum total bond of $2,000. The Director has set the bond at $400 per acre. The reclamation must be completed within 12 months after the permit has expired. Section 350.250 of the Kentucky law allows for some citizen complaints. The hearing process takes 30 days to complete. The operator is not allowed another permit unless reclamation is completed without cost to the state. It requires 60 days for the exhaustion of appeals. The findings of the agency, if supported by substantial evidence, are conclusive. The total value of bonds held by the state is $8.4 million. In the past year, about $53,000 worth have been forfeited. The involved land amounts to 21,000 acres. (Title XXVIII, Chapter 350)

Maryland: The Maryland law is administered by the Director of the Bureau of Mines. The enforcement section has seven employees with a yearly budget of $127,954. The bond is $400 per acre with a minimum total bond of $3,000. Maryland offers some state aid for reclamation. Maryland will not issue a new permit to anyone who has previously forfeited a bond in any state. They estimate the time necessary for the exhaustion of appeals as being from 30 to 60 days. (Article 66C, Maryland Code)

Ohio: The Chief of the Division of Forestry and Reclamation of the Department of Natural Resources administers the Ohio law with 36 employees and a biennium budget of $959,700. The amount of bond per acre is to be determined by the Chief. The minimum total is $5,000. Reclamation (except planting) must begin within three months. Citizens are given some remedies in Section 1513.15. The hearing process takes two to three weeks. After the third violation under the statute, the operator's license will be revoked and no new license will be issued for five years. Before going to the courts, there is an appeal to the Reclamation Board of Review. They estimate the time necessary to complete all appeals as up to one year. For the period from April 12, 1973, through April 12, 1974, the total value of bonds posted was $45,828,387. The acres licensed were 18,667. No bonds were forfeited. (Chap. 1513: Strip Mining and Reclamation of Land)

Pennsylvania: The Secretary of the Department of Environmental Resources is the administrator of the Pennsylvania law. The amount of bond per acre is to be determined by the Secretary with a minimum total bond of $5,000. Section 18.3 allows for remedies for private citizens under the statute. No new license shall be issued to any operator who has been found in violation of any terms of the license. The first step in the appeals process is to the Environmental Hearings Board. The total value of bonds put up in 1972 was $7,848,511. The total value of bonds forfeited in that period was $10,000. Permits to mine 10,000 to 12,000 acres are issued each year. (Surface Mining Conservation and Reclamation Act)

Tennessee: The Department of Conservation headed by the Commissioner is in charge of the Tennessee statute. The section involved with enforcement of this statute has 17 employees and a budget of $264,600. The minimum per acre bond is $600. There is a fine of from $100 to $5,000 for each day in which a violation of the statute continues. Appeals from agency rulings are heard first by the Board of Reclamation Review. In 1972 there were 5,181 acres permitted. This involved about sixty-five operators. There were three bonds forfeited that year with a total value of $10,350. (Title 58, Chapter 15)

Virginia: The statute is administered under the Department of Conservation and Economic Development by 25 employees with a biennium budget of $465,615. The bond is set at $200 to $1,000 per acre with a minimum total of $2,500 except where there are less than five acres involved when the minimum is $1,000. The Department has set the bond at

$200 per acre. The hearing process takes five weeks. The appeal is first to a Board of Surface Mining Review. There are 8,424 acres now under permit. Under the 1973 law there have been $33,060 in bonds forfeited. (Chapters 16 & 17, Code of Virginia)

West Virginia: The statute is administered by the Department of Natural Resources. The enforcement section has 56 employees with a budget in excess of $500,000. The bond is set by law at $600-$1,000 per acre. The Director has set it at $750 per acre. The minimum total is $10,000. All reclamation must be finished within 12 months after the permit expires. Possible citizen remedies are found under Sections 20-6-11 and 20-6-28 & 29. The agency decision on a violation may be appealed to the Reclamation Board of Review. The total appeals process may take several years. A 1968 case is still pending [1974]. In 1972, 246 permits were issued, 24,508.13 acres were bonded, and the total bond was $18,381,097.50. From July 1, 1961, to September 30, 1973, the total value of bonds forfeited was $1,358,363.50. (Chapter 20, Article 6 and 6A, Code of West Virginia, as amended)

Comparison of Statutes and Enforcement Procedures

In this section the provisions of state and federal legislation are highlighted and compared. In each state statute, there appears at least an outline of the manner in which the statute shall be administered. The administration of the statute is given to one of the state's agencies and the responsibility for the proper enforcement of the statute is usually placed specifically on the director of the designated agency. In five of the Appalachian states—Ohio, Pennsylvania, Tennessee, Virginia, and West Virginia—the statute is administered by the departments of conservation or resources. Of these, Ohio and Virginia have special divisions within the departments which deal with these statutes. In Ohio, it is the Division of Forestry and Reclamation and in Virginia it is the Division of Mined Land Reclamation. The Alabama law is administered by the Department of Industrial Relations. In Kentucky, it is the Department of Environmental Protection, Division of Reclamation, and in Maryland the Bureau of Mines. The federal bills included in this study all give the responsibility for administration to the Department of the Interior. It should be noted that in one of the federal bills, the states can become the real administrators, and the federal agency will step in only in the event that the states fail to conform to the standards set up by the law or by the federal agency acting in accordance with the federal law.

At times, difficulty in enforcement of a statute of this type is due to lack of funds or personnel in the agency with this responsibility. For this reason, as well as to find out the type of emphasis the states were placing on these statutes, the states were asked the question (above) concerning their personnel and budget. Alabama did not respond to this question. Pennsylvania did not initially respond, but it was found that they have a staff of eighty-seven people. We have no figures on Pennsylvania's mine inspection budget. Of those states which did respond fully to this question, Kentucky has the largest mine inspection bureau with seventy employees and the largest budget with a yearly figure of $1.4 million. They range from these figures down to Maryland's seven employees and $127,954 budget. The figures for the other states are as follows: West Virginia—56 employees with a budget in excess of $500,000; Ohio—36 employees, budget (biennium) $959,700; Virginia—25 employees, budget (biennium) $465,615; Tennessee—17 employees, budget $264,600.

Of central concern to any discussion of bonding requirements is the dollar amount of the bonds which must be filed. These are per-acre bonds and depend directly on the size of the area to be stripped. The amount of the bond varies from state to state. The usual practice is for the statute to set up a range for the bonds or a minimum total requirement or both. For example, in Kentucky the amount per acre is $200 to $1,000 and the minimum total bond is $2,000. In Ohio and Pennsylvania, there is no specified amount per acre, only a minimum total of $5,000 in each state. On the other hand, Tennessee has a minimum per acre of $600 with no ceiling and no minimum total requirement. The basis for the amount of the bond is stated in the statute in all the states except Tennessee and Alabama. The basis in all the other states may be generally stated as the cost of reclamation. For example, in the Pennsylvania statute (S 4(2) -K-c), it is "based upon the total estimated cost to the Commonwealth of completing the approved reclamation plan." It would appear from this that the amount of bond required would vary with the type and location of the land being stripped. This often does not appear to be the case. To return to the example of Kentucky where the range on the amount per acre is $200-$1,000, the basis of the bond according to the statute (350.06) is "the character and nature of the overburden, the future suitable use of the land involved and the cost of backfilling, grading, and reclamation to be required." In spite of this specific basis for variation in the amount of the bonds, Kentucky strip mine operators put up $400 per acre without any regard to the type, location, or uses of the land involved. The same is true of Virginia where the bond required is $200 per acre which is the minimum required by law. The statutory basis (S 45.1-206) for the amount should be the "estimated cost of reclaiming the land to be disturbed and the quality and quantity of coal."

In the federal legislation, House Bill 3 requires a minimum per acre bond of $500 with a total bond of at least $5,000 which is based on the amount necessary to assure the reclamation. Senate Bill 425 does not provide for a specific per acre amount, but it does provide for a $10,000 minimum total bond to be based on a sufficient amount to assure the completion of reclamation by a third party.

There seems to be no doubt that the legislative intent was for the agencies to base the amount of the bonds required on each individual mining operation on the specific types of land, etc., involved. Although not all the figures on this were available for this study, it appears that the agencies have not conformed to the will of the legislatures in this area. The reasons for this are not known. One agency employee interviewed in the course of this study stated that "we figure it averages out." This may be accurate, but it seems doubtful that that was the original purpose of the act. In other words, an operator who is mining land which is difficult to reclaim may prefer to forfeit the bond rather than go to the additional expense of reclaiming the land. If that is the case, the cost of reclamation to the state is going to be greater than the amount of the bond held by the state for reclamation of that land. On the other hand if the reclamation of the land is not a difficult or expensive project, the operator will probably do whatever is necessary to prevent forfeiture of his bond. The state will be left with nothing to make up the difference. Most states have set up a fund into which they place a forfeited bond which is to be used to reclaim those lands not properly reclaimed by the operators.

The length of time which the strip mine operator has to reclaim the stripped area varies considerably from state to state. The state statutes are considerably more specific on this point, however, than the federal bills. The federal bills apparently leave the setting of time limits entirely to the agency in charge of administering the proposed law. The only specific limitation is that the reclamation must proceed along with the mining operation. In House Bill 3, the backfilling, grading, and resoiling, must be completed before the equipment is removed. Senate Bill 425 requires that it be "contemporaneous."

The state statutes, as mentioned above, vary considerably in their time allowances for reclamation. The best way to illustrate this is to give some examples of these requirements. Alabama allows three years from the date the permit to mine the area expires. Ohio requires that reclamation must begin within three months and must be completed within twelve months after the end of the license year or the end of the operation, and the planting must take place in the next appropriate season. The Virginia statute requires that grading be completed prior to removal of the operator's equipment, and that vegetation must be planted in the first proper planting season following the grading. Maryland is the only state whose law does not specifically place some time limitation on the reclamation. Pennsylvania requires that the time for reclamation meet the plan required of the operator and approved by the Secretary of the Department of Environmental Resources. Kentucky allows twelve months for completion. West Virginia law requires that reclamation must be completed within twelve months also, but further requires that grading, backfilling and water control must be kept current. Tennessee requires that grading must be completed within three months after removal from any given acre, but the statute allows two acres to be temporarily abandoned and does not specify how long this "temporary" abandonment might be.

Under the state statutes and also the federal bills, reclamation appears to be considered a private undertaking. Only two state statutes, Maryland and Alabama, make any mention of available state aid in the process of reclamation by the surface mining operator. The findings of this study do not show whether or not there may be further aid available in other states, but it seems likely that there would be at least in the form of seedlings and certain other planting materials which are often available from the state for private individuals.

In the event that the operator does not complete the reclamation of the mining site and is therefore in violation of the statute, we arrive at the question of possible bond forfeiture and other possible sanctions. Generally it is the responsibility of the agency to initiate a complaint against any operator who is in violation of the statute. Four states—Kentucky, Ohio, Pennsylvania, and West Virginia—will allow some citizen complaints. Both the proposed federal bills allow citizen complaints.

Once the complaint has been initiated, the controversy proceeds to the hearing stage. The hearings take place within the agency which has jurisdiction over the administration of the statute. The initial hearings are conducted by whatever type hearing structure has been set up by the agency. Of the states that replied to this question, the average time required for the initial hearing is about once a month. The finding of the agency may lead to forfeiture of the bond or to other sanctions provided by the statute or both. Other sanctions are generally fines for each day of the violation. These range from $100 to $5,000 with Tennessee allowing the fine to be anywhere within that range. These may also include revocation of the permit or license to mine and limitations on if and when a new permit or license may be issued. Maryland will not issue a permit to any operator who has previously forfeited a bond in any state.

Following the agency hearing on the question, the question may be appealed. Five states—Ohio, Pennsylvania, Tennessee, Virginia, and West Virginia—have boards set up for this purpose. The membership of the boards is set up by the states' surface mining statutes with the exception

of Pennsylvania. The Pennsylvania board is the Environmental Hearing
Board and is not set up under this statute but proceeds under the Administrative Agency Law. In those states which do not have appeals boards set
up, the appeal goes directly to the courts unless there is some type of
further decision-making within the department.

The membership of the board varies from state to state with the one
consistency being that the coal mining industry always has at least one
representative on the board. The Ohio reclamation board of review has
seven members. They are appointed by the governor with the advice and
consent of the senate. The board consists of one member representative
of the coal industry, three members from the public, one forestry expert,
one agronomy expert, and one expert on earth grading. The Tennessee
Board of Reclamation Review has five members including the Commissioner of Public Health, two representatives of the coal mining industry,
and two other members who must have no financial interest in coal mining. The Virginia Board of Surface Mining Review has, besides the Director of the Department of Conservation and Economic Development, three
members appointed by the governor. These three are two mining operators
and only one landowner. The five-member Reclamation Board of Review
for the state of West Virginia consists of one representative of the coal
mining industry, one forestry expert, one agriculture expert, one engineering expert, and one water conservation expert, all appointed by the governor. Most states either did not respond to the question concerning the
length of time necessary for exhaustion of the appeals process or they had
no experience with appeals under the current law. Ohio, although having
no actual experience, estimated that the time necessary would be about
one year. West Virginia responded that they would estimate the time as
several years since they have a 1968 case still pending. Kentucky reported
that the time necessary was only sixty days and Maryland estimated the
necessary time as being from thirty to sixty days.

As to what presumption might be given to the agency decision in the
event of an appeal, most states felt that the agency would be given some
deference. The only state which specifically felt that there would be no
presumption in the agency's favor was West Virginia. Their letter of September 5, 1973, stated that "Any case on appeal to the court of competent
jurisdiction is based upon the record of that case and the applicable law."

18
The Strip Mining Dilemma: The Case of Virginia

James E. Rowe

The surface coal mining industry is presently the focus of an intense public debate. Surface mining obtains coal by less expensive techniques than deep mining, and its use has greatly increased in recent years. Barring legislation or court intervention, this trend may be expected to continue because stripping is more profitable than underground methods. On an average, strip mining recovers up to 90 percent of a coal deposit, while deep mining removes 60 percent. By stripping, production per man-day approximates thirty-five tons versus eleven tons by deep mining.[1] In Virginia and Tennessee, where coal seams range from eighteen to sixty inches thick, contour stripping yields 2,400 to 8,100 tons per acre.[2]

A growing concern is being expressed about the amount of land disturbed by surface mining; it focuses on the need for the industry to develop and apply effective reclamation procedures.[3] In response to this concern, there is pending federal legislation that would require a significant upgrading of current reclamation standards. A dilemma develops with respect to regulations: Should surface mine operators be restrained with stringent reclamation standards that may possibly hinder the efficient and economical recovery of coal at the least cost? Or, should the environment be protected at all costs? If the environment is protected, the price of coal would be raised in order to cover the added cost involved with reclamation standards. This article discusses a weakness in the pending federal legislation in regard to alternative uses of reclaimed land in Virginia.

From *Land: Issues and Problems*, no. 22, October 1976. Reprinted with permission. Illustrations adapted from *Surface Mining . . . A Changing Industry* (Charleston, W. Va.: West Virginia Surface Mining and Reclamation Association, n.d.), p. 18.

James E. Rowe is Community Development Planner for the Mount Rogers Planning District Commission, Marion, Virginia. This article is based on research supported by the National Science Foundation/RANN grant No. S1A72-91535. The author thanks Eddie L. Stamper and Donald L. Henderlite for their assistance in preparing the graphics used with this article.

The "Back-to-Contour" Issue

The proposed federal regulations would make complete back-filling necessary in all cases in Virginia and other states; the land would have to be restored to its original contour. At the same time, the legislation would require the "utility and capacity of the reclaimed land to support a variety of alternative uses."[4] If and when such Federal legislation is enacted, mine operators, regional planners, and those responsible for post-mining reclamation would be required to consider alternative uses.[5] The relationship of these uses to existing land-use policies and plans would have to be approved in the post-mining plans required before a stripping operation may start or continue.

Back-to-contour reclamation greatly restricts the options available for post-mining planning. The actual process of recontouring the land to its before-stripping form not only increases the cost of mining but precludes, in some instances, many post-mining land uses. The writer believes that it would be less expensive and more productive to the regional economy not to reclaim *all* sites back to contour. Some sites would benefit from alternative post-mining land uses, which require unique reclamation methods. For example, head-of-hollow fill and mountain top removal are two strip mining methods that would increase the amount of level and, therefore, potentially developable land. However, the pending legislation would make these methods illegal because the land would not be returned to the original contour.

To be effective, restoration must be planned at the earliest stages of the mining operation. The process must take into account such factors as economics, aesthetics, revegetation, erosion, water conservation, and soil type. Long term benefits, future land use, conservation of natural values and beauty, as well as many other objectives must be considered. Where a proposed re-use would require reclamation procedures that would make the stripping economically unfeasible, the area, in most instances, would not be mined. Schlottman and Abrams studied the regional impact of a ban on mining coal on slopes in excess of 20 degrees.[6] They concluded that, if the steep sloped areas, such as those common in Virginia, were banned as strip-mine sites, then other areas, such as southern Illinois, would increase production in order to meet the demand for coal. If this occurs, an important Virginia industry would be reduced in size.

The potential exists, however, for residential, commercial, and industrial land uses on surface-mined areas in situations where access, load bearing capabilities, and water and sewer availability meet residential and industrial requirements. The highest potential for commercial or industrial uses would be in close proximity to the urban areas in the region and on strip-mined sites close to interstates and other major access routes.

Reclamation Methods

Reclamation procedures are related to the method of surface mining employed. The nine most frequently used reclamation methods utilized in central Appalachia are slope-reduction, parallel-fill, box-cut, head-of-hollow fill, mountain-top-removal, block-cut, lateral-movement, long-wall-stripping, and multiple-seam. High-walls are associated with most of these methods. Many studies have been conducted on how to reduce highwalls and how to improve their aesthetic qualities.[7] The nine methods listed here have all been developed in response to either pending or actual legislation and are used to minimize adverse effects on the environment and to maximize the recovery of coal.

Most coal surface mining operations in Virginia do not use the slope-reduction and parallel-fill methods because the overburden is placed on the downslope. However, the parallel-fill method may prove useful as an emergency measure in case of slides.[8]

The box-cut is a conventional method, which prevents most of the overburden from being placed downslope (Figure 1). It, too, would become illegal, if the pending federal legislation is enacted into law.

The head-of-hollow fill improves aesthetics, reduces landslides, allows for more coal recovery, and creates flat land in mountainous areas. This method is useful in reclaiming orphan mines (Figure 2). (The term orphan mine is defined as any abandoned surface mine operated prior to the enactment of comprehensive reclamation laws.) When properly constructed, such a fill substantially reduces erosion and minimizes the probability of land slides.[9]

Mountain top removal is being used near Charleston, West Virginia, by Cannelton Industries. This method makes available a number of alternative land-use considerations in the limited number of places where it is applicable. All the overburden is excavated and the spoil and top soil are replaced and graded into flat to rolling topography. A new town is planned for the site being mined by Cannelton. This method is also being utilized in several other locations, including one in Tazewell County, Virginia.

The block-cut method eliminates downslope overburden and integrates the reclamation process with the process of mining. The overburden is hauled back and deposited in the pit of the previous cut. The aesthetics of the area are improved because the tree line is undisturbed by this method. The lateral movement, sometimes called the haul-back method,

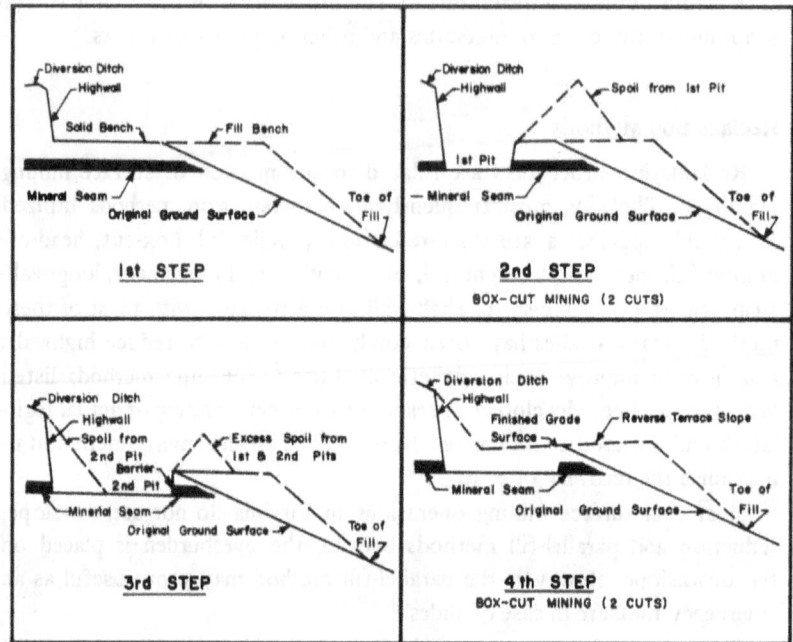

Figure 1. Box-cut mining (two cuts).

is a modification of the block-cut method.

Longwall-stripping is an experimental method being examined as an alternative or modification that is in a combination of the conventional shaft and surface mining methodologies.[10]

Multiple-seam mining is being tested by TVA at its Massengale site. TVA has shown that several coal seams can be simultaneously mined by this method without long-range environmental damage.[11] This experimental method of mining and reclamation meets all the proposed federal laws and returns the site to its original contour.

Virginia Surface Mining Law

The extraction of bituminous coal by surface mining was unregulated in Virginia until the enactment of legislation by the 1966 session of the General Assembly.[12] This legislation was amended in 1972. The Assembly determined that unregulated strip mining of coal caused soil erosion, stream pollution, and, in general, created conditions inimical to life, property, and the public welfare. After determining that the public welfare

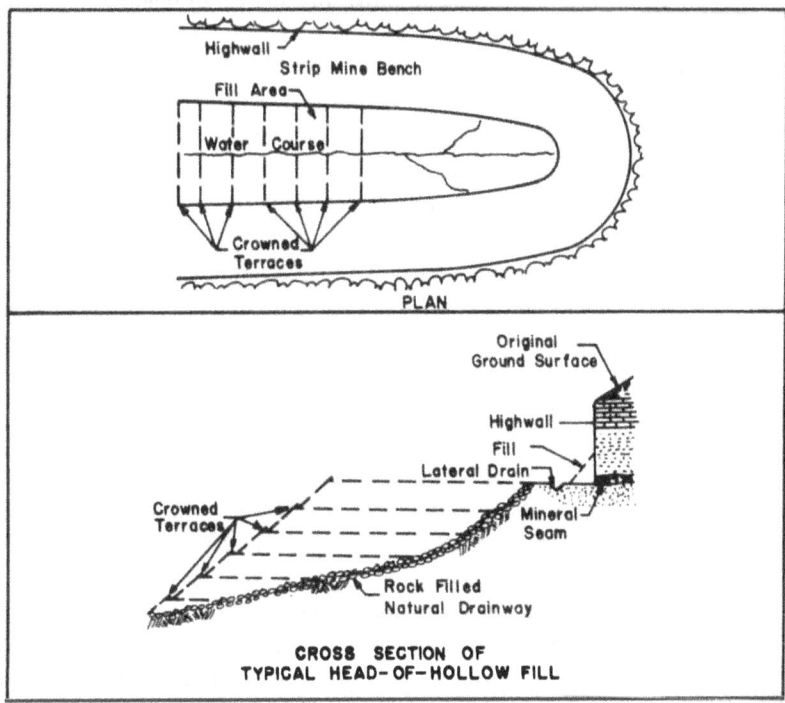

Figure 2. Head-of-hollow fill.

was infringed upon by strip mining, the Assembly decided to exercise the state's police power and regulate the industry.

The Act vested the Director of the Department of Conservation and Economic Development with the authority to require and prescribe conditions for the issuance of permits for strip mining, to collect bonds, and to set the conditions for forfeiture of the bonds. The Act mandated the Director to create a Strip Mining Fund to be used for the reclamation of orphan mines. The Director was also granted authority to enforce the Act and its associated regulations and to determine penalties for violations. The Director's authorized agent is the Commissioner of the Division of Mined Land Reclamation, who is officed at Big Stone Gap.

The Act requires permits to start or to continue any surface mining operation. The permits, issued on an annual basis, have to be obtained from the Department. The operators are charged a basic fee of $12 per acre for a permit, applicable to an area not to exceed 250 continuous acres.[13]

The Virginia law has a notable provision, similar to one in the pending federal legislation, that encourages adoption of more productive land uses. Since enactment of the law, operators have to prepare a mining and re-clamation plan for the area affected by their operation. The plan must be approved by the Commissioner before a permit is granted. Each plan describes all mining operations, gives detailed estimations of water and soil problems, and contains the post-mining reclamation and revegetation pro-cedures.[14]

Each operator must post a bond with the Commissioner, guaranteeing faithful performance of the plan and the provisions of the Act. Bonds range between $200 and $1,000 an acre.[15] The total bond cost is the estimated cost of reclamation. Bonds are released when the Commissioner determines that all reclamation is completed.[16] Appeals from the Division's rulings are heard before the Board of Conservation and Economic Develop-ment.[17] Surface mining in Virginia has disturbed 75,363 acres, of which 25,000 need reclamation.[18] These lands consist of mining debris and are often detrimental to the health and safety of any person who may be on or near them.

Forfeiting the bond results in the operator not being legally able to receive future coal surface mining permits unless authorized by the Board of Conservation and Economic Development. A Bond Forfeiture Fund was set up as a receiving account for the forfeited bonds. Such funds are utilized to carry out the reclamation and revegetation of the lands unreclaimed by an operator.

An Orphaned Land Fund was established in 1972 in order for the state to establish a systematic schedule to be carried out for the reclamation and revegetation of orphan mines. As of November 1, 1976, only about 250 acres of orphan lands had been reclaimed because of lack of funds.

In contrast to the law in Tennessee, Virginia law prohibits the State from exercising the power of eminent domain to purchase any orphan land(s) affected or disturbed by coal surface mining.

Pending Federal Legislation

The pending federal legislation, directed toward controlling the mining and reclamation methods employed by the surface mining industry, would supersede Virginia law in all instances where the federal regulations were more stringent than the state's law and regulations.[19] If and when the legislation is enacted, the Department of the Interior would administer the regulations. A national reclamation fund would be established similar to the Orphaned Land Fund in Virginia.[20]

There are major differences between H.R. 25, the proposed federal bill, and the Virginia law. The performance bond would be at least $10,000 per surface mining operation, and the proposed reuse of an area would have to be consistent with adjacent land use and existing state and local land-use plans or programs. Instead of allowing a reasonable highwall, depending upon the slope, the proposed federal legislation would require completed backfilling with spoil materials that would completely cover any highwall and return the area to the approximate original contour in all mining operations.[21] This reclamation standard is being demonstrated on Massengale Mountain in Tennessee by the Tennessee Valley Authority. The procedure meets the requirements of the pending federal regulations, but the costs are considerably increased.[22]

The pending legislation would permit states to assume exclusive jurisdiction over the regulation of surface coal mining and reclamation operations, except as provided in Section 521 and Title IV of the bill. States also would have to meet all Environmental Protection Agency standards pertaining to strip mine operations. The federal government would retain the right to enforce any section of the new strip mine controls if it finds that the state is not enforcing them.

The President vetoed the strip mine bill that was passed by Congress on May 20, 1975, on the grounds that the energy crisis and potential unemployment made the bill undesirable. An attempt to override the veto failed. Another effort, H.R. 9725, a bill of the same substance, was tabled by the House Rules Committee on March 23, 1976. A final effort to bring this subject before the House as H.R. 13950, essentially the same bill as H.R. 9725, was again tabled by the Committee on September 15, 1976. This bill would provide that, in special circumstances, reclamation to the original contour would not apply to some sites.

Conclusions

The economy of southwest Virginia is dependent upon the growth of the surface coal mining industry. The energy crisis has and will precipitate further strip mining on the already scarred landscape. Effective reclamation is dependent upon both the *quality* and the *enforcement* of the surface mining laws in existence and those enacted in the future. Both state and federal laws should allow for the employment of alternative post-mining uses of land without making the costs prohibitive.

Virginia surface mining regulations presently allow the use of several different methods of extraction and reclamation. Pending federal legislation would supersede the state's reclamation laws, in some instances, and possibly hinder the employment of some alternative post-mining land uses. The bank-to-contour reclamation required by the proposed federal legislation has proven to be expensive and, in some instances, it might prove detrimental to the regional economic growth and development of southwest Virginia.

Notes

1. H. R. Wagstaff, *A Geography of Energy* (Dubuque, Iowa: W. C. Brown, 1974), p. 18.

2. Tennessee Valley Authority, *An Appraisal of Strip Mining* (Knoxville, Tennessee, 1963), pp. 4-5.

3. Surface or strip mining is a method of extraction whereby overlaying materials, that is, the overburden, are removed to expose the mineral being mined. Reclamation is the process of recontouring surface mined land, to either its original slope or to slopes suitable for productive uses other than the original use.

4. H.R. 25, 508 (a) (3).

5. For examples, see: J. E. Rowe, *An Inventory of the Unique Uses for Reclaimed Strip Mined Land in the Appalachian Region*, ARP Pub. No. 33 (Knoxville, Tenn.: The U. of Tenn. Environment Center, June 1975).

6. A. Schlottman and W. Abrams, *The Regional Impact of a Ban on High Slope Coal Surface Mining*, ARP Pub. No. 34 (Knoxville, Tenn.: The U. of Tenn. Environment Center, March 1975).

7. T. Plass, "Highwalls—An Environmental Nightmare," in *Proceedings of the Symposium on Revegetation and Economic Use of Surface-Mined Land and Mine Refuse at Pipestem State Park, West Virginia.*

8. E. C. Grim and E. Hill, *Environmental Protection in Surface Mining of Coal*, EPA-670/2-74-093 (Washington, D.C.: 1974), p. 60.

9. Ibid.

10. *Surface Mining . . . A Changing Industry* (Charleston, W. Va.: W. Va. Surface Mining and Reclamation Assoc., n.d.), p. 26.

11. P. Pitts, "Surface Mine Reclamation in East Tennessee," in *Proceedings of the Symposium on Surface Mine Reclamation and Land Use Planning* (Lee's College, Ky.: Scientist and Engineers for Appalachia, Oct. 26-27, 1973), p. 42.

12. Va. Code, Title 45-1, Chapter 17, amended.

13. Ibid., 45.1-202 (e).

14. Ibid., 45.1-203 (a).16. I(a).

15. Ibid., 45.1-206 (a).

16. Ibid., 45.1-212 (a).

17. Ibid., 45.1-206 (b).

18. *An Action Program for Virginia's Abandoned Coal Mines*, T.V.A. and Virginia Department of Conservation and Economic Development, 1975.

19. H.R. 25, Sec. 505 (a).

20. Ibid., Sec. 401 (a).

21. Ibid., Sec. 515 (d) (2).

22. "What Can You Do With Strip-Mined Land?" *Tennessee Valley Perspective* (Knoxville, Tenn.: Tenn. Valley Authority, Fall 1975), p. 10.

19
Surface Mining Control and Reclamation Act of 1977 (P.L. 95-87)

James R. Wagner

Contents

From *Congressional Quarterly Weekly Report*, July 23, 1977: 1596-1600. Reprinted with permission.

Provisions

As sent to the President, HR 2, the Strip Mining Control and Reclamation Act of 1977:

Title I—Findings and Policy

Found that surface mining operations adversely affect commerce and the public welfare by diminishing or destroying land use, polluting water, damaging natural beauty and habitats, and creating hazards to life and property.

Found that expanded coal mining to meet the nation's energy needs required establishment of protective standards.

Found that the primary responsibility for developing and enforcing regulations for surface mining and reclamation should rest with the states because of the diversity of terrain and other physical conditions.

Recognized the need for national standards in order to eliminate competitive advantages or disadvantages in interstate commerce among sellers of coal.

Called for reclamation of mined areas left unreclaimed before enactment of the act.

Title II—Office of Surface Mining

Established an Office of Surface Mining Reclamation and Enforcement in the Interior Department; provided that its director be subject to Senate confirmation.

Identified the specific duties of the office, including administering the act's regulatory and reclamation programs, approving and disapproving state programs, and providing grants and technical assistance to the states.

Stipulated that there was to be no use of coal mine inspectors hired under the Federal Coal Mine Health and Safety Act of 1969 for strip mining inspection unless the director published a finding in the Federal Register that such activities would not interfere with inspections under the 1969 act.

Directed the office to develop and maintain an Information and Data Center on Surface Coal Mining, Reclamation and Surface Impacts of Underground Mining to provide information to the public and other government agencies.

Prohibited any federal employees who performed functions under the act from having a direct or indirect financial interest in coal mining operations and made violators subject to fines of up to $2,500, imprisonment of up to one year, or both.

Title III—State Mining Institutes

Authorized each state to establish, or continue to support, a state mining and mineral resources research institute at a public or private college that would conduct research and train mineral engineers and scientists.

Authorized for each participating state $200,000 in fiscal 1978, $300,000 in fiscal 1979, and $400,000 for each fiscal year thereafter for five years, to be matched by nonfederal funds, to support the institutes.

Authorized an additional $15-million in fiscal 1978, to be increased by $2 million in each fiscal year for six years, for specific mineral research and demonstration projects of industry-wide application at the mining institutes.

Specified that the use of federal funds for the institutes did not authorize federal control or direction of education at any college or university.

Established a center in the Interior Department for cataloging current and projected research on mining and mineral resources.

Established an Advisory Committee on Mining and Mineral Research composed of representatives from the Bureau of Mines, National Science Foundation, National Academy of Sciences, National Academy of Engineering, U.S. Geological Survey and four other persons knowledgeable in the field, at least one of whom represented working coal miners.

Title IV—Abandoned Mine Reclamation

Established within the U.S. Treasury a trust fund called the Abandoned Mine Reclamation Fund consisting primarily of amounts derived from the sale, lease or use of reclaimed land and from a reclamation fee of 35 cents per ton of surface mined coal and 15 cents per ton of underground mined coal (or 10 percent of the value of coal at the mine, whichever was less), except that fees for lignite or brown coal were set at 2 percent of the value of the coal or 10 cents per ton, whichever was less. Such fees were to be paid by 30 days after each calendar quarter ended.

Provided that the fund be used to acquire and reclaim abandoned surface mines and deep mines, including sealing off tunnels and shafts. However, up to one fifth of the fund would be transferred to the Agriculture Secretary for a rural lands reclamation program; up to 10 percent, but not more than $10 million annually, was to be used for hydrologic planning and core drilling assistance on behalf of small mine operators.

Provided that up to 50 percent of the fees collected annually in any state or Indian reservation were to be allocated to that state or reservation for abandoned mine reclamation. But, after reclamation was completed, the Secretary could allow use of the remainder of the 50 percent for construction of public facilities in communities impacted by coal development—if certain specified federal payments were inadequate to meet the needs.

Provided that the balance of reclamation funds could be spent in any state at the discretion of the Secretary for mine reclamation.

Required the Secretary to set rules and regulations for state reclamation programs within 180 days of enactment. States having approved regulatory programs could submit reclamation plans for funding, including grants of up to 90 percent for the cost of acquiring lands to be reclaimed.

Authorized the Secretary of Agriculture to enter into agreements with small rural landowners of abandoned mines for land stabilization, erosion and sediment control, and reclamation. Landowners were to furnish conservation and development plans, and agree to effect the land uses and treatment outlined in the plans. Federal grants to carry out the plans were not to exceed 80 percent of costs on not more than 120 acres, or lower amounts on up to 320 acres in certain instances, unless justified to enhance off-site water quality or to enable a landowner of limited income to participate.

Gave the Interior Secretary and the states broad authority to study reclamation sites, acquire lands not already owned by the public, reclaim the land according to a cost-benefit analysis for each project, and determine use of the land after reclamation. For work done on private lands, the Secretary and the states were directed to establish a lien on the property after reclamation to the extent that the market value of the land was enhanced. Restored land could be sold by competitive bidding or added to the public lands.

Authorized the Interior Secretary to construct public facilities necessary to a reclamation project that created public outdoor recreation areas.

Provided for the filling of voids and the sealing of abandoned tunnels, shafts and entryways, and reclamation of other surface impacts of mining—not limited to coal mine impacts.

Gave the Secretary power to use the fund for emergency abatement or prevention of adverse coal mining practices and gave him access to land where any such emergency existed.

Authorized the transfer of abandoned mine reclamation funds to other federal agencies in order to carry out reclamation activities.

Title V—Environmental Control of Surface Mining

Required the Interior Secretary to issue interim regulations for environmental standards within 90 days of enactment, and waived provisions of the National Environmental Policy Act of 1969. Permanent regulations were to be issued one year after enactment.

Required approval of the administrator of the Environmental Protection Agency for regulations concerning air and water quality standards.

Required all new mines within six months of enactment and all existing mines within nine months to comply with the interim standards. However, an exception was made for operators whose surface and underground mines combined produced no more than 100,000 tons per year; they were given until Jan. 1, 1979, to comply.

Established a federal enforcement program within six months of enactment, to include at least one inspection for every mining site every six months.

Set interim standards requiring surface mine operators to keep waste materials off steep slopes, return mined lands to their approximate original contour, preserve topsoil for reclamation, stabilize and revegetate waste piles, minimize disturbances to water tables, notify the public about blasting schedules and take certain prescribed safety precautions.

Provided that within two months after approval of a state regulation plan, or after implementation of a federal regulatory plan, all mine operators within a state had to apply for a permit to mine lands they expected to be working on eight months later. The state regulatory authority or the Secretary had to grant or deny a permit within eight months.

Directed states that wished to assume jurisdiction over surface mining to submit state regulatory programs to the Secretary within 18 months of enactment, demonstrating that they had the legal, financial, and administrative ability to carry out the act. Among the other requirements, the state program was to provide sanctions for violations of state laws and establish a process for the designation of areas as unsuitable for surface mining.

Directed the Secretary to approve or disapprove a state program. States were allowed 60 days to submit a new program if their first attempts were unsuccessful; the Secretary was required to rule on the resubmitted program in 60 days.

Authorized the Secretary to implement a federal strip mining regulation program in any state that failed to submit a program within 18 months, or failed to resubmit an acceptable program within 60 days of federal disapproval, or otherwise failed to implement, enforce and maintain an approved program. Federal programs were to be implemented no more than 34 months after the enactment, and following a public hearing in each affected state. A state could apply for approval of a new state program any time after implementation of a federal program.

Allowed state programs to include more stringent environmental protection regulations than required by the act.

Required surface mine operators to obtain a permit no more than eight months after approval of a state program or implementation of a federal program. Permits were to be issued for a period of five years, but could be extended if necessary for an operator to obtain financing. If mining operations did not begin within three years, under normal circumstances, the permit expired.

Required that permit applications be accompanied by fees as determined by the regulatory authority.

Required mine operators to submit detailed information with their applications, including the following: identification of all officials and corporations involved; history of the applicant's experience with past mining permits; a demonstration of compliance with public notice requirements; maps of the proposed mining area and land to be affected; description of the mining methods; starting and termination dates of each phase of the mining operation; schedules and methods for compliance with environmental standards; description of the hydrologic consequences of mining and reclamation; results of test borings; soil surveys if the mine might include prime farmlands; a blasting plan.

Provided, for mining operations not expected to exceed 100,000 tons annually, free hydrologic studies and test boring analyses performed by qualified public or private laboratories designated and paid by the regulatory authority.

Required proof of public liability insurance, or evidence of other state or federal self-insurance requirements, as part of a permit application.

Required operators to submit a reclamation plan as part of their permit application.

Required that reclamation plans submitted with permit applications must include the following information: identification of the area to be mined or affected; condition of the land prior to mining, including a description of the uses, topography and vegetation; the use to be made of the land following reclamation and how that use is to be achieved; description of the steps taken to minimize effects on renewable resources; engineering techniques for both mining and reclamation; consideration given to maximum recovery of coal to avoid reopening the mine later; estimated timetable for each reclamation step; measures to be taken to protect surface and ground water systems, and the rights of water users; confidential results of test boring.

Prescribed the requirements for obtaining a performance bond of at least $10,000 covering the area to be mined within the term of the permit. Bonds, payable to federal or state authority, had to cover the full cost of reclamation. States were permitted to establish alternative systems in lieu of bonding programs, subject to federal approval.

Provided that a mining permit could not be approved unless the regulatory authority found that all requirements of the act would be met, reclamation could be accomplished, damage to the hydrologic balance would be prevented, and the area to be mined was not one designated as unsuitable for mining—unless the operator showed that substantial legal and financial commitments were made before Jan. 1, 1977.

Required findings that mining operations would not interrupt, discontinue or preclude farming on alluvial valley floors west of the 100th meridian, nor materially damage the quality or quantity of underground or surface water there. Exempted undeveloped rangeland and farmland of small acreage where mining would not adversely affect agricultural production. Also provided that the requirement did not apply to mines in commerical production or for which state permits had been granted during the year preceding enactment.

Authorized the Secretary to lease other federal coal deposits in exchange to operators who had made "substantial financial and legal commitments" before Jan. 1, 1977, to alluvial valley mining outlawed under the act.

Required, in cases where the private surface ownership and private mineral ownership were separate, the written consent of the surface owner for strip mining the property, or a conveyance that expressly granted the right. Disputes were to be settled under state law and in state courts.

Permitted the mining of prime farmlands, as defined in the act, if the regulatory authority finds in writing that the mine operator has the technological capability to restore the area within a reasonable time to equal or higher farm productivity.

Set terms for revision of permits when there was to be significant alteration in the permit plan.

Prescribed standards for coal exploration, including reclamation requirements, and restricted exploration operations to removing 250 tons of coal.

Established procedures for public notice and hearing of an applicant's intention to mine. Required the regulatory authority to hold public hearings if operators requested them or if serious objections were filed.

Required the regulatory authority to rule on a permit application within 60 days if a public hearing is held, or "within a reasonable time" under other circumstances. Set procedures for appeals.

Established in section 515 the performance standards for environmental protection, to apply to all surface coal mining and reclamation.

Required operators to regrade mining sites to their approximate original contour in most instances and to eliminate highwalls, spoil piles and depressions. Regraded slopes had to assure mass stability and prevent surface erosion and water pollution.

Directed operators to preserve, segregate and reuse topsoil taken from the mine site, protecting it from erosion and contamination. If prime agricultural areas were mined, operators had to provide in the regraded soil a root zone of comparable depth and quality to that of the natural soil.

Allowed water impoundments as a part of reclamation if they met certain water quality and dam safety standards and if embankments were graded properly.

Required operators to minimize disturbances to the hydrologic balance and to the quality and quantity of surface and underground water by avoiding acid and toxic mine drainage, preventing suspended solids from entering the stream flow, cleaning out and removing temporary settling and siltation ponds, preserving hydrologic functions of alluvial valley floor in arid regions, and avoiding channel enlargement in operations having a water discharge.

Permitted permanent disposal of surplus spoil in areas other than mine workings (but within the permit area) if certain standards were met to stabilize the spoil mass, control surface erosion, provide internal drainage and take other precautions.

Required operators to revegetate mined lands with cover native to the area, and to assume responsibility for revegetation for five years after the last seeding or planting. In areas having less than 26 inches of annual precipitation, the responsibility period was extended to 10 years.

Prohibited surface mining within 500 feet of active or abandoned underground mining to protect the health or safety of miners. However, variances could be permitted if the mining efforts were coordinated and if they improved resource recovery.

Prescribed conditions and standards for blasting, including advance notice of schedules.

Permitted the mining practice known as mountaintop removal without regrading to its approximate original contour although no highwalls were permitted in certain cases when the proposed post-mining use of the land was an equal or better economic use, or when the applicant presented specific plans for the post-mining use.

Required complete backfilling of all highwalls, but permitted a variance for certain operations that left a very wide and stable bench for post-mining land uses.

Set standards for mining on slopes steeper than 20 degrees, including a prohibition against placing any spoil or other mining debris on the downslope below the bench (mining cut), and a requirement that highwalls be covered completely by backfilling and the land be returned to its approximate original contour.

Set minimum environmental standards to control the surface impacts of underground mining operations. These included protection of surface land uses from subsidence hazards and protection of surface waters from mine discharges and drainage from mine waste piles. Required coordination in this effort between the Interior Secretary and the Administrator of the Mine Enforcement Safety Administration.

Required inspections of each surface mining operation, without prior notice, to occur on the average of at least one partial inspection a month and one complete inspection every three months. Provided for rotation of inspectors and public availability of inspection reports.

Established environmental monitoring procedures, with special procedures for operations that remove or disturb strata that serve as aquifers affecting the hydrologic balance either on or off the mining site.

Placed tight restrictions on the financial interests any employee of a regulatory authority performing functions under the act might have in coal operations.

Set civil penalties of up to $5,000 for each violation under Title V, and provided that each day of continuing violation could be deemed a separate violation for purposes of penalty assessments.

Provided that any person who knowingly violates a condition of a permit, or makes a false statement on an application, could be fined up to $10,000 or imprisoned for one year, or both.

Provided that civil and criminal provisions of state programs be no less stringent than such provisions in the act.

Established procedures for the release of performance bonds.

Set forth the standing and procedural rules to be applied to lawsuits brought under the act. Allowed citizens to bring suit against the United States or other government instrumentalities under the act, or against any person for violations of rules, regulations, orders or permits issued under the act—including violations that resulted in injury.

Gave primary responsibility for enforcing state programs to the states, but allowed the Interior Secretary to reinforce that authority with federal action following public hearings.

Gave the Secretary authority to stop a mining operation immediately if it posed an imminent danger to public health or safety or might cause irreparable damage to the environment.

Required states to establish plans for designating lands unsuitable for surface mining. The designation was required if land could not be reclaimed under requirements of the act. All other designations were discretionary with the regulatory authority, but lands could be deemed unsuitable if: strip mining would be incompatible with government objectives; the lands are fragile or historic; the site is a natural hazard area where development could endanger life or property; the area contains renewable resources where development would result in a loss of long-range productive capacity.

Exempted unsuitable lands on which surface mining operations were being conducted on the date of enactment, or where substantial legal and financial commitments in such operations were made prior to Jan. 4, 1977.

Directed the Secretary to review federal lands to determine which were unsuitable for surface mining. Existing operations on federal lands were allowed to continue until completion of the review.

Prohibited strip mining operations in the National Park System, National Wildlife Refuge System, National System of Trails, National Wilderness Preservation System, Wild and Scenic Rivers System, or Custer National Forest.

Permitted surface mining in other national forests if the Secretary found there were no significant recreational, timber, economic or other values that might be incompatible with such operations, and where surface operations were incident to underground mines and where the Agriculture Secretary determined that surface mining in areas west of the 100th meridian having sparse forest cover would be in compliance with existing law.

Required the Interior Secretary to promulgate a program for federal lands within one year of enactment, incorporating all requirements of the act.

Authorized the Secretary to enter into a cooperative agreement with a state for state regulation of strip mining on federal lands, provided that the Secretary retained authority to approve or disapprove mining plans and to designate federal lands as unsuitable for surface coal mining.

Required all public agencies, public utilities and public corporations to comply with the environmental protection standards of Title V.

Provided for judicial review in the appropriate U.S. District Courts of the Secretary's decisions regarding approval or disapproval of state programs.

Authorized separate regulations for bituminous coal mines in the West which were in production prior to Jan. 1, 1972, and which met special criteria.

Authorized the Secretary to issue separate regulations for anthracite coal surface mines if such mines were regulated by environmental protection standards of the state.

Title VI—Lands Unsuitable for Non-Coal Mining

Permitted the Secretary of Interior, if requested by a state governor, to review any federal land within a state to assess whether it was unsuitable for mining for minerals other than coal.

Authorized designation of any area as unsuitable if it were predominantly urban or suburban or if a mining operation would have an adverse impact on lands used primarily for residential purposes. The provision would not apply to any lands already being mined.

Permitted any person with an interest which might be adversely affected to petition for exclusion of such an area from mining activities.

Title VII—Miscellaneous Provisions

Defined technical terms and descriptions used throughout the act.

Exempted coal owned by the Tennessee Valley Authority from the surface owner and federal lessee protection requirements applicable to other federal coal, but authorized the Secretary to set guidelines for mining TVA-owned coal.

Made it unlawful to discharge or discriminate against an employee for filing suit or testifying under provisions of the act.

Made it a criminal violation to resist or impede investigations carried out by a regulatory authority under the act.

Authorized the Secretary to make grants to the states to develop and implement state regulatory programs. Grants could be up to 80 percent the first year, 60 percent the second year, and 50 percent each year thereafter.

Required the Secretary to report annually to the President and Congress on activities under the act.

Authorized the Secretary to modify application of environmental protection provisions of the act to Alaskan surface mines for up to three years if he decided that was necessary to continue operation of the mines. Required the Secretary to make a study of strip mining conditions in Alaska and report to Congress within two years and authorized $250,000 for the Study.

Mandated a study within 18 months of enactment concerning surface and open pit mining and reclamation technologies for minerals other than coal, with emphasis on oil shale and tar sands deposits in western states. Authorized $500,000 for the study.

Required a special study of surface mining regulation on Indian lands, and authorized $700,000 to assist the Indian tribes in the study.

Permitted departures from environmental performance standards for mining and reclamation on an experimental basis, in order to allow post-mining land use for industrial, commercial, residential or public use, including recreation.

Authorized $10-million a year for fiscal 1978 through fiscal 1980 for initial regulatory procedures and administration of the program; $10-million each year for 15 years beginning in fiscal 1978 for hydrologic studies and test borings for small mine operation; $20-million in fiscal 1978 and $30-million each in fiscal 1979 and 1980 for grants to the states in preparing their regulatory plans. It also authorized up to $2-million in fiscal 1977 for the Secretary to begin implementing the act.

Provided that surface owners, as defined by the act, must give their written consent before the Secretary could lease federally owned coal beneath the land they lived and worked on.

Provided, in cases where the surface above federally owned coal was subject to a federal lease or permit, that there must be either written consent of the lessee or permittee, or evidence of bonding for payment of damages to the lessee or permittee.

Title VIII—University Coal Research Laboratories

Authorized the administrator of the Energy Research and Development Administration to designate 10 institutions of higher education for establishing university coal research laboratories.

Authorized the administrator to make grants of up to $6-million for initial costs, and up to $1.5-million annually for operating expenses for each institution.

Established an 11-member Advisory Council on Coal Research to help administer the title.

Authorized appropriations of $30-million for fiscal 1979, and $7.5-million annually for fiscal years 1980-1983.

Title IX—Energy Resource Graduate Fellowships

Authorized the Administrator of the Energy Research and Development Administration to award up to 1,000 graduate fellowships annually in fiscal years 1979-1984 for study and research in applied science and engineering related to the production, conservation and use of fuels and energy.

Set terms and conditions of the fellowships.

Authorized appropriations of $11-million for each of the six fiscal years.

Authorized the Interior Secretary to conduct and promote research and demonstration projects of alternative coal mining technologies.

Authorized annual appropriations of $35-million for fiscal year 1979-1983.

Part 5
The Future of Surface Mining

20
Coal and the Future
of the Appalachian Economy

William H. Miernyk

There is an ancient association between coal and economic backwardness in Appalachia. Generations of Appalachian miners labored in "dogholes" for little more than a subsistence wage. Even in large, unionized mines wages lagged behind those of industrial workers until recently. Aggressive competition in coal markets kept the price of coal low, and low prices were reflected in low wages. Add to this the historic instability of the coal industry, the ever present danger of mining as an occupation and the environmental damage it has caused, and one can understand the ambivalent attitudes of many Appalachians toward their most important natural resource.

Coal Production

Coal production in the United States hit an all-time peak of nearly 631 million tons in 1947. This was followed by a 14-year decline. In 1961, only 403 million tons were produced—36 percent less than in the peak year. During this period the nation's railroads were dieselized, and there was a massive shift from coal to cheap imported oil by electric utilities.

Employment fell even more rapidly than production, especially after the mechanization agreement between John L. Lewis's United Mine Workers and the major operators was reached in 1950.

Between 1947 and 1961, total employment in coal production declined by 63 percent—from 497,782 to 166,415 workers. As the nation's major coal-producing region, Appalachia bore the brunt of the decline in jobs and production. Coal employment in the Region fell from 427,600 in 1947 to 144,914 in 1961, a decline of 65 percent. Although consumer prices went up 32 percent during this period, the average price of coal fell from $4.99 in 1948 to $4.48 in 1962, a drop of 10 percent.

From *Appalachia*, vol. 9, no. 2, Oct.-Nov. 1975, pp. 29-35. Reprinted by permission.

The long, slow and painful decline of coal ended in the early 1960's however. Between 1963 and 1973, national coal production increased from about 459 million tons to 591 million tons—a gain of almost 29 percent. Appalachia's share of national coal production declined, but the Region enjoyed a healthy 16-percent increase in output during this period. Table 1 shows Appalachian coal production by state, and the relative contribution of each state to national output.

In 1963, West Virginia was the largest coal producer in the nation. It is the only Appalachian state which registered a decline in production during this period, however, and by 1973 it ranked behind Kentucky. West Virginia still accounts for more than a fifth of the nation's total coal output. The accelerating search for energy, which started during the early 1960s, has opened up new sources of supply in the Midwest and the Far West. But it has also stimulated increases in production in Appalachia. At present, Appalachia remains by far the most important coal-producing region in the nation.

Table 2 gives the 1973 distribution of coal production by method of recovery. In that year surface-mined coal accounted for slightly less than half of the total U.S. production (it now accounts for more than half). Sixty percent of Appalachia's coal came from deep mines, however.

Table 1

Coal Production by Appalachian States, 1963-1973
(thousands of tons)

State	1963 Amount	1963 Percent of U.S. Total	1973 Amount	1973 Percent of U.S. Total	Percent Change
Alabama	12,359	2.7%	19,793	3.3%	60.1%
Kentucky	77,350	16.9	127,000	21.3	64.2
Maryland	1,162	*	1,754	*	50.9
Ohio	36,790	8.0	45,345	7.7	23.3
Pennsylvania	71,501	15.6	76,646	13.0	7.2
Tennessee	6,121	1.3	8,993	1.5	46.9
Virginia	30,531	6.7	33,869	5.7	10.9
West Virginia	132,568	28.9	115,239	19.4	-13.1
Appalachia	368,382	80.3%	428,639	72.0%	16.4%
United States	458,928	100.0%	595,386	100.0%	29.7%

*Less than .05 percent.

Source: United States Bureau of Mines as reported in *Coal Facts 74/75*, Bituminous Coal Operators Association.

Underground mining is by far the most important method of recovery in West Virginia, Virginia and Pennsylvania. The other states rely heavily, and increasingly, on surface mine techniques.

Coal Prices and Revenues

Between 1963 and 1973 the average price of U.S. coal at the mine mouth increased from $4.39 to $8.42, or 92 percent. In Appalachia, coal prices rose from $4.50 to $9.77, or 117 percent. There are wide variations in the price of coal among the Appalachian states, as Table 3 shows. These variations are due to differences in market conditions and differences in quality. Alabama, which ranked sixth in production in 1973, ranked first in the average price of coal. West Virginia, which was second in terms of production, ranked third in terms of prices.

Table 4 shows the gross revenue generated by coal production in 1963 and 1973, and the relative changes during this decade. Appalachian coal revenue increased by 153 percent. The largest absolute increase came in West Virginia, which was the first state to produce more than a billion dollars worth of coal in one year (1970). By 1973, the gross revenue of the Appalachian coal industry was well over $4 billion.

Table 2

Distribution of 1973 Bituminous Coal Production in Appalachia by Mining Method, by State
(thousands of tons)

State	Underground	Surface	Auger	Total	Underground as Percent of Total
Alabama	7,892	11,836	65	19,793	39.9%
Kentucky	63,400	55,100	8,500	127,000	49.9
Maryland	64	1,611	79	1,754	3.6
Ohio	16,205	28,130	1,010	45,345	35.7
Pennsylvania	46,255	30,043	348	76,646	60.3
Tennessee	4,785	4,053	155	8,993	53.2
Virginia	23,339	5,703	1,827	33,869	68.9
West Virginia	95,448	17,575	2,216	115,239	82.8
Appalachia	257,388	154,051	14,200	428,639	60.0%
U.S. Total	301,500	275,300	142,000	591,000	51.0%
Appalachia as Percent of U.S.	85.4%	57%	10%	72.5%	

Source: *Coal Facts 74/75.*

Information gathered from *Moody's Government and Municipal Manual* (1975), and from the West Virginia Tax Department, shows that four of the Appalachian coal-producing states have some sort of tax on coal. Alabama levies a tax of 13.5 cents per ton, and the state collected an estimated $3.7 million in coal taxes in 1973. Ohio, with substantially more production than Alabama, collects a tax of four cents per ton. This yielded an estimated $1.8 million in coal-tax revenue in 1973.

Kentucky levies a tax of 4 percent of gross value, but not less than 30 cents a ton. In 1973, Kentucky collected an estimated $53.5 million from coal. West Virginia levied a 3.5 percent business and occupations tax on coal in 1973, and collected more than $44 million that year. This tax subsequently was raised to 3.85 percent. The two major coal-producing states in Appalachia—Kentucky and West Virginia—derive substantial sums from their most important natural resource. Coal production in West Virginia declined between 1972 and 1973, but because of a substantial rise in prices total tax collections increased.

Throughout the present recession, personal income in major Appalachian coal-producing states has increased more rapidly than that of the nation as a whole. This has been particularly true in Kentucky and West Virginia. While other factors might be partly responsible for this favorable showing, rapid increases in coal revenues are undoubtedly the most important cause.

Table 3

Average Price of Bituminous Coal per Ton, by State
(F.O.B. Mine)

State	1963	1973	Percent Change
Alabama	$7.38	$12.43	68.4%
Kentucky	3.82	7.36	92.7
Maryland	3.73	7.20	93.0
Ohio	3.70	7.37	99.2
Pennsylvania	4.90	10.73	119.0
Tennessee	3.71	7.95	114.3
Virginia	3.96	11.59	192.7
West Virginia	4.79	11.39	137.8
Appalachia	$4.50	$ 9.77	117.1%
United States	$4.39	$ 8.42	91.8%
Appalachia as Percent of U.S.	103%	116%	

The parts of Appalachia that were hardest hit by the decline of coal from 1947 until 1961 have been exhibiting more robust recovery than the rest of the Region. Recent Census reports show that Central Appalachia—long regarded as Appalachia's more seriously depressed subregion—is one of the four fastest-growing areas in the nation.

Projected Coal Production

What of the future? Is the latest burst of coal prosperity likely to last? Many old-timers are skeptical. As one put it to me: "If you had been around coal towns as long as I have, you would know it's either feast or famine. These good times have got to come to an end." What those who feel this way don't realize, however, is that the demand for and supply of energy have undergone drastic changes in the past few years on a world-wide basis. The world's supply of fossil fuels is dwindling, but the demand for energy continues to grow as population and income increase.

Those who are incurable technological optimists feel there really is no problem. Nuclear fusion, solar energy—or something—they say, will come along to solve the energy problem. The Energy Research and Development Administration (ERDA) is somewhat less optimistic, however. For the immediate future, ERDA reports have shown, we will have to rely on fossil fuels. The agency has earmarked substantial sums for coal research and development.

The Federal Energy Administration also has projected large increases

Table 4

Estimated Gross Revenue of Coal Sectors, by State*
(millions of dollars)

State	1963	1973	Percent Change
Alabama	$ 91.2	$ 246.0	169.7%
Kentucky	295.5	934.7	216.3
Maryland	4.3	12.6	193.0
Ohio	136.1	334.2	145.6
Pennsylvania	350.4	882.4	151.8
Tennessee	22.7	71.5	215.0
Virginia	120.9	392.5	224.6
West Virginia	635.0	1,312.6	106.7
Appalachia	$1,656.1	$4,186.5	152.8%

*Average price x total production.

in coal production between now and 1990. The FEA's energy experts have made two sets of coal projections with regional breakdowns (Project Independence Report, November 1974, pp. 102-108). One of these, the "business as usual" projection, shows national output of coal rising to about 1.3 billion tons by 1990. The "accelerated development" projection however, shows coal production reaching 2.8 billion tons by 1990. The first is probably too low, and the second may be too high. The midpoint of these two projections yields a number that will no doubt come closer to actual production in 1990 than either of the extremes. This projection comes to slightly more than two billion tons.

It is generally conceded that Appalachia will account for a declining share of the national coal production, as output in other regions expands. In 1973, for example, Appalachia produced 72.5 percent of the nation's coal. The midpoint projection shows Appalachia's share dropping to 60 percent by 1977, and to 57 percent by 1990. But even with Appalachian coal production increasing at a slower rate than the national average, the Region is expected to produce more than 1.1 billion tons of coal by 1990. This would represent an increase of 167 percent over the 1973 level of production (see Table 5).

Are these projections technically feasible? Is there enough coal in Appalachia to permit production increases of the magnitudes given in Table 5? First, these projections are well within the bounds of technical feasibility. Indeed, the 1990 production levels could be achieved as early as 1985 if demand (and price) conditions warranted such a rapid expansion.

Table 5

Projection of Coal Production by State to 1990
(thousands of short tons)

State	1973	1990	Increases	
			Tons	Percent
Alabama	19,793	47,700	27,907	+141.0%
Kentucky	127,000	366,500	239,500	188.6
Maryland	1,754	4,000	2,246	128.0
Ohio	45,345	130,000	84,655	186.7
Pennsylvania	76,646	177,300	100,654	131.3
Tennessee	8,993	21,200	12,207	135.7
Virginia	33,869	90,400	56,531	166.9
West Virginia	115,239	308,100	192,861	167.3
Appalachia	428,639	1,145,200	716,552	167.2%

The capital requirements will be enormous—industry spokesmen claim it will have to raise between 20 and 25 billion dollars over the next decade—but with high prices and high profits the coal sector will be able to do this. And it is inconceivable that short-term bottlenecks in equipment, including transportation equipment, could hold up expansion of output for any appreciable time.

Coal Reserves

The answer to the second question raised above is also encouraging. It is given in Table 6, which shows the relationship between estimated reserves and projected output to 1990. The second column shows the cumulative production that would be required to reach the target levels given in Table 5. The years of production left—assuming that output levels off at the 1900 rates—are given in the final column.

Even at the high rates of annual production forecast for 1990 and subsequent years, Appalachia's coal stocks will not be exhausted until 2336. On the basis of the assumptions behind Table 6, Virginia will be the first to exhaust its supply of coal (in 2141). At the other extreme, Alabama's coal stocks would not be exhausted until the year 2884. These should not be considered as actual "projections"—the assumptions behind them are too rigid. They are, however, indicators of the extensive coal reserves in

Table 6

Appalachian Coal Production and Reserves

State	Total Reserves[1] (millions of short tons)	Cumulative Production 1973-1990 (millions of short tons)	Years of Production Remaining at 1990 Rate
Alabama	41,432	607.4	854
Kentucky	116,842	4,441.5	307
Maryland	1,558	51.8	377
Ohio	43,358	1,578.1	321
Pennsylvania	87,269	2,285.5	479
Tennessee	4,572	271.7	203
Virginia	14,787	1,118.4	151
West Virginia	100,628	3,810.1	314
Appalachia	410,446	14,164.5	346

[1] Source: "Coal Reserves in the United States," *Coal Facts, 1974-75*, p. 74.
Note: Details might not add to totals because of rounding.

Appalachia, and rough guides to the useful lives of the Appalachian coal fields over the next few centuries.

Projected Coal Prices and Revenues

Forecasting average coal prices is a hazardous statistical exercise. Average prices of steam coal may be less meaningful than other price averages because of wide variation. At present, for example, a single utility might pay $6.00 a ton for part of its coal, $12 for another part and $25 for the remainder. The first two prices were determined by long-term contracts, and the last is the spot, or open-market, price. To calculate the firm's average price, we would have to know the proportions of purchases to which each price applies, and this information is not published. As old contracts expire and new ones are negotiated, however, the average price moves up. It is possible to project this average. Such projections have been made for the Appalachian states. These have been applied to the cumulative production figures of Table 6 to obtain estimates of the cumulative revenue to be derived from coal in this Region. These are very crude projections, and the results, given in Table 7, should be considered only as rough approximations.

If prices and coal production rise as projected, the coal industry in Appalachia would generate about $277 billion of revenue between 1973 and 1990. The annual average comes to slightly more than $16 million.

Table 7

Projected Cumulative Appalachian Coal Revenue, 1973-1990

State	Cumulative Production 1973-90 (tons)	Projected Average Price, 1982	Cumulative Revenue (millions of dollars)
Alabama	607,436,968	$24.80	$ 15,064
Kentucky	4,441,499,960	14.72	65,379
Maryland	51,786,048	14.40	746
Ohio	1,578,105,016	14.74	23,261
Pennsylvania	2,285,514,064	21.46	49,047
Tennessee	271,737,024	14.94	4,060
Virginia	1,118,421,008	23.18	25,925
West Virginia	3,810,051,040	22.78	86,793
Appalachia	14,164,551,128	$19.54	$270,275

Note: Details might not add to totals because of rounding.

The amounts that individual states will derive from coal taxes will depend, of course, on the extent to which they tap into this rich new source of revenue. At present tax rates, for example, Kentucky's coal industry would yield about $26 billion in taxes (on the assumption behind Table 7). West Virginia would do somewhat better, with a gross tax yield of more than $33 billion. It is likely that other Appalachian states will follow the lead of Kentucky and West Virginia in levying taxes based on the value of coal during the coming period of expanding coal production.

Projected Productivity and Employment

The last two tables deal with productivity and employment. Productivity in the Appalachian coal sector dropped between 1970 and 1972 (the latest year for which detailed data are available). But this was a temporary decline engendered in part by new health and safety legislation. The long run upward trend in productivity will be resumed as new and better mining machinery and methods are introduced.

Table 8 shows annual output per worker projected to 1990 on the basis of two sets of assumptions. The first is that productivity will increase at an average rate of 1 percent per year; the second is a somewhat more optimistic rate of 3 percent per year. The observed variations from state to

Table 8

Annual Output per Worker, 1972, and Projections to 1990

| | | Projected Output Per Worker | |
State	1972[1] (tons)	Assumption A[2] (tons)	Assumption B[3] (tons)
Alabama	4,112	4,958	7,001
Kentucky	4,388	5,290	7,471
Maryland	4,970	5,997	8,456
Ohio	5,360	6,466	9,127
Pennsylvania	3,137	3,778	5,340
Tennessee	4,419	5,329	7,525
Virginia	2,745	3,305	4,671
West Virginia	2,620	3,155	4,461
Appalachia	3,410	4,109	5,807

[1] Calculated from data in *Coal Facts, 1974-75*, pp. 81 and 87.

[2] Assuming productivity increases of 1 percent annually.
[3] Assuming productivity increases of 3 percent annually.

state, in 1972, are largely due to differences in mining methods. The states which obtain large proportions of total output by surface mine methods had the highest output-per-worker figures. The projections of output-per-worker assume that these proportions will be maintained. This may or may not be the case, depending in part on the course of future surface mine legislation.

The productivity and production projections of Tables 8 and 9 have been used to project coal employment in 1990. On the basis of the low productivity assumption, coal employment in Appalachia would more than double, with almost 147,000 new jobs in the industry. More rapid productivity increases would, of course, reduce manpower requirements. Even on the assumption of annual increases of 3 percent in productivity, however, more than 66,000 new jobs will be created in the industry.

These projections relate only to the jobs that would be created directly in the industry. But as coal production expands in Appalachia, other sectors of the economy will grow as well. If we take a weighted average of the deep-mine and surface-mine employment multipliers calculated from the West Virginia Input-Output Model, with a slight downward adjustment for broadened areal coverage, we obtain an employment multiplier for the Region of 1.9. Direct and indirect employment in the Region, on the low-productivity assumption, would then come to almost 279,000 new jobs. Based on the high-productivity assumption, the corresponding figure is over 125,000 new jobs.

There are likely to be other indirect effects. Historically, only a few types of manufacturing activities have been attracted to relatively low-cost power sites. Examples are aluminum reduction and other electrolytic processes. But as the price of coal, petroleum and electric energy continue to increase, the cost of transporting energy will also rise. This could lead several types of manufacturing activities to consider relocation to sources of energy, just as textile mills shifted to sources of low-cost labor in the 1940s and 1950s. The likelihood of such relocations will increase as low-Btu coal gasification becomes a reality. Low-Btu gas cannot be shipped long distances, but as it becomes available in significant quantities, industries that need gas as a boiler fuel might consider locating some of their plants in this Region. These and other energy developments could complement the locational pull of the highway network that will exist in Appalachia when the Interstate and Development Highways are completed.

Is Coal the Panacea?

Will coal solve Appalachia's problems? It is not likely to. The employment and income generated by the revival of coal will have a major impact

Table 9

Employment, 1972, and Projected Employment, 1990

		1990					
		Assumption A[2]			Assumption B[3]		
			Increase over 1972			Increase over 1972	
State	1972[1] Number	Number	Number	Percent	Number	Number	Percent
Alabama	5,062	9,620	4,558	90.04%	6,813	1,751	34.59%
Kentucky	27,616	69,281	41,665	150.87	49,050	21,434	77.61
Maryland	330	667	337	102.12	473	143	43.33
Ohio	9,509	20,105	10,596	111.43	14,243	4,734	49.78
Pennsylvania	24,216	46,929	22,713	93.79	33,202	8,986	37.11
Tennessee	2,548	3,978	1,430	56.12	2,817	269	10.56
Virginia	12,398	27,352	14,954	120.62	19,353	6,955	56.10
West Virginia	47,223	97,654	50,431	106.79	69,065	21,842	46.25
Appalachia	128,902	275,586	146,684	113.79%	195,016	66,114 51.29		51.29%

[1] *Coal Facts 1974-75*, p. 87.
[2] Assumes annual increase in productivity of 1 percent.
[3] Assumes annual increase in productivity of 2 percent.

Assumes annual increase in productivity of 3 percent.

on West Virginia, Kentucky and the Appalachian portions of other coal-producing states. But the economic scars left by the contraction of coal following World War II and the other economic disadvantages of rural Appalachia will not be easily eradicated. There will be a need for continued investment in educational and health facilities if rural Appalachia is to be raised to approximate parity with the rest of the nation. But there is no doubt that the job of stimulating general economic development will be much easier than it would have been without the revival of coal and the new prosperity which this has engendered.

Finally, coal is an exhaustible natural resource. Table 6 shows that the coal fields of Appalachia have potentially long useful lives in terms of known coal reserves. But the day will come when it will no longer be feasible—because of either physical or economic constraints—to depend upon coal as a source of revenue. Appalachia's economic future beyond the coal age will depend upon the extent to which the Region's political and business leaders acted to broaden its economic base at a time when funds were available to do so.

21

Surface Mined Land Reclamation in Germany

William S. Doyle

To help design effective mining and land conservation measures, it is useful to examine programs which have been adopted by other industrial nations faced with similar problems. ORNL-NSFEP-16 describes the planning, technological, and regulatory procedures which are used in the Federal Republic of Germany (West Germany) to ameliorate the harmful environmental consequences of large-scale surface mining of brown coal.

In general, the land restoration program has been largely successful and strip-mining is no longer a controversial public issue in Germany. Of particular interest are the institutional arrangements that have been worked out to assure comprehensive planning for land restoration before the start of mining. Many features of the German approach could be applied to strip-mining problems in the United States.

The Importance of Lignite in West Germany

The Federal Republic of Germany is fortunate to possess large reserves of brown coal (lignite) to serve in electric power production and possibly as a raw material for producing synthetic fuels. The total brown coal reserves of West Germany are estimated to be about 60 billion tons. Of these, some 55 billion tons are located in the Rhineland coal fields alone, making them the largest continuous deposit of lignite in Europe. (The lignite reserves of the world are estimated to amount to some 2,100 billion tons, as compared to 155 billion tons for all of Europe.) Using

From William S. Doyle, *Strip Mining of Coal: Environmental Solutions*, 1976, pp. 268-286. Published by the Noyes Data Corporation. Reprinted by permission.

This material was excerpted from E. A. Nephew, *Surface Mining and Land Reclamation in Germany*, May 1972. The original article is available from the National Technical Information Service, U.S. Department of Commerce, 5285 Port Royal Road, Springfield, VA 22151, and is cataloged as ORNL-NSFEP-16.

present-day mining techniques, approximately eight to ten billion tons of Rhineland lignite lie close enough to the surface for economical recovery.

As improved mining methods will almost certainly be developed in the future, it is clear that brown coal will continue to play a vital role in the West German economy far into the next century. The total West German production of brown coal in 1970 amounted to some 108 million tons, with 92.6 million tons being mined in the Rhineland alone. About 81 million tons of lignite were burned in thermal power stations to produce 60 billion kilowatt-hours of electricity; 24.7 million tons were used for briquette manufacture; and the remainder was used for miscellaneous purposes.

In view of the declining market for briquettes, new applications for brown coal are under study. It is believed that favorable physical properties of brown coal, such as high chemical reactivity, porosity, and low sulfur content, may make it suitable for a number of different production processes. Some of these potential processes are the reduction of iron ore, the gasification of lignite using nuclear process heat, and the production of metallurgical coke.

About one-third of all power generated in German thermal electric power stations comes from lignite-fueled plants. The present and potential uses of brown coal, both as an energy source and as a raw material, assure that it will eventually be mined to depletion. In Germany, the need to treat effectively the environmental brown coal mining has been fully recognized. Furthermore, this necessity will remain even if brown coal is eventually replaced by an alternative fuel for electric power production.

The vital role of lignite in the German economy can be seen from the intensive efforts which have been made to modernize the mines. This modernization process has required an enormous financial investment, more than one billion dollars since the end of World War II, and has led to a consolidation of the brown coal mining industry. The many shallow surface mines which were formerly common in the Rhineland have been gradually replaced by larger, more efficient open-pit mines. (While the number of active mines in the Rhineland decreased from 23 in 1950 to 6 in 1970, the annual production of lignite increased from 64 to 93 million tons.)

The post-war lignite industry in the Rhineland was shared by four major companies: Rheinische Braunkohle AG, Roddergrube AG, Braunkohleindustrie AG (BIAG), and Neurath AG. In 1959, these companies merged to form the present-day Rheinische Braunkohlenwerke AG, which today dominates the brown coal mining industry. This single company, a subsidiary of Germany's largest electric power utility—the Rheinisch Westfälische Elektrizitätswerke AG (RWE)—employs nearly 16,000 workers and produces 85 percent of the German lignite. Its large size and many resources have helped the Rheinische Braunkohlenwerke AG to institute enlightened land reclamation practices which have received worldwide recognition.

Brown Coal Mining Technology

Open-Pit Mines

The Rhineland brown coal field lies in flat, plains country in the 1,000 square mile triangular area formed by the cities of Aachen, Cologne, and Mönchen-Gladbach. The lignite is deposited in highly faulted seams that are from 65 to 350 feet thick, with varying overburden up to 650 feet. The coal bed lies on a slightly-folded, inclined plane, with the shallower seams located at the base of the triangle, in the vicinity of Cologne and Aachen. Mining began in shallow surface mines near Cologne and has moved steadily northward, becoming progressively more complicated as the deeper coal deposits were reached. Prior to World War II, the mining of lignite in the Rhineland area was confined to sites where the brown coal lay sufficiently close to the surface (<100 ft) that conventional surface mining equipment could be used. Since the number of such suitable locations was limited and would be exhausted in the foreseeable future, it was clear that new mining methods needed to be developed to exploit the deeper lignite deposits.

Attempts to develop suitable new mining techniques began in 1938 when tests were made to explore the possibility of using deep mines to extract the lignite. These experiments were delayed by the war and were finally abandoned in 1953 when it became apparent that the high groundwater level of the region and the unconsolidated nature of the overburden precluded the large-scale introduction of deep mines on an economical basis.

At about this time, however, modern, massive excavating machines were developed that made the deeper brown coal seams accessible for the first time to surface mining techniques. Because the capacity and efficiency of these bucket-wheel type excavators were significantly greater than those of the older equipment, it became possible to design new open-pit lignite mines on a scale not possible heretofore.

Thus, in the years 1953 to 1955, a new epoch in the history of the Rhineland brown coal industry began. The result is such mines as the Fortuna-Garsdorf open-pit mine located near Bergheim. This mine is the largest material handling operation on earth, nearly twice as large as its closest competitor—the Kennecott Copper Bingham pit in Utah. In 1970, a total of 86.8 million cubic yards of spoil material, together with 36.2 million metric tons of lignite, was taken from the Fortuna-Garsdorf open-pit mine alone.

Mechanization and Automation of the Lignite Mines

The modernization of the Rhineland brown coal mines, beginning with the introduction of giant wheel-excavators in 1955, has greatly increased productivity and helped to make brown coal the cheapest source (next to hydroelectric power) of energy in Europe.

The process of mechanization did not end with the massive, new digging machines. A transportation system, capable of matching the prodigious capacity of the wheel excavators, had to be developed to haul away the spoil material and lignite from the mines. Since full restoration of the land disturbed by the mining operations was planned, giant spreader machines—similar to the wheel excavators—were designed and built to spread the overburden from active mines back into mined-out pits.

In addition, because the depth of the new surface mines extends well below the groundwater level, methods had to be developed to prevent flooding of the pits. That these problems were solved successfully is shown by the fact that lignite has not only held its position with respect to competitive fossil fuels, but has actually expanded its market.

Bucket Wheel Excavators: Each of the large, open-pit, brown coal mines is equipped with several wheel excavators for stripping off the overburden and extracting the lignite. A single machine costs up to ten million dollars and consumes as much as 10.4 Mw of electrical power in operation. Excluding maintenance personnel, only two operators are needed to operate the huge excavator. The lignite or loose overburden is carried by conveyor belt from the excavator wheel to the discharge boom, where it falls into waiting railroad cars. Because of their ability to excavate selectively and deliver the loose overburden, lignite, or topsoil to a separate, interfacing transportation system, the wheel excavator is especially suitable for use in areas where land reclamation is planned.

Transportation of Bulk Materials: Transporting the massive amounts of spoil material and lignite from the mine is accomplished with a specially designed system consisting of conveyor belts, heavy-duty trains, and slurry pipelines. Much of this equipment is automated or remotely controlled, thereby contributing to the high productivity of the overall mining operations. Seven-foot-wide conveyor belts moving on steel rollers at speeds up to 15 miles per hour are used to transport the lignite out of the mine pit. These conveyor belts can be installed in a straight line and operate satisfactorily on relatively steep inclines, thus eliminating ramps which would be required if trains were employed in the pits.

The total installed length of the conveyor belt network comes to about 70 miles. The belts are used to haul both lignite and spoil material. A crawler, equipped with special handling devices, is used to move the skid-mounted conveyor belt sideways as mining progresses. This can be done very rapidly, with little interruption of the mining operations.

Bulk material from the mines which must be transported over long distances is hauled on a company-owned railway system. The rail network consists of 310 miles of tracks connecting the active brown coal mines, the mined-out areas undergoing restoration, the briquette factories, and the power generating stations. The spoil material is hauled in eight-axle-gondolas, each with a capacity of 125 cubic yards and a gross weight of 240 metric tons. The gondolas can be emptied in a matter of seconds by a hydraulic system which tips the cars sideways.

Raw lignite is transported in four-axle, ninety-ton rail cars. Specially profiled, heavy-duty rails, with a linear density of 43 pounds per foot, have been designed to accommodate the enormous, 30-ton-per-axle loads which are encountered. Electric locomotives weighing up to 139 tons are used to pull loads of as much as 2,000 tons. This private railroad network carries a larger annual tonnage (not ton-miles) than the whole German Federal Railway system.

Ground and Surface Water Control

To extract the lignite using surface mining techniques, it is necessary to lower the groundwater level to prevent flooding of the mine pits. This is accomplished by pumping water from some 1,850 deep wells which have been drilled in the Erft river basin to an average depth of nearly 600 feet. Submersible motor pumps, some nearly 33 feet long and weighing more than 12 tons, pump water at rates as high as 33 tons per minute. At the present time, 1,100 wells in continuous operation provide sufficient water removal capacity for the six surface mines in the Erft region. On the average, 14 tons of water must be pumped out of the ground for each ton of lignite mined.

The water from the deep wells is discharged into the Erft, Inde, and Merzbach waterways and into a special drainage canal which connects the coal fields to the Rhine River. The canal can also be used to provide supplementary flood control benefits to the region north of Cologne. During periods of high water, a pumping station can divert 10 cubic meters per second from the Erft into the Rhine River. The Rheinische Braunkohlenwerke AG mining company is also studying the possibility of converting some of the huge mine pits into a freshwater reservoir to improve the supply and quality of water for industry in the region. Thus, the brown coal mining activities provide unforeseen community benefits.

Land Rehabilitation

Extent and Costs of Land Reclamation

All lignite mining operations to date have affected less than one-tenth of the 620,000 acre Rhineland brown coal area. The total land area disturbed by brown coal surface mining from the turn of the century to January 1, 1969, amounted to 36,750 acres. Of this, 16,480 acres were not yet restored, representing either active mine sites or depleted pits being reclaimed. The remainder, 20,270 acres, has been restored for forestry (10,290 acres), agriculture (7,390 acres), and artificial lakes (2,590 acres). Restoring land to full agricultural productivity is the most expensive type of reclamation, costing from $3,000 to $4,500 per acre.

In recent times, rising land prices and lower reclamation costs, due to improved, more efficient methods, have resulted in arriving at an economic break-even point. Today, the market value of the restored farmland compares favorably with the expenditures for reclamation. In the United States, the cost of full land restoration would, in most cases, greatly exceed the value of the land. However, it is interesting to note that land reclamation was required in Germany long before it became marginally profitable.

A Panoramic View of Land Reclamation: The steady, northward progression of mining operations during the past fifty years occurred as the shallow, southern lignite deposits were gradually exhausted. Because of this, today the various stages öf the mining and land restoration cycle are open to view, spread out in sequential order. At the active mines in the northern and central regions of the lignite field, the giant bucket-wheel excavators selectively strip off and save the top layer of loess. (Loess is an extremely fertile type of loam which covers most of the Rhineland region and has gradually come to be regarded as an important mineral in its own right.)

The excavators next peel off thick layers of sand, gravel, and clay overburden before extracting the loose, black layers of exposed lignite. Some commercial exploitation of the sand and gravel has begun, thus turning the extraction of brown coal into a total mining operation.

Further southward, near Quadrath-Ichendorf and Berrenrath, the huge surface mines have been exhausted of lignite, and restoration of the land is underway. Brought in by trains from the north, the discarded spoil material is filled back into the mined-out pits by mammoth spreader machines and leveled off by bulldozers. The leveled areas are subdivided into 5- to 10-acre tracts, or polders, by six-foot-high dikes of loam. These polders will eventually be filled with a loess slurry, which leaves behind a one- to two-meter-thick top layer of loess when it dries. Still further to the south, near Berrenrath, fields of grain and hay can be seen thriving on restored land which is less than 5 years old. The sequence in the forested areas is similar: To the north are newly planted stands of young trees, and in the south are forested areas reclaimed fifty years ago. The latter are nearly indistinguishable from natural forests and are superior to the stands of scrub timber which originally grew there.

Meticulous Advance Planning: The German land restoration program actually begins, with the preparation of detailed plans for the relocation of populated settlements and for the restoration of the land, years before the first shovel of brown coal is mined. Land-use patterns are proposed in advance, and the new landscape is designed accordingly—the topography, the drainage system, lakes, and the designation of areas to be restored for forestry and for agriculture. This comprehensive early planning enables the mining operations to be coordinated with concurrent land restoration work. New towns for the displaced people are designed according to modern urban requirements and are more compact than the former, unplanned settlements.

The basic resettlement costs are borne by the mining company with local and state governments providing supplementary funds to pay for the incremental costs of better schools, sewer systems, and other community services than those which existed at the former town site.

This comprehensive approach reflects an acceptance of the fact that surface mining affects not only coal, but also trees, buildings, people, and the land itself. In Germany, the State of North Rhine Westphalia and the lignite mining industry have accepted the challenge of finding acceptable solutions to the entire set of social and environmental problems created by brown coal surface mining. This approach makes it possible to treat the overall problem as a whole rather than dealing with separate aspects of the problem on a piecemeal basis.

Forests and Lakes

The restoration of brown coal mining lands began in the Rhineland some fifty years ago. Today, extensive tracts of both first and second generation forests (and some thirty-nine lakes of varying size) can be seen in the area. Historically, the forestland reclamation program divides naturally into three main periods. These are the "greening" action of the 1920s, the extensive planting of poplars and alders after the second World War, and the current phase of reforestation which began about 1958. The current program is distinguished by the planting of commercially valuable trees, with some poplars planted merely to provide a measure of protection against the weather. The planting program is also concerned with replanting areas which did not take well in earlier actions and the planting of trees and shrubs on slopes to reduce erosion.

The planting of commercial trees directly—without first having to prepare the way by establishing hardy but worthless quick growth tree types—has been made possible by applying a loess-improved topsoil to the areas which are to be reforested. Instead of simply grading the spoil banks and planting them to pioneer trees, a special layer of loess and overburden mixture is applied in depths of 3 to 5 meters. This mixture forms a loose, porous soil with good physical properties for tree growth. Because of the presence of limestone in the loess, the pH value of the soil ranges from 6.8 to 7.4. At first, the humus and nitrogen are at rather low levels, but this condition is improved by sowing lupine at the time of, or prior to, tree planting. Later, fallen leaves and organic debris resulting from forest thinning activities provide a rapid buildup of the humus content.

The Rheinische Braunkohlenwerke AG Mining Company maintains some 16,000 acres of forests, extending from Bonn in the south to Grevenbroich in the north. During 1970, some 670 acres of land were reforested—three million trees planted—and 9,500 cubic meters of wood harvested. The mining company has established a Forestry Division, with a staff of about 40 foresters and technicians, to plan and supervise these sizeable operations. During peak periods in the planting season, the mining company supplements its forestry division staff by contracting with local firms.

Much of the effort of the Forestry Division has been devoted to establishing the large forests in the southern part of the coal field. The focus is gradually shifting as mining operations continue to move northward—into areas primarily used for agricultural purposes. In the future, activities of the Forestry Division will center increasingly on forest maintenance, thinning, and harvesting.

A special problem to be dealt with is the upgrading of nearly 5,000 acres of poplars and alders planted during the second reforestation period. Extensive thinning and planting of more valuable trees are needed. More recently, there is a notable diversity in the kinds of trees planted. The tree types selected for a particular location are chosen in accordance with soil conditions and the expected exposure to sun and wind. To promote efficiency and economy in harvesting, the tree mixture in a given section is limited to one or two types, but the mixture varies strongly from section to section.

In all, the reforestation program employs twenty-two different types of deciduous trees (including basswood, oak, ash, beech, willow, alder, locust, poplar, elm, and maple), eleven types of conifers (including pine, fir, and hemlock), and eighteen different varieties of shrubs, such as hazelnut, dogwood, mountain ash, and wild roses. This mixture of trees and shrubs serves to provide not only an ecologically sound forest, but also the diversity appealing in a recreational forest. Forests in Germany have traditionally been used for recreation by the general public as well as for timber production.

In the past, most public and many private forests were open to all for biking, hiking, picnicking, and other recreational uses. In recent times, the accessibility of forests for relaxation and enjoyment has been extended by the 1969 Federal Forestry Law which requires that all private forests be open to the public. Commonly, all agricultural lands as well are open to people wishing to take walks along field paths and farm roads.

Winning New Farmland

In 1970, the Rheinische Braunkohlenwerke AG Mining Company restored some 470 acres of mining wasteland to agricultural productivity and an additional 670 acres as forest land. In view of the high costs of farmland reclamation, it is interesting to examine the considerations which have led to the adoption of this policy. Before improved reclamation techniques were introduced in 1960, the land restoration costs were far higher than the worth of the reclaimed land. Even today there are few economic incentives for rehabilitating the land. It is difficult for German agriculture to compete with neighboring European Common Market countries, and in many instances direct government subsidy is required to keep it viable.

Furthermore, the surface mining industry is not even the principal destroyer of farmland in North Rhine Westphalia. While the brown coal mining industry consumes about 600 acres of farmland annually, some 300 acres of farmland are lost each week to city growth, highway construction, and industrial expansion. These considerations clearly demonstrate that the decision to reclaim mined-out areas as farmland was not at all a "foregone conclusion."

The incentives for restoring mining areas to agricultural productivity in the Rhineland are diverse. First, the brown coal deposits are located in the richest, most fertile farm country of Germany which is favorably situated near the large population centers of the Ruhr valley. Thus, a ready, nearby market is on hand for the agricultural produce, affording a significant savings in transportation costs. Second, the problem of dispossessing and resettling farmers who are in the path of the mining operations is greatly eased by having restored land available as an acceptable substitute. Many of these farmers would be loath to part with long established family farms were not satisfactory restored areas, fully commensurable in fertility and productivity with their former holdings, available as compensation.

Thus, the farmland restoration policy aids greatly in reducing social and political tensions and contributes to public acceptance of the temporary disruptions caused by the mining industry. Probably the most compelling reason for farmland restoration, however, is the prevailing conviction that to allow valuable soil to be irrevocably destroyed by a strictly temporary land use—mining—would represent extreme folly. Saving the loess has become one of the highest priority items of land planning in the Rhineland, reflecting a basic land ethic which cannot be evaluated or explained in purely economic terms.

Structuring the New Landscape: The spoil material and soil from active mines is transported to the site selected for restoration and is used to fill in the deep pits left by earlier mining operations. There, a completely new landscape, with a topography specified by prior design, gradually takes shape. Mammoth spreader machines, quite similar to the wheel excavators used in extracting the brown coal, are used to distribute the spoil material evenly over the area being reclaimed. Each spreader weighs 2,300 tons and requires only two operators.

A single machine is capable of handling up to 200,000 cubic yards of material daily. The transportation network of conveyor belts and trains on heavy duty tracks described earlier brings a steady stream of spoil material to the spreaders, enabling continuous operation. Ordinary bulldozers are used to level off and compact the overburden in preparation for applying the top layer of loess.

Before the loess is applied, the surface of the prepared area is deliberately furrowed, or deeply roughened, to prevent the formation of a clear interface between the top layer of loess and the substrata. In the past, the same spreader machines used to distribute the spoil material were also used to apply the loess. Again, final leveling of the surface is performed with bulldozers, which produce some undesirable compacting of the top layer to depths of about one foot. Harrow disks are employed to break up the compacted loess before planting. In recent times, an alternative method of applying the top layer of loess has been developed. After the spoil material has been distributed, the surface is divided into small diked areas, or polders. The dikes consist simply of loosely piled loess and are about two yards in height. Loess and water are mixed in a one-to-one ratio and pumped through pipes into the polders, which are from 5 to 10 acres in size. The application of slurry is carried out in successive steps; the polder is flooded with slurry to a depth of about 2 feet and allowed to dry before more slurry is applied. This process is repeated until the final six-foot-thick top layer of loess has been obtained. Normally, a month or two is required for the drying process, but in unfavorable weather up to 8 months may be required.

The slurry technique of applying loess, adapted from methods developed in Holland to reclaim land from the sea, is faster and more economical than mechanical methods. Furthermore, it has been found to possess other superior attributes as well. For example, the pore volume of land prepared by the slurry method is higher than that of either virgin soil or mechanically prepared land.

The method has found increasing application in Germany, and much practical experience has been gained. To a large extent, the ease with which the slurry is prepared depends upon the condition of the loess. If the loess is strongly wetted, it does not readily form a suspension and loess-to-water ratios of 1-to-3 may be required.

On the other hand, if the loess was mined and stored during relatively dry weather, ratios as low as 1-to-0.7 suffice. The slurry discharge pipe must be relocated several times during the filling of a polder to prevent the formation of zones with different loess particle sizes. During the first slurry application the loess settles out and partly seals the bottom. In subsequent applications, less water penetrates into the subsoil and consequently the drying process can be greatly speeded by simply draining off the water after the loess has settled.

Once the land has received the top layer of loess, it is important to establish a vegetation cover as quickly as possible. Such a cover prevents hardening of the ground from heavy rains and siltation and makes it difficult for undesirable weeds and other plant growth to gain a foothold. Working land which has been formed by the slurry method is initially difficult because of its softness.

At first, the land cannot sustain the heavy loads of normal farm machinery, and special lightweight equipment must be used to work the ground and sow the first crop of alfalfa. Once established, the vegetation assists in drying out the polder, and the root system penetrates deeply into the fresh land. If reclamation occurs in late fall or early winter so that crops cannot be planted at once, special measures must be taken to prevent surface hardening due to winter precipitation. Experience has shown that the land should be deeply furrowed after the loess is applied. The ground then tends to freeze in rough clumps during the winter, promoting the drying process and allowing early spring planting.

Initial Cultivation and Interim Management: The newly reclaimed farmland is retained by the company for an interim five-year period. During this time it is subjected to intense management by agricultural experts of the brown coal mining company. The land improvement methods are based on experience and the results of scientific research which has in part been carried out in cooperation with the Agronomy Institute of the University of Bonn. The field work is conducted by trained personnel, working from centrally located company farms to assure that uniform methods of soil preparation and fertilization are used. The company farms also serve for conducting diverse experiments in soil physics and agricultural chemistry.

A primary objective of the interim management program is to build up the humus and nutrients in the soil. The nutrient level can be regained rather quickly by the use of fertilizer, or by planting leguminous crops such as alfalfa or lupine. Restoring the humus level which prevailed before mining, however, is a much slower process.

The humus content of the newly restored land averages about 0.5 percent while that of undisturbed areas in the same vicinity ranges from 1.3 to 1.8 percent. For this reason, alfalfa is prized not only for its ability to fix nitrogen, but also because of its deep root system which contributes to the buildup of humus when the plant is harvested. Research has shown that normal crop rotation increases the humus level by only 0.04 to 0.05 percent annually.

Therefore, other methods are employed to speed the process of soil conditioning. The stubble from winter rye, which follows the alfalfa crop, is disked and plowed under. Sewer sludge, composted garbage, and other organic wastes have been used experimentally to increase the humus level. Rape, a plant of the mustard family used for fodder, is sometimes sown after the grains have been harvested and the straw and stubble disked into the soil. The plant is fertilized to achieve rapid growth and the plant foliage is then plowed under as a form of green manure. Since rape possesses a flexible planting date and is relatively inexpensive, the method has proved attractive.

The reclaimed land requires higher than usual applications of fertilizer for at least the first 10 years of cultivation. Experiments have shown that the optimal amounts on the newly restored land are: 135 to 180 pounds of P_2O_5 and K_2O per acre, and 180 pounds of nitrogen per acre. Controlled experiments were carried out by the Agronomy Institute of the University of Bonn to determine the crop yields attainable on restored mining lands during the 5 year period of interim management.

Experimental plots of land which had been restored by the slurry technique were selected near Inden, Germany and control experiments were conducted on nearby, similar land which had not been disturbed by surface mining. The restored land was first conditioned by planting a crop of alfalfa. In succeeding years, identical crops were planted on the restored and the undisturbed land. The results shown in Table 1.

Social and Economic Improvements

Agriculture

The principle of primogeniture—or the exclusive right of the eldest son to inherit his father's land—is not embodied in the inheritance laws of North Rhine Westphalia. As a result, once-large farms have gradually become splintered and subdivided over the centuries, as the property passed

Table 1

Crop Yields on Restored

and Undisturbed Land*

| Year | Crop | Yield (pounds per acre) | |
		Old Land	New Land
1962	Winter Rye	3,822	410
1963	Rape (dry mass)	3,349	3,456
	Oats	3,411	3,367
1964	Sugar Beets		
	Roots	44,025	54,652
	Sugar, %	16.7	16.7
	Foliage	34,291	43,489
1965	Sugar Beets		
	Roots	46,168	46,168
	Sugar, %	15.4	16.5
	Foliage	65,010	60,278
1965	Winter Wheat	5,537	5,572
1965	Winter Wheat	4,688	5,019

*Number of pounds to the bushel:
rye (56), wheat (60), and oats (32)

from one generation to the next. Today, it is not uncommon for a farmer to own, or to have to lease, numerous small parcels of land which may be widely separated from one another. Such a land-holdings pattern is highly inefficient because it precludes the application of modern, mechanized methods of farming.

Consequently, a consolidation of the many small holdings to form larger, economically viable units has long been a prime objective of governmental planning. Obviously, the handling of such a sensitive issue requires much care, if the reform is to be accomplished in an equitable manner. Hence, in settled areas, the land consolidation program must proceed slowly and cautiously to avoid arousing dissatisfaction. On the other hand, in the resettlement of reclaimed brown coal lands, it was recognized at the onset that a unique opportunity existed for quickly accomplishing the desired land consolidation.

Resettlement consists of two distinctly separate transactions: the indemnification of the farmer for property confiscated and the purchase of new, reclaimed land from the mining company. In the sale of restored land, the mining company favors buyers who were dislocated by the mining operations. The purchaser is granted a $190 per acre rebate by the mining company to compensate for the extra fertilizer and seeding costs initially required in cultivating reclaimed land.

An intricate method has been worked out to estimate fairly the worth of farms confiscated by the mining company. State assessors appraise the property and judge the value of the land not only on the basis of its area, but also according to its fertility as established by past records of crop productivity. The farm buildings and such external improvements as woods, orchards, and wells are appraised by taking into account their current replacement cost and their present depreciation based on age. The value of each item is added to obtain the final settlement sum. If the farmer is dissatisfied with the negotiations, he may appeal through the courts. However, this happens in less than one out of six cases. The settlement payment, supplemented by savings and loans, enables the farmer to purchase new land and buildings. How high the replacement costs may be is revealed by a recent survey, which found that the total investment in buildings on an average farm in North Rhine Westphalia amounts to about $70,000.

The nearly 3,000 acre agricultural community developed on reclaimed land near Berrenrath, in the southern sector of the brown coal field, illustrates the new socioeconomic structure which is being attained. The new community will comprise some 27 separate farms, each with 40 to 80 acres in a single tract of land. The community is being built according to plans specified by the winning entry to a landscape design contest sponsored by the state of North Rhine Westphalia. The plan envisages 70 percent of the land being used for cultivation, 20 percent for forested tracts, and the rest for village growth and industry.

The farmers are clustered together in small hamlets of about six to eight families each. Contrary to usual German practice, the farm dwellings are located amidst the cultivated fields, reducing unnecessary traveling time. The hamlet pattern reduces the isolation of an individual farm family and makes it possible for several farmers to pool resources in purchasing expensive farm equipment since the machinery can then be used communally. The size of the farms, the spacious modern buildings, and the economies afforded by cooperative endeavor make it likely that the Berrenrath agricultural community will remain economically resilient for many decades to come.

Villages

Whole villages, caught up in the path of the brown coal mining operations, must be torn down and relocated. Thus, the mining juggernaut uproots people and institutions as well as the landscape itself. The number of people evacuated from their homes and resettled in fully new locations reached a total of 19,552 at the beginning of 1971 and is expected to grow to 30,000 by the end of this century. In all, nearly 5,000 homes, farms, and places of business have been forced to yield to the brown coal industry and move elsewhere. Moving an entire village requires extensive preparations and takes a relatively long time.

For this reason, the decision must be reached and made known well in advance of the planned start of mining. Most of the villages which have been relocated are fairly small in size, with a population of about 350 to 2,000 inhabitants. In some few cases, particular villages have been spared because of their historic interest or because they are located close to the edge of the brown coal field.

The villagers participate directly in selecting a new town site, choosing one from a dozen or so possibilities which have been presented to them. All of the sites being considered have previously been approved by the regional planning commission. Past experience has shown that the villagers prefer to move en masse, retaining some portion of the community identity. This practice tends to avoid excessive social upheaval and the creation of a condition of rootlessness. The village may simply be reconstructed at another site, or the community may decide to join another already existing town. The latter is desirable because the connecting arteries of transportation already exist, and the consolidated population pattern which results makes it possible for the new, merged community to afford improved public services. The new community invariably provides better schools, parks, sports facilities, and playgrounds than the old one. Because of unsatisfactory conditions in the old village, the resettlement in many cases amounts to nearly a complete urban renewal action.

Mining company representatives negotiate directly with the individual villagers to determine a fair settlement payment for the confiscated property. The money paid by the mining company enables the villager to replace his home at the new town site. Generally the new homes are substantially improved—larger than before, and with central heating and indoor plumbing. The added improvements are paid for privately, not by the mining company. The new village is designed according to modern principles by professional architects and planners commissioned by the town council. By-pass roads separate the residential section from areas of heavy traffic, and more space is allotted for playgrounds, ball parks and kindergartens.

The businesses are centrally located, the village is provided with a sewer system, more public green areas, and is generally more compact and efficient than the former one. Given this type of choice, the villagers apparently prefer to trade the picturesqueness of their old towns for a new environment with more modern conveniences.

A sociological survey of some resettled communities reveals that the end result has been generally satisfactory. According to the survey 85 percent of the resettled people are happier in their new homes, and 57 percent likewise prefer the new location of the town. Some 83 percent of those interviewed believe that the advantages outweigh the disadvantages in resettlement. This does not imply, however, that the resettlement proceeds without a certain quota of worry and concern. The typical evacuee experiences feelings of anxiety about his future job, income, the new school and community accommodations, his social position in the new community, and the unavoidable problems involved in constructing a new house. On the plus side, however, he gains the following:

1. He receives a desirable building lot located on an improved residential street away from areas of heavy traffic.
2. His modern, new home is equipped with central heating and indoor plumbing and is considerably more spacious, nearly one-third larger than his former dwelling.
3. His new house is worth more, will last longer, and has lower maintenance costs than his old one.
4. In building a new home, he receives substantial federal tax advantages which are provided in Germany to encourage the construction of residential housing.

Recreation

The recultivation of mined lands furnishes a unique opportunity to sculpture the new landscape to meet the recreational needs of the people as well as to serve the interests of commerce, industry, and agriculture. The growing population and changing life styles in Germany have combined to create an unprecedented demand for outdoor recreational facilities at the same time that ever greater amounts of land are being consumed for highways, cities, industry, and military uses. For this reason, land reclamation in the Rhineland brown coal fields is planned in such a way as to provide for multiple land usage. The forests are planted not only to meet the commercial need for lumber production, but also to provide restful settings for weekend relaxation.

The agricultural landscape as well is prized for its potential recreational value. Trees and hedges are planted by the mining company on reclaimed farmland to divide the fields and protect the land against moisture loss from action of the sun and wind. These windbreaks and groves of trees amidst the cultivated fields provide a scenic variety which attracts countless city dwellers on Sunday afternoons. The well maintained farm roads, with side paths and benches, enable the elderly as well as the young to undertake leisurely strolls and enjoy a closer contact with nature than would otherwise be possible in their urban life.

The oldest and most impressive restored region lies in the southern section of the coalfield near Brühl, where a forest-and-lake landscape of distinctive charm has been created. The numerous ponds and lakes, formed from deep pits left over from the lignite mining, have been artfully fitted into the landscape and are scarcely distinguishable from natural bodies of water. The lakes have sloping shores and shallow edges which are planted to reeds and grasses to provide wildlife habitats. The restored region comprises some 3,000 acres and serves both as a recreation park and a wildlife preserve.

Because of the many opportunities for water sports and hiking, the park attracts some 20,000 visitors each weekend from the greater Cologne area. It is maintained jointly by the Forestry Division of the mining company and the State Forestry Service in cooperation with civic organizations of neighboring communities. The achievements here, and elsewhere in the Rhineland brown coal area, demonstrate that, with proper planning and effort, the needs of industry for raw materials and power can be met without producing excessive environmental damage.

Government Regulation and Supervision

Historical Development

The present form of the brown coal industry and the adoption of enlightened land restoration practices in Germany evolved gradually over the past several decades. Although reforestation of land areas disturbed by lignite surface-mining has been carried out in Germany since the early 1920s, the practice of restoring land for agricultural use was not instituted until much later.

The location of the Rhineland lignite deposits, in the midst of rich, fertile farmland, provided a strong incentive for this type of land reclamation. Public concern over the large tracts of unreclaimed land left over from World War II began to appear after the war, particularly in the vicinity of Cologne, where the proximity of the mining operations made the disturbed lands highly visible to great numbers of people. As a result, new surface-mining control legislation was enacted in the year 1950 to assure orderly, well-planned mining practices.

On March 11, 1950, the state legislature of North Rhine Westphalia passed Germany's first Regional Planning Law. This law, modified in May 1962, established a Land Planning Commission charged with the responsibility of developing overall guidelines for land use within the state. The main purpose of the commission is to help coordinate the diverse social, economic, and industrial activities of the region. The commission designates specific land areas for use by agriculture, forestry, and industry and sets the boundaries of populated settlements. It develops long range plans for transportation arteries and networks, the preservation of historic sites, and the construction of recreational facilities to serve the entire region.

On April 4, 1950, the state legislature enacted two additional laws applying specifically to the brown coal producing areas of the region. These are The Law for Overall Planning in the Rhineland Brown Coal Area (Gesetz Uber die Gesamtplanung im Rheinischen Braunkohlengebiet— GS NW. S. 450) and another law establishing a Community Fund to finance land restoration. The first of these laws created the Brown Coal Committee which develops detailed plans for exploiting the lignite resources of the state within the framework and spirit of the overall regional planning law.

The Brown Coal Committee

The basic responsibility of the Brown Coal Committee is to safeguard land areas temporarily used for brown coal mining from long-term damage and from being rendered unsuitable for more lasting uses. This responsibility encompasses more then merely preventing the creation of desolate areas by requiring that the land be restored for forestry or agriculture. Rather, in light of the general objectives of the overall regional planning, the Brown Coal Committee assures that the land is restored in such a way as to harmonize with the social, cultural, and industrial interests of the rest of the region.

The Brown Coal Committee is composed of 27 members especially selected to represent the societal interests most affected by the impact of the mining operations. Members of the committee are the district governors in Cologne, Aachen, and Düsseldorf; the head of the state mining agency; the Rhineland land planning commissioner; the minister of agriculture; a representative of the Rhineland Agriculture Association; the director of the State Land Settlement Office; three representatives from the brown coal mining industry; three representatives of mining unions; five representatives from county governments, three representatives of the farmers; one representative of Crafts and Trades; one representative from the power industry; one representative of the stoneware industry; one representative from the industrial union for chemistry-paper-ceramics; and one representative from the Erft Basin Conservation Club.

This broad base of representation on the committee affords an opportunity to resolve conflicts long before actual mining activities begin. The committee formulates land restoration requirements based on the future use of the land. These requirements are determined as early as possible to enable the mining company to design its mining operations accordingly.

The primary function of the committee is to act as a review body to consider proposals for extending mining operations to new land areas and to make appropriate recommendations to the minister-president of North Rhine Westphalia. As can be expected in view of the composition of the Brown Coal Committee, the final recommendation to the state government is based on considerations of overall land use, conflicting local issues, and national coal requirements. The Brown Coal Committee has gradually emerged as a powerful force, defining the conditions under which the brown coal industry must operate. Its existence subjects the brown coal industry to public scrutiny and has been instrumental in bringing about the conservation practices of the industry. The Brown Coal Committee serves as a quasi-public forum where the divergent interests of society can be considered before mining commences. Public hearings and the signature of the state chief executive are required before the recommendation of the committee become legally binding.

The adoption of requirements that a certain portion of the land disturbed by the surface mining of brown coal be restored to agricultural productivity illustrates the importance of such a planning and review body. Shortly after World War II, a coalition of agricultural groups within the Brown Coal Committee became concerned over the destruction of fertile farmland by the mining operations.

In the late 1950s, this coalition of agricultural interests, known as the "green front," successfully campaigned within the committee to require that the valuable top layer of loess, often 15 to 20 feet thick, be saved, and that a portion of the land disturbed by surface mining be restored to agricultural productivity.

Implementing the Mining and Reclamation Plan

After the mining and land rehabilitation plan is adopted, the State Mining Office is responsible for supervising its implementation and assuring that the mining and land restoration activities are carried out in accordance with its stipulated provisions. The mining company is required by law to cooperate by providing all information which the state enforcement agency needs to carry out its regulatory function.

For example, the brown coal mining company routinely submits aerial survey photographs every six months to document its mining and land restoration progress. The planning and enforcement process, with participation of nonmining interests, affords flexibility in resolving the social and environmental problems posed by surface mining. The recommendations of the Brown Coal Committee serve as a living law which changes in accordance with the requirements of specific situations. Since the deliberations take place well in advance of actual mining, amply sufficient lead time is available for full consideration of all of the issues and problems.

By virtue of its representation on the Brown Coal Committee, the state enforcement agency is fully cognizant of the spirit and intention behind provisions of the operations plan and is able to draft supplementary regulations accordingly.

State Mining Office

The State Mining Office (Bergamt) of North Rhine Westphalia is the agency which oversees mining operations and enforces the provisions of the land restoration plan. Since most of the German lignite deposits are located within this single state, nearly all government control of brown coal surface mining is on the state, rather than the federal, level. The task of setting adequate reclamation standards is facilitated by the lack of significant economic competition from neighboring states. The legal authority to regulate the extraction of minerals in West Germany derives from a general mining law based on an older Prussian model written June 24, 1865. The law reserves nearly all mineral rights to the state which may grant mining concessions to private companies. The concession confers on these companies the right to commercially exploit the mineral resources by state-approved methods. Traditionally, the concession has generally been granted to the discoverer of new mineral deposits in order to encourage prospecting.

The State Mining Office of North Rhine Westphalia has thirteen regional offices to oversee the mining activities in the state, including the mining and reclamation operations of the brown coal industry, with its annual production of nearly 100 million tons. Each regional office is staffed with about ten mine inspectors, usually trained mining engineers, who spend most of their time in the field observing the progress of operations and checking for compliance with regulations.

The cost of administering and enforcing a strip-mine reclamation bill can constitute a substantial part of the overall restoration costs. In spite of the expense, however, a well-supported enforcement agency is vital to the success of any reclamation program. The agency not only enforces the explicit provisions of the law, but also, by writing supplementary regulations as part of its interpretive role, largely sets the tone of the program.

In Germany, the brown coal mining industry was at first reluctant to undertake the highly expensive reclamation of farmland. The regulatory agency, backed up by a strong law, played a highly important role in bringing about an acceptance of the practice, today an even slightly profitable venture.

Relevance to U.S. Surface Mining

Some elements of the German surface mining and land reclamation techniques are applicable to U.S. strip-mining in spite of important differences in the climate, terrain, and geological features of the coal-bearing regions of the two countries. Bucket wheel excavators of the type used in the Rhineland have already been used in North Dakota and Illinois to remove soft, unconsolidated overburden. These machines can operate continuously and deliver the loose overburden by conveyor belt to a separate, interfacing transportation system. Because of its digging selectivity, the bucket wheel excavator is especially suitable for separating the fertile topsoil from the remaining overburden material and saving it for later use in land reclamation.

It is not well-suited for hard rock digging or for the handling of drilled or blasted materials. Similarly, the extremely heavy weight and limited mobility of the bucket wheel excavator make it wholly unsuitable for Appalachian contour mining. Nevertheless, the coalfields of the United States, especially those in the interior, northern great plains, and western provinces, contain vast expanses of gently rolling or flat land where the German technology could be applied if there were sufficient incentives for full land restoration.

Whether or not legislation exists, requiring that quality land restoration be integrated into the mining cycle can change the economics of mining and thereby influence the selection of a specific excavation technology. If the digging machines do not have to be coupled to a separate transportation system, to haul away and save the topsoil and to transport massive amounts of spoil material to refill areas, the operational costs of excavating would tend to dictate choosing the giant shovels and draglines used in southern Ohio and Illinois. If full restoration of the land after mining is planned, it may well be that some adaptation of the excavation and transportation system used in the Rhineland is more economical and efficient. In other words, different conclusions are reached depending upon what portion of the overall mining cycle is included in the optimization process.

In the main, the methods of land reclamation which are adaptable from those employed in the Rhineland lignite fields apply to the rolling plateau country and the flat lands described earlier. In such topography, the slurry method of spreading loam on graded, filled-in areas is technically feasible.

Of course, as in the case of choosing an excavation technology, economic considerations may favor some alternative method of accomplishing the same purpose. The large amount of directed research carried out at the University of Bonn to determine suitable plants and trees for revegetation of the mined lands, and the factors affecting their growth rates, should be valuable and useful in the United States.

Much of the environmental degradation from surface mining can be prevented by making a conscientious effort in land reclamation. Nevertheless, it is still too early to exclude the possibility of long-range, adverse effects from surface mining. For example, the altered ground strata and mineral content of mined lands could unfavorably affect groundwater movements or percolation characteristics. These could conceivably lead to undesirable long-range results such as increased soil salinity in the mined land areas or elsewhere.

Although there is no reason to suspect that such events are actually occurring in the Rhineland, the possibility of subtle, but ultimately harmful changes cannot be dismissed. For this reason, it is important to gain a better understanding of possible geological effects and to continue to develop improved mining techniques. The restoration achievements in the Rhineland clearly demonstrate that a meticulously planned, well-funded program can produce impressive results in land reclamation.

In summary, there is much that can be learned from the German experience in restoring surface-mined lands. Their program has been in effect for some twenty years and has helped to minimize social dislocations and environmental damage from brown coal surface mining.

The land restoration program in North Rhine Westphalia embodies four main principles which have made it viable and effective. First, the regulation of surface mining is incorporated within an overall regional development plan. This makes it possible to protect the larger interests of the whole region. Second, a planning body composed of representatives from diverse interest groups participates in formulating detailed requirements for mining and land restoration long before the actual mining begins.

Thus, a broad spectrum of society is consulted and untimely haste is avoided. Third, the recommendations of the planning body are submitted for public review before being adopted and implemented. This provides a political pressure relief valve as well as a mechanism for detecting possible adverse side-effects which had escaped consideration. Fourth, an enforcement agency is empowered to enforce the plan which is finally approved and adopted. The German program offers visible evidence that, with detailed advance planning, striking achievements in reducing environmental damage from surface mining are possible at a price that can be borne by the consumer.

22
A Regional Analysis of Air Quality Standards, Coal Conversion, and the Steam-Electric Coal Market

Alan Schlottmann

Introduction

In the United States coal-fired electric power plants generate the major portion of sulfur emissions, accounting for most of the approximately 57 percent of emissions attributable to electric utilities.[1] Certain standards for sulfur emissions to be imposed either at the stack or directly on the fuel being burned have been proposed for air quality controls. Most states now have plans to reduce sulfur emissions in the near term in conventional steam electric generation by regulating the type of fuel used. In the long run, nuclear power is an important factor. Also in the long run sulfur standards could provide incentives to develop techniques, such as stack gas scrubbing, for controlling sulfur emissions. However, the major effect of alternative sulfur emissions policies through the mid-1980s would be to promote the substitution of lower sulfur fuels.

In recent years, serious concern has been expressed over the United States' reliance on imported oil as a major energy source. The demand for energy in the United States has been increasing at an annual average rate of 3.5 percent per year. Much of this increase is being met by petroleum, mostly imported petroleum products.[2] This trend has been most noticeable in the past few years, however. Imports of foreign oil products, both crude and refined, have increased by 82 percent from 1970 to 1973

From the *Journal of Regional Science*, vol. 16, no. 3, 1976, pp. 375-387. Reprinted by permission.

This work was supported in part by the National Science Foundation/RANN at the Appalachian Resources Project, University of Tennessee. The author is indebted to Lawrence Abrams, University of California, Santa Cruz, for assistance in early development of the model. The comments of John Moore and Sidney Carroll on an earlier version of this paper are gratefully acknowledged. Any errors, however, are the author's alone.

Alan Schlottmann is Assistant Professor of Economics at the University of Tennessee.

while output of domestic petroleum products has decreased by approximately 2 percent. The implications of this increasing reliance on foreign sources were clarified by the 1973 Arab oil embargo and subsequent increases in the price of imported oil. As a result proposals have been made to prohibit conventional power plants from using oil or natural gas as their primary generating fuel, with the explicit purpose of increasing the utilization of domestic coal resources. Given the concern over sulfur emissions from electric power plants, there has been serious debate on the impacts on air quality of reconversion to coal.

Since the coal used in electric generation, often termed steam-electric coal, provides not only the major market for coal but the only one which has not declined in recent years, national coal and sulfur emissions policies and their related effects would have a major impact on the coal industry. In this paper we first investigate the possible impacts that such standards for existing power plants would have on coals with different sulfur contents and consequently, the regional effects on the entire coal industry. Estimates of the impacts of these standards on the coal industry vary greatly, ranging from declines in coal use of 50 to 200 million tons per year.[3] The feasibility of placing sulfur limits directly on coal use is then considered. With these results, a final purpose of our analysis will be to investigate the implications for public policy emphasizing power plant conversions to coal. The regional impacts of sulfur-reducing technology, which would permit current high sulfur coal production, are analyzed for their interaction with coal conversion and air quality.

Table 1 shows the level of shipments and the average sulfur content of steam electric coal by supply district. For our purposes, this classification has been surprisingly useful. Districts One and Two, for example, are both in Pennsylvania. The coal fields of District Two supply the industrial centers of southern Pennsylvania, while District One is the primary source of shipments to electric utilities in the northeast. Similarly, Districts Eight and Nine split Kentucky into eastern and western coal fields. The western fields of Kentucky are a major source of surface mined, high sulfur coal, while the eastern fields are an important area for lower sulfur underground coal. The districts which exceed the average sulfur content of 2.9 percent are the major producers in total tonnage of steam electric coal. Further, these districts produce almost entirely for electric utilities, because their coals have no coking or related properties suitable for steel production and other uses.

The Regional Model

Coal resources are nonrenewable stocks with fixed locations. The study of a resource which has a spatial dimension fixed by the nature of its

TABLE 1: Shipments and Average Sulfur Content of
Steam Electric Coal by Supply Districts[a]

District of Origin	Average Sulfur Content (percentage by weight)		Quantity Shipped (thousands of tons)	
	1972	1973	1972	1973
1 Eastern Pennsylvania	2.2	2.2	34,346	34,362
2 Western Pennsylvania	2.1	2.1	8,923	9,403
3 and 6 Northern West Virginia and Panhandle	2.7	2.7	36,836	34,818
4 Ohio	3.5	3.5	41,731	38,926
7 Southern Numbered 1	—	—	746	789
8 Southern Numbered 2	3.9	3.9	71,313	65,747
9 West Kentucky	4.6	4.6	49,374	52,894
10 Illinois	3.4	3.4	53,137	49,705
11 Indiana	3.4	3.4	20,286	20,454
12 Iowa	3.4	3.4	703	618
13 Southeastern	1.7	1.7	13,049	11,628
14 Arkansas-Oklahoma	—	—	—	—
15 Southwestern	4.9	4.9	7,430	12,665
16 Northern Colorado	.5	.5	561	492
17 Southern Colorado	.6	.6	2,566	2,974
18 New Mexico	.6	.6	10,184	11,008
19 Wyoming	.6	.6	10,637	14,113
20 Utah	.7	.7	1,611	1,903
21 North-South Dakota	.9	.9	6,032	6,098
22 and 26 Montana and Washington	1.5	1.5	10,797	13,567

[a] 1972 data from Bureau of Mines [19, pp. 65, 71]. The 1973 data are from National Coal Association [14, pp. 83, 85].

available deposits must consider the regional implications of the analysis. Because of the impact of coal extraction on land use, these regional considerations are relatively more important for coal resources than for other fossil fuels.

We have attempted to model these fundamental structural characteristics of regional production capabilities and electric utility demand for coal in power plant generation in a spatial programming model of steam-electric coal production, distribution and use.[4] The model's basic activity is the delivery and utilization in a demand region of coal which has been extracted and shipped from one of the coal supply districts. The programming model consists of 52 electric power producing demand regions and the 23 coal supply districts. Coal shipments from any one supply district are differentiated by type of mining (surface and underground) and by source (new or existing mines). This is an important distinction because surface mining has grown dramatically in recent years.[5] In 1950, 24 percent of production was from surface mines. By 1972, surface mines accounted for 49 percent of total production. The relative importance of these two mining methods varies across coal supply districts. Surface mining is predominant in Western states. With the exception of the Eastern Kentucky fields, which are a center for underground mining, the two methods are used almost equally in the majority of Appalachian districts. Surface mining is more important in the Midwest than underground operations. Missouri, for example, has only surface production. Approximately three-fourths of total surface mine production is delivered to steam-electric plants, which is 61 percent of coal delivered to utilities.

The utilization of any coal in steam-electric generation is dependent on its energy value, measured in B.T.U.'s. These energy values can differ by extraction method and by sulfur level. The average heat content by supply district of underground coal, for example, is about 8 percent higher than that for surface mined coal. This difference is greater in the lower sulfur coals, generally from 12 to 14 percent more.[6] On a price per million B.T.U. basis, this helps to offset the initial disadvantage of underground mining where the extraction cost per ton is greater than surface production.

The sulfur content of coal can differ from one region to another. As we have seen, sulfur is the chief chemical property of coal which is subject to air pollution regulations. By most classifications of coal reserves by sulfur content, the West has the most low sulfur coal. The Midwest has relatively small low sulfur deposits. The Appalachian area, contrary to popular belief, does contain a significant amount of low sulfur reserves. The programming model differentiates coal shipments by five levels of sulfur, ranging from those emitting 0-.6 pounds of sulfur per million B.T.U. to those emitting 3 pounds or more of sulfur per million B.T.U. Thus, differences by method of extraction, sulfur content and B.T.U. value for each supply district are considered separately.

Solutions are characterized as determining the most efficient (minimum cost) network of production, distribution and use for coal in steam-electric generating plants. The basic programming model used to obtain this solution is linear.[7] Regional air quality standards which limit average sulfur emissions from coal-fired power plants can have important interactions with the spatial location of production activities if a premium is placed on low sulfur emissions. As public policy alters the delivered price of steam-electric coal to a demand region, two types of substitution can occur. First, since there are nonuniform impacts on coal supply regions, particularly from the simulation of various sulfur emissions policies, substitution between shipments from alternative domestic coal producing regions can be made, resulting in significant interregional production shifts. Second, substitution of alternative fuel sources, particularly lower sulfur oil, can be analyzed in the model by interfacing demand equations for coal use and their elasticities with the linear programming model. This results in a heuristic programming procedure to approximate the spatial equilibrium values for coal production and use as an alternative solution technique to the classical nonlinear models.[8] As we have discussed, our main purpose here is to consider the feasibility and implications of possible interregional domestic coal substitution rather than the increased use of alternative fossil fuels to coal.

Air Quality Standards: Existing Coal-fired Plants

The regional impacts on coal shipments of alternative sulfur emissions standards at existing coal-fired power plants for 1977 are shown in Table 2. We see that the sulfur emissions standards can significantly affect the utilization of alternative regional production. The effect on Western production is particularly strong. With a 1.65 sulfur standard, Western shipments increase by 53 million tons, replacing almost 25 million tons of Midwest surface production. Since Western coal has a lower heating value per ton than the higher sulfur coal it replaces, total coal production increases as a larger tonnage of Western coal is needed to replace Midwestern shipments. An increase in Western production as a substitute, particularly for Midwestern coal, is a definite result of public policy which attempts to restrict total sulfur emissions.[9]

TABLE 2: Regional Shipments with Sulfur Emissions Standards at Existing Coal-Fired Power Plants
(Millions of Tons)

Region[a]	Sulfur Emissions Standards (pounds per million B.T.U.)		
	2.00	1.85	1.65[b]
Northern Appalachia	152.7	146.3	131.8
Central Appalachia	121.8	127.5	127.5
Southern Appalachia	18.9	18.9	18.9
Total: Appalachia	293.4	292.7	278.2
Midwest	134.3	116.4	106.7
West	53.1	74.6	106.0
UNITED STATES	480.8	483.7	490.9
Total Sulfur Emissions (Millions of Tons)	10.00	9.33	8.74
Total Sulfur Emissions (Millions of Tons) Without Standards	10.72		

[a] The regions are identified as: Northern Appalachia = Districts 1, 2, 3, 4, 5, 6; Central Appalachia = Districts 7, 8; Southern Appalachia = District 13; Midwest = Districts 9, 10, 11, 12, 14, 15; West = Districts 16, 17, 18, 19, 20, 22, 23.

[b] This standard represents the lowest uniform source standard for electric power plants which is feasible, assuming there is no coal-oil substitution or stack gas scrubbing technology.

The generally accepted hypothesis that Eastern coal producers would be hurt most severely is not supported by our results.[10] Central Appalachia, for example, is the largest single source of coal which can meet proposed air quality standards. Appalachian producers face difficulties to be sure, but it is coal production in the Midwest which is most severely affected.

As shown in Table 2, although total sulfur emissions do decrease significantly with the sulfur emissions standards, these reductions are moderate reductions rather than extreme declines in overall emissions. At a 1.65 standard, for example, emissions are reduced by 19 percent, dropping to 8.74 million tons from 10.72 million tons with no standard. The increased reliance on coal shipments from more distant coal supply regions in Midwestern and Eastern utility markets resulting from the sulfur emissions standards is reflected in the increases in the delivered prices of steam-electric coal. In Eastern markets and in the Midwest, the average cost increase is 24-28 cents per million B.T.U. at the 1.85 standard and 31-35 cents at the 1.65 standard. The current average cost of coal to Midwestern and Eastern utilities is 80-90 cents per million B.T.U. Western utilities generally have no cost increases.

The increases in Western surface production indicated in the results bring into focus the debate over surface mine reclamation in the West. The environmental issues for surface mining in the West differ from other producing regions. Adequate reclamation to insure soil stability, particularly on steep slopes in the mountainous coal fields, has been the primary cause of concern in the producing areas of Appalachia. In the Midwest the environmental issues in strip mining deal mainly with the elimination of unreclaimed land from alternative uses in the future and the disruption of wildlife habitats. The coal industry often points out the potential of reclaimed Midwestern surface mined lands for grazing, farming, orchards, and possibly new wildlife and lake areas.[11]

In the relatively dry areas of the West, however, there is some question of whether proper revegetation is possible as a method of reclamation. Soil conditions in the West are extremely fragile, and direct rainfall and other sources of water are limited. A study undertaken by the National Academy of Sciences [1] concludes that reclamation does not appear feasible where the rainfall is less than 10 inches annually and where soils have difficulty in retaining the moisture. A similar conclusion is reached in the more recent analysis of the Northern Great Plains coal study [15]. At current levels of production, these minimum conditions are met in the West's two most heavily mined areas, the Northern Great Plains and the Rockies, which also contain 60 percent of the region's surface mine reserves of coal. However, these studies generally conclude that, although water requirements for mining and reclamation in these areas can be met, "there is not enough water available there for large scale operations like gasifying and liquefying coal or generating electric power [1]." The implication of the reports for reclamation in other Western areas is not clear, however, since revegetation issues are emphasized where substantial original ground cover exists.

The remaining area of significant surface production in the West is found in the generally contiguous fields of northeastern Arizona and northwestern New Mexico. This area supplies mostly Western utilities, particularly those in Arizona. In parts of these states, annual rainfall can be less than five inches and the surface soil is generally alkaline. As a result, little or no vegetation exists under normal circumstances. Since most of the deposits currently under extraction are on Navajo or Hopi Indian lands, the effort at present is to determine which vegetation will produce the best grazing crop for sheep and cattle.[12]

A Ban on the Highest Sulfur Coals

A major effect of the sulfur emissions standards was to reduce the use of the highest sulfur coals. We define the highest sulfur coals as those emitting three pounds of sulfur or more per million B.T.U. At the 1.65 standard for example, the use of these coals was reduced by 74 percent, from 97 million tons at a 2.00 standard to 26 million tons. It has been suggested that one way to reduce the use of such coal as a generating fuel is simply to ban it. These proposals generally suggest that alternative coals, rather than oil, be substituted. This has occurred in some urban situations. For example, in recent years Chicago has used some Montana coal instead of relying solely on Illinois production, as does most of the state.

But what would be the effect of such a policy as an overall regulation? The shortage of lower sulfur coals and the variances granted for utilities to burn high sulfur coal even in urban areas would lead us to believe that such a policy over the near term would not be feasible. In our model, a ban on the use of the highest sulfur coals without oil substitution simply cannot be upheld, since there is not enough other mining to compensate for the production lost. Over 17 million tons of high sulfur coal produced mostly in the Midwest cannot be replaced by substitution of other coals under present capacities. The basic solution in our model shows that the highest sulfur coals comprise 20 percent of all coal used. One might ask if this figure is perhaps too high; that actual use is so much less than in the basic model (97.0 million tons) that we overestimate the importance of high sulfur output. If we aggregate the data available from the F.P.C., which lists all coal deliveries to utilities by sulfur content, we find that, over the last half of 1972 and through 1973, high sulfur coal comprised approximately 24 percent of total steam-electric coal shipments.[13] An overall ban of such production for coal-fired utility boilers over the short run would not be feasible.

Table 3 shows the simulated increases in local production required to replace high sulfur coal in the basic solution without changing the relative regional production levels. The greatest expansion would be required in the Midwest, where failure by Midwestern producers to supply this additional tonnage would require the use of Western coal. This would add approximately 25 to 30 cents per million B.T.U. to the average delivered costs of coal in the Midwest utility market. Net sulfur emission levels would decrease by 20 percent, almost 2.135 million tons, if the use of all high sulfur coal by utilities was eliminated and other coal substituted. This is a reduction comparable to that under the 1.65 sulfur emissions standard, if oil substitution is not allowed. However, new sources for 97 million tons of coal would have to be found, a difficult task considering the current time lag for expanding existing underground and surface mining or for establishing new mines.[14]

Coal Conversion

The main public policy tool to increase coal use in conventional power generation is the Energy Supply and Environmental Coordination Act of 1974, which allows the Federal Energy Administration to prohibit power plants with a capability to use coal from relying on natural gas or oil as their main generating fuel.[15] The regional distribution of the maximum feasible power plant capacity conversions of existing and planned units is shown in Table 4. The majority of the additional demands for coal would be in the Western states as a substitute for natural gas and in the East to replace oil. Significant reductions in oil use particularly would result.

TABLE 3:   Relative Regional Increases in Coal Production
to Compensate for the Highest Sulfur Coal Losses

Region	Percentage Increase in Other Capacity	Required Additional Tonnage (millions of tons)
Appalachia	9.68	27.1
Midwest	83.21	69.9
West*	—	—
United States	23.82	97.0

* No increase required.

As shown, a uniform feasible source standard for sulfur emissions with the coal conversions, given current regional production capabilities for coal and planned mine additions, would be a standard in the range of 1.75 for 1977 and 1.50 for 1980. Most of the planned additions to regional coal capacity by 1980 are in lower sulfur seams in Appalachia and the West, which would allow the additional power plant demands for coal to be met at a reduced overall sulfur emissions standard.

The Environmental Protection Agency, which is required to certify the FEA actions and essentially is given a veto power, has been less than enthusiastic about coal conversion. This has resulted because of difficulties in existing coal-fired power plants as well as additional conversions to meet the suggested primary air quality standards of 1.2 pounds of sulfur oxide per million B.T.U. in the amended Clean Air Act of 1971. Our results in Table 4 tend to support their position that, although coal conversions would definitely allow both a greater availability of natural gas for interstate shipments to industrial markets and a reduction in oil use and imports, overall sulfur emission levels at the converted power plants would indeed exceed the 1.2 standard.

Stack gas scrubbing, which removes sulfur from the fuel gases after

TABLE 4: Regional Power Plant Conversion Potential

Region[a]	Conversion Capacity (Megawatts)	Decreased		Additional Coal (millions of tons)[b]
		Oil Use (million barrels)	Natural Gas (billion CF)	
Northeast	10,862	84.67	30.76	24.9
East Central	2,366	5.95	69.76	4.9
Southeast	3,948	28.50	25.62	9.1
West Central	730	5.04	.93	1.4
South Central	3,067	27.65	2.11	7.8
West	3,418	29.22	10.18	8.6
Total, United States	24,392	181.04	139.38	56.6

	1977	1980
Capacity Conversion Potential[c]	22,080	24,392
Lowest Feasible Sulfur Emissions Standard (pounds per million B.T.U.)	1.75	1.50

[a] Regions correspond to the FPC National Power Regions.

[b] Additional demand based on a uniform coal ton equivalent of 22.2 million B.T.U. per ton as described in Federal Energy Administration [5], actual coal shipments by region from the model in Table 5 differ as a result.

[c] As estimated by F.E.A. based on the FPC Form 67, Schedule 9-A data.

combustion, could affect the use of higher sulfur coals and allow low sulfur emissions standards to be met.[16] Efforts to reduce the sulfur level of coal before shipment and combustion have not had much success. The sulfur is found in coal in two forms, organic and pyrite. Pyrite sulfur, which varies among coal fields from four percent to 60 percent of total sulfur, can be reduced by mechanical cleaning and crushing. At present, around 50 percent of all coal produced is cleaned, but, since organic compounds are unaffected by cleaning, these operations usually remove less than half of the sulfur.[17] The only possibility of eliminating sulfur from coal appears to be through the development of a method of producing synthetic fuels from coal, particularly synthetic gas.

In the metallurgical coal market, cleaning is a routine step in preparing high grade coal for coking and related uses. Other cleaning operations are undertaken explicitly to reduce the sulfur levels in coal for the electric utility market. In Illinois, Indiana, and Western Kentucky, for example, 94 percent of all coal produced in 1972 was for electric utilities. The average sulfur content, shown in Table 1, was 3.4 percent, a high level. Yet production records indicate that almost all of this coal had been first mechanically cleaned at the mines.[18] Thus, the removal of pyrite sulfur alone, while it improves the quality of coal somewhat, cannot reduce the sulfur content to a sufficient degree.

Table 5 shows the increases in regional shipments required to satisfy the demands resulting from coal conversion as well as to meet low primary air quality emission standards at alternative levels of stack gas scrubbing costs. These results clearly indicate that Western coal use would be competitive in Midwestern and Appalachian markets given the current levels of stack gas scrubbing costs. It is important to consider what level of surface mining public opinion will allow in the West before local and state action hinders its development. Most surface mining reserves occur in remote areas, where the alternative value of grazing is relatively low. Yet the controversy over surface mine reclamation shows that public opinion can be an important variable in the West. It could significantly affect the Midwestern utilities' use of Western coal for meeting sulfur emissions standards. Sulfur emission policies have boosted the competitive position of Western coal over local coal production in Midwestern markets. Whether Western producers could expand production into new reserve areas might well depend on local response.

It is true that "the large scale installation of stack gas scrubbers will minimize the need for heavily increased low sulfur coal producing capacity" (Federal Energy Administration [7, p. 94]). Our analysis shows, however, that this could be expected only if the current costs of scrubbing technology are reduced by at least half or if large scale development of Western surface mining is prohibited.

TABLE 5: Regional Shipment Increases in Coal Production to
Meet Coal Conversion Demands (1980)

Region	Stack Gas Scrubbing Costs (cents per million B.T.U.)	
	20–25 cents[a]	70–80 cents[b]
Appalachia	34.17	24.84
Midwest	13.30	5.20
West	12.03	36.50
United States	59.50	66.54

[a] The figures here represent the extreme lower end of the range for possible scrubbing costs.

[b] One of the major supports for scrubbing technology is contained in a government report of an interagency committee considering scrubbing systems. (See U. S. Sulfur Oxide Control Technology Assessment Panel [21].) A major observation was the "promising" work of Commonwealth Edison. Commonwealth Edison's research experience with scrubbers has not made the company optimistic over their potential. The dual scrubbers installed at their Chicago plant have had a simultaneous availability of only 8.1 percent. Component failures of such systems have been lessened through redesign as in the newer systems, but the problems of plugging and scaling can still exist. In terms of our main interest concerning costs, however, research by Commonwealth Edison indicates a cost range for new plants of 50 to 63 cents per million B.T.U., rising to 75 to 85 cents in the later years of a plant's life. Existing plants attempting to retrofit scrubbers due to conversion will face these higher costs also. It is of interest that current delivered coal prices are approximately equal to the higher level of scrubbing costs.

A similar conclusion occurs when considering coal gasification.[19] Coal gasification could make these higher sulfur coals almost sulfur free. Unfortunately, it is difficult to use the high coking coals of the Eastern fields in the most advanced technique presently available for gasification—the Lurgi process. Assuming a 15 percent cost of capital and coal at 40 to 50 cents per million B.T.U., the cost estimates for gasification would range from $1.63 to $1.70 per million B.T.U. Any changes in the price of coal utilized for gasification are significant because the efficiency of conversion of only 67 percent magnifies any rise in coal prices. In terms of domestic energy independence, this conversion factor means that any long-term reliance on gasification would deplete coal reserves at a faster rate than if they were mined and used directly.

Conclusions

The use of Western coal in particular has been shown to be significantly affected by sulfur emissions policies. The issue of adequate surface mine reclamation in the West would receive new emphasis if public policies were seriously undertaken to lower current levels of sulfur emissions at coal-fired power plants. These results would only be intensified with coal conversion policies at power plants currently using oil and natural gas as their main fossil fuel. Midwestern high sulfur coal cannot compete with Western coal with the probable operating costs for sulfur reducing technologies.

Perhaps the main point to be learned through this analysis is that public policy towards alternative energy-environmental issues can have important interactions. Air quality policies which decrease sulfur emissions can stimulate surface mining and regional land use controversies. Simply banning the use of highest sulfur coal may not be feasible given current regional production capabilities; coal conversion at power plants does reduce oil use but overall cannot meet low sulfur emission standards; pressure by government agencies for stack gas scrubbing or similar technology to be installed at power plants without consideration for the prices of substitute fuels may aggravate the surface mining controversy even further. In short, the interrelationships among energy policies in the United States concerning coal use and its environmental impacts should be recognized.

Appendix: The Parametric Procedures

The basic activity in the model is denoted as X_{ijklm}—coal extracted in the ith coal supply district by the jth mining method (with the kth sulfur level) in new or existing mines (mth) and delivered to the lth demand region, measured in 1,000 ton units.

The main behavioral assumption of the model is contained in the objective function

$$(1) \qquad \text{Minimize} \ \sum_i \sum_j \sum_k \sum_l \sum_m (C_{ijm} + t_{il}) X_{ijklm}$$

where C_{ijm} represents the per unit extraction costs of the respective activities and t_{il} represents the per unit transportation costs. The optimal solution to (1) minimizes the total cost of the production, delivery, and utilization of steam electric coal subject to the constraint system. If a supply district can provide the requirements of a demand area at a lower delivered price than a competing district, either because of an extraction cost advantage or a transport cost advantage, the model attempts by the specification in (1) to ship from that district. Utilities in each state are seen as attempting to minimize the costs of their fuel input deliveries, and producing districts as shipping where they have the greatest relative cost advantage.

Minimum cost increases in altered regional production capacities to meet the impacts of increased demands resulting from coal conversion and banning the highest sulfur coals involved modifying the regional capacities, K_{ijkm}, by increments Δ in the regional capacity constraints based on minimum cost vectors in the original solution

$$\sum_l X_{ijkm} \leq (K_{ijkm} + \Delta_{ijkm}) \qquad \text{for all} \quad i, j, k, m$$

Given the increased demands resulting from coal conversion, the impacts of stack gas scrubbing were analyzed by modifying the objective function to:

$$(1') \quad \sum_i \sum_j \sum_k \sum_l \sum_m \sum_s [(C_{ijm} + t_{il}) X_{ijklm} + \phi(b_{ijkms}) X_{ijklms}]$$

where ϕ is the level of the stack gas scrubbing costs. The terms X_{ijklms} represent the scrubbing of coal in the model. This technology alters the user cost of a ton of coal from any region by the emitted pounds of sulfur per million B.T.U.'s in that ton times the appropriate scrubbing costs. The regional constraints on the average allowable sulfur emissions from coal-fired plants, S_l', were modified to represent the lower levels of sulfur emissions per million B.T.U. from scrubbed coal, S_{ijklms}, relative to untreated coal, S_{ijklm}

$$\sum_i \sum_j \sum_k \sum_m \sum_s [S_{ijklm} X_{ijklm} + \tilde{S}_{ijklms} X_{ijklms})] \leq S_l' \qquad \text{for all} \quad l$$

Notes

1. The figure for electric utilities' share of sulfur emissions is from U.S. Environmental Protection Agency [20]. Natural gas is essentially a sulfur-free fuel, since the technology to remove its sulfur compounds during routine preparation prior to shipment is well advanced. Oil is generally low sulfur due to alternative blending and refining techniques. The highest monthly sulfur content of oil deliveries to electric utilities in 1974 for example was .9 percent (based on preliminary data supplied by the Federal Power Commission).

2. The following discussion and figures, except as noted, are based on FEA [5, pp. 1-6].

3. The Bureau of Mines had estimated lost coal production of 100 to 200 million tons. In contrast the estimates by the EPA were a loss of only 50 million tons. See FEA [7, pp. 36-41].

4. A formal algebraic description of the model where the impacts on interfuel substitution from public policy are the main concern is in Schlottmann and Abrams [18]. For a description of the parametric procedures used in this paper, see the Appendix.

5. Following figures computed from data in National Coal Association [14, p. 15].

6. Based on data reported on Federal Power Commission Form 423, "Cost and Quality of Fuels for Steam Electric Plants," calendar year 1973.

7. A complete discussion of this model is contained in Schlottmann [17, Chaps. 4, 5].

8. A similar programming procedure is described in FEA [6]. Such a procedure allows greater econometric flexibility in estimating the form of the demand equations compared to estimation explicitly designed to place the results in a quadratic programming model.

9. Attaining lower standards over the short run requires increased oil use. If we allow interfuel substitution in order to reach a stringent standard of 1.25 pounds of emitted sulfur per million B.T.U., total coal shipments are 371 million tons. The overall decline of 110 million tons from a moderate standard of 2.00 lends support to the pessimistic conclusions of the Bureau of Mines' estimates, and represents additional oil usage of 1.14 million barrels per day.

10. Our discussion here is based on the study by Schlottmann [17, Chap. 6].

11. This type of reclamation is extensively described, for example, in Mined Land Conservation Committee [12].

12. By contrast, average rainfall in Appalachia can be up to 45 inches and in other Western areas, 14 to 16 inches, as reported in Council on Environmental Quality [3, p. 16].

13. FPC Form 423 data.

14. For a discussion of constraints on expanding coal production. See FEA [7, pp. 43-63].

15. The main reference for the data used here is an unpublished report by FEA [5]. Further details are available upon request from the author.

16. For a general discussion of this technology and cost estimates, see Gordon [9, pp. 109-125].

17. A summary discussion of coal cleaning is in Congressional Research Service [2, pp. 31-33].

18. Yearly data on coal cleaning is available from the National Coal Association [14].

19. A general discussion of the technology of coal gasification appears in Federal Energy Administration [6].

References

1. Box, T. et al. *Rehabilitation Potential of Western Coal Lands.* National Academy of Sciences. Cambridge: Ballinger, 1974.
2. Congressional Research Service. *Factors Affecting the Use of Coal in Present and Future Energy Markets.* Washington, D.C.: U.S. Government Printing Office, 1972.
3. Council on Environmental Quality. *Surface Mining and Reclamation.* Washington, D.C.: U.S. Government Printing Office, 1973.
4. Denis, S. *Some Aspects of the Environment and Electric Power Generation.* Santa Monica, Cal.: Rand Corporation, 1972.
5. Federal Energy Administration. "Coal Conversion Program," unpublished draft environmental statement, Office of Coal, 1975.
6. ___. "Documentation of the *Project Independence* Assessment Model." Forthcoming, 1977.
7. ___. *Project Independence: Coal.* Washington, D.C.: U.S. Government Printing Office, 1974.
8. Federal Power Commission. *Monthly Report of Cost and Quality of Fuels at Steam Electric Generating Plants.* Washington, D.C.: U.S. Government Printing Office, December 1973.
9. Gordon, R. *Coal and the U.S. Electric Power Industry.* Baltimore: Johns Hopkins Press, 1975.
10. Henderson, J. *The Efficiency of the Coal Industry: An Application of Linear Programming.* Cambridge, Mass.: Harvard University Press, 1958.
11. M.I.T. Energy Policy Study Group. "Energy Self Sufficiency: An Economic Evaluation," *Technology Review,* 76 (1974), 22-58.
12. Mined Land Conservation Conference. *What About Strip Mining.* Washington, D.C.: Mined Land Conservation Committee, 1964.
13. National Petroleum Council, Coal Task Force. *U.S. Energy Outlook.* Washington, D.C.: National Petroleum Council, 1973.
14. National Coal Association. *Bituminous Coal Data, 1973.* Washington, D.C.: National Coal Association, 1975.
15. Northern Great Plains Coal Study Group. *Northern Great Plains Coal.* Minneapolis: Upper Midwest Council, 1975.
16. Office of Emergency Preparedness, Executive Office of the President. *The Potential For Energy Conservation: Substitution for Scarce Fuels.* Washington, D.C.: U.S. Government Printing Office, 1973.
17. Schlottmann, A. "Environmental Regulation and the Allocation of

Coal," unpublished Ph.D. dissertation, Washington University, 1975.

18. ___ and L. Abrams. "Sulfur Emissions Taxes and Coal Resources," *Review of Economics and Statistics* (forthcoming).

19. U.S. Bureau of Mines. *Coal—Bituminous and Lignite 1973.* Washington, D.C.: U.S. Government Printing Office, 1975.

20. U.S. Environmental Protection Agency. *Nationwide Air Pollutant Emission Trends 1940-1970.* Research Triangle Park, N.C.: Environmental Protection Agency, 1973.

21. U.S. Sulfur Oxide Control Technology Assessment Panel. "Final Report on Projected Utilization of Stack Gas Cleaning Systems by Steam Electric Plants," unpublished, 1973.

23
A Suitability Matrix
for Selecting Land Use Alternatives
for Reclaimed Strip Mine Areas

James E. Rowe

Introduction

The occurrence of surface coal mining is increasing and, as a result, a need has developed for a logical method for choosing post-mining land uses. Growing concern has been expressed about the amount of land disturbed by strip mining and the need to develop effective reclamation procedures (see Figure 1).

What to do with the land after the coal is extracted is the central issue of this paper. A process for choosing an economically justifiable use for the reclaimed strip mined site is presented.

Pending federal legislation will require mine operators to reclaim their site for the "utility and capacity of the reclaimed land to support a variety of alternative uses."[1] If and when such federal legislation is enacted, mine operators, regional planners, and those responsible for reclamation will be required to consider alternative uses (Rowe, 1975).

Potential exists for residential, commercial and industrial land uses in situations where access, water and sewer availability, and load bearing capabilities meet industrial and residential requirements. The highest potential for commercial or industrial uses would be in close proximity to the urban areas in the region and in strip mined sites close to interstate and other major access routes. In areas where a proposed re-use would require reclamation procedures that would make the stripping economically unfeasible, the area in most instances would not be mined.

The Extent of Mining in the Study Area

The study area, the Tennessee counties of Anderson, Campbell, Claiborne, Morgan, and Scott, was chosen because its physical characteristics

From *Landscape Planning*, vol. 4, no. 3, September 1977, pp. 257-271. Reprinted by permission of the Elsevier Scientific Publishing Company.

Research supported by National Science Foundation/RANN Grant No. 51A72-03525 A04. The author acknowledges George Zufall for his graphics assistance.

Figure 1. An active strip mine in Campbell County, Tennessee, on 25° slope.

are similar to those of the rest of Central Appalachia, which includes central and southern West Virginia, eastern Kentucky, southwest Virginia, and the Cumberland Plateau portion of Tennessee (Figure 2). The five counties contain approximately 80 percent of the total strip mined acreage in Tennessee. The area is situated in the northeast corner of the Tennessee portion of the Cumberland Plateau. In Claiborne and Anderson Counties, only the western section is on the plateau and these sections contain the mined sites. The total area occupied by the five counties is 1,480,400 acres. Total area disturbed by strip mining of coal as of July 12, 1973, was 41,320 acres, ranging from a low of 5,155 acres in Claiborne County to a high of 14,229 acres in Campbell (Tennessee Bureau of Mines files). More stripping is expected in the future because of the large amounts of desirable high-quality low-sulfur coal underlying the region. Although properly reclaimed land can be used for many purposes, including recreation, agriculture, forests, housing, and industrial development, only 42.9 percent of the disturbed land in Central Appalachia has been reclaimed (Paone, 1974) (Table I; see Figures 3 and 4).

Development which occurs on poor soils, steep topography, or in areas subject to periodic flooding will be significantly more expensive and restricted than in areas which have favorable soils. Land suitable for agricultural purposes is also utilized. Urban development on such unsuitable land

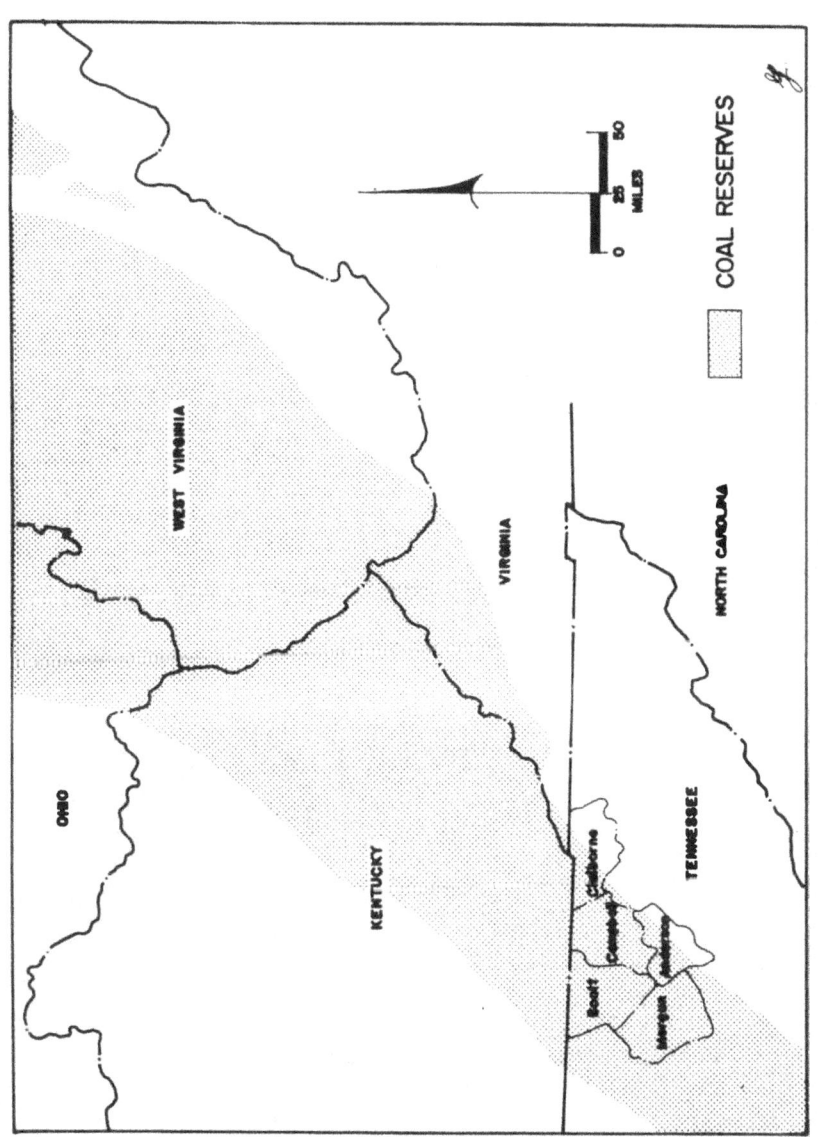

Figure 2. Central Appalachian study area.

TABLE I

Acreage strip mined and reclaimed: Central Appalachia, 1930—1971

Area	Acres disturbed	Acres reclaimed	Percent reclaimed
Eastern Kentucky	108,000	42,000	38.8
Tennessee	67,800	23,400	34.5
Virginia	78,800	28,900	36.9
West Virginia	210,000	105,000	50.0
Total	464,600	199,300	42.9

requires estensive landscaping work, heavy reliance on sewer systems, and demands more expensive site preparation. Industrial and commercial development normally require large tracts of level land which are rare in the study area.

Strip mining can create level land that can later be employed for industrial or commercial development (Rowe, 1976). Residential areas offer greater flexibility in design but still must bear the additional, often costly, expense of site improvement. Most of the land in the study area is subject

Figure 3. An abandoned (orphan) strip mining pit in Anderson County, Tennessee. This particular pit was never reclaimed due to low bonding requirements in 1967.

to severe development limitations due to either slope or accessibility. Except for small, isolated areas around Oneida in Scott County and between Wartburg and Petros in Morgan County, the remainder of these two counties is not suitable for major urban-industrial development due to the terrain. Campbell County, with the exception of the southeast quarter, is likewise not suitable for major urban-industrial development. Major portions of Claiborne and Anderson Counties also have severe development restrictions.

When determining land use, many factors must be considered, including competing uses for the same land and the problem of providing for the use of marginal and sub-marginal land for urban-type development because of population and commercial pressures. For example, it is frequently economical for one to risk building on a floodplain which also happens to be along a major highway or in a resort area.

Post-mining redevelopment for certain uses demands accessibility, a population base, an economically viable area, and the physical characteristics essential for the various proposed re-uses. Naturally, certain forest and wildlife management areas do not require maximum accessibility. The criteria selected and the use groups formed are based upon the five-county study area's characteristics. This paper develops a process for selecting land

Figure 4. An orphan mine in Scott County, Tennessee. A very poor example of natural reclamation.

use alternatives by combining the traditional physical factors with spatial parameters developed for strip mined sites in Appalachia.

Alternative Uses

An early study by Limstrom and Merz (1949) found that most strip mined lands in Ohio were adaptable to three general uses: forestry, recreation and wildlife, and agriculture. Other uses, such as industrial or residential purposes, were considered to be of minor importance, except near municipalities and along railroads and highways. Limstrom and Merz found that, in addition to the production of wood, forests help control erosion, improve soil, conserve water, and provide recreational opportunities. They felt that most of the stripped land in Ohio was best suited for forestry.

More recently, Buckner and Kring (1967) have expanded forestry opportunities by determining that Christmas trees are a profitable, short-rotation crop suitable to the extreme site conditions in the study area. University of Tennessee studies found that Christmas tree crops grown on reclaimed strip mined land have three advantages over other uses: they are not subject to vandalism; mine roads provide year-round use and access; and, due to different site characteristics, exposure, slope and altitude, several species can be utilized (see Figures 5 and 6).

West Germany's largest strip mining operator, Rheinische Braunkohlenwerke AG, is world renowned for its reclamation efforts in the Ruhr. In a recent news magazine article, Rheinbraun was praised for its successes in totally relocating several villages in its reclamation efforts. Their postmining plans called for wildlife preserves and recreational areas (Newsweek, 1975).

A strip mine site can be re-used for many purposes. The alternative uses which fit the regional needs and, at the same time, are environmentally compatible are naturally the most desired. This analysis does not quantify variables such as scenic features and landscape characteristics. Fisher et al. (1972) have presented an analytical framework for evaluating the benefits of environmental resources that could be appropriately modified for strip mined areas. Their research was based upon recreational areas. Such analysis would remove some of the subjectivity of existing methods.

Physical Factors

Land morphology, soils, stream patterns, plant association, wildlife habitats, and land use can be examined through the concepts of the physiographic region (McHarg, 1971). Nature often performs work for

Figure 5. A regraded strip mine bench in Morgan County, Tennessee.

Figure 6. A post-mining alternative use for the same bench as that in Figure 5. In this instance Christmas trees were being grown.

man and in many cases this is best accomplished in a natural condition. Many strip mined areas are best reclaimed and revegetated to the original status and left in a wilderness state. Many areas are intrinsically suitable for certain uses while other are less suited. If we accept this proposition and modify it, one can develop families of land uses that can be utilized in strip mining reclamation.

To arrive at an alternative, one can chart an area's underlying design by analyzing major natural factors such as the water table and geologic elements. One by one, each element is traced to determine where, all other factors being equal, it would be best to build or develop and where development would be unwise. The result is an accumulative plan, the charts of each element are overlaid on top of the others, and the natural design emerges.

The natural and ecological features of certain sections of the study area dictate that they would be protected. These areas should be placed in low density and open space type uses. The major areas are: groundwater intake areas, major water run-off areas, agricultural and forest soil areas, mineral resource areas, steep land, areas affected by certain air drainage patterns, floodplains, and areas of scenic features and regional landscapes. These physical factors were mapped onto overlays and the resultant natural pattern emerged (Figure 7).

Certain lands serve man best if left in an undeveloped state. These include floodplains, water run-off areas, areas which are highly erodible due to slope and/or soil characteristics, and major aquifers and aquifer recharge areas. These considerations are modified and applied to developing a suitability matrix for determining alternative land use for reclaimed mined sites.

Spatial Parameters

A survey of the various state regulatory agencies indicates that surface mined lands have been utilized for unique post-mining uses only since 1971 and then only as an alternative to minimum requirements of existing state reclamation laws. The data reported here represent a 90 percent sample of all such post-mining uses in Appalachia from 1971 through 1974.

Table II presents a summary of the 150 cases of post-mining land uses observed by type and location. Urban related uses are predominant, with 45 percent of the examples. The recreation category, 24 percent, is liberally defined to include forest areas. This permits land involved in timber production and forest uses that include hunting and camping to be considered as potentially multirecreational in nature. Overall, the data do not support

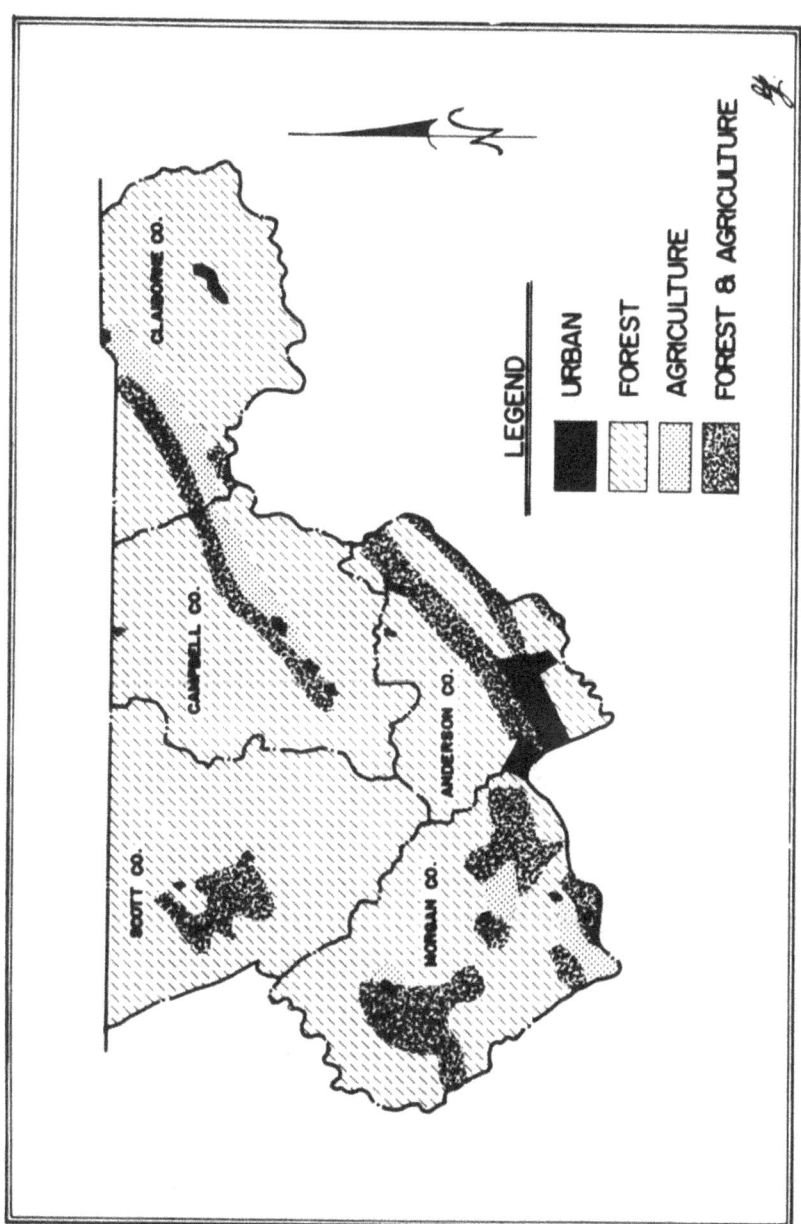

Figure 7. Land use suitabilities.

TABLE II

Spatial relationship of post-mining uses to urban centers and interstate highways

Use	Number of observations by miles to interstate			Number of observations within 10 miles of a city of 20,000	Number of observations by miles to SMSA			Total observations
	0—5	5—15	15+		0—25	25—50	50+	
Urban development	14	20	34	45	43	20	5	68
Agricultural uses	3	4	14	10	7	8	6	21
Recreational uses	21	6	10	19	22	11	4	37
Other uses	9	4	11	10	11	4	9	24
Totals	47	34	69	84	83	43	24	150

Source: Rowe (1975, p. 9).
SMSA, standard metropolitan statistical area.

claims that post-mining use of surface mined land is widespread.

The spatial pattern of productive uses for surface mined areas in Appalachia for selected uses suggests that accessibility to a major highway system and proximity to a regional population center are major determinants of the location and types of post-mining activity.[2] Inaccessible surface mined areas are often left unreclaimed or are reclaimed only to the minimum required by law.

Since urban related and recreational uses represent over 70 percent of the post-mining activities at present, it seems likely that most future development will also be in these two categories. Therefore, the effects of accessibility and distance to urban centers on these uses in particular were considered.

A general interstate highway/major arterial system variable was used as a measure of accessibility for potential users of a site. A variable measuring distance from a standard metropolitan statistical area (SMSA) was constructed as an indication of a population center's potential market. Grouped data in 5-mile intervals up to 50 miles represent the most reliable distance information available.

It is well known that ordinarily least squares estimation using grouped data does not satisfy the classical assumption of homoskedasticity of the disturbance terms. Thus, generalized least squares procedures were applied to estimate the relationship between urban related, recreational and all other land uses and the distance variables (Johnson, 1972). Urban related/recreational use was treated as a qualitative dependent variable (Kmenta, 1971). That is, the following regressions were estimated.

$$y = a + b_1 x_1 + b_2 x_2 + e$$

where $y = \begin{cases} 1 \text{ if observation was an urban related/recreational use} \\ 0 \text{ otherwise} \end{cases}$

x_1 = distance from interstate highway/major arterial system
x_2 = distance from SMSA
e = disturbance term

The estimated regression equation was as follows.

$$y = 1.225 - 0.024x_1 - 0.012x_2 \qquad r^2 = 0.57$$

standard errors: (0.50)   (0.03)   (0.04).   (1)

This result shows that accessibility and proximity to a regional population center is important in determining the degree to which urban related/recreational sites are developed. In particular, the estimates are significant at the level of 5 percent. Accessibility is apparently the most important variable in recreational uses. If one were to disaggregate the recreational uses and redefine the qualitative dependent variable as having a value of 1 for a recreational use and 0 for all others, including urban related uses, it is estimated that

$$y = 0.800 - 0.028x_1 \qquad r^2 = 0.57$$
$$\text{standard errors:} \quad (0.07) \quad (0.03).$$

(2)

The SMSA variable is statistically insignificant, primarily because urban related observations were considered separately from the recreational uses. Similarly, the analogous estimated regression equation for the disaggregated urban related uses shows that nearness to a regional population center is the most important variable. Note that the classification of urban related uses was independent of their spatial relationship to an urban area. The dependent variable is redefined as having a value of 1 for an urban related use and 0 for all others, including the recreational uses:

$$y = 0.871 - 0.016x_2 \qquad r^2 = 0.51$$
$$\text{standard errors:} \quad (0.08) \quad (0.01).$$

(3)

The above regressions support the hypothesis that the accessibility and SMSA variables are significant for urban related and recreational uses. Although this result was expected, it apparently has not influenced public policy toward surface mining in Appalachia.

The economic viability of a proposed re-use is estimated by the percentage of existing examples being located within the same general distance from an interstate highway or an SMSA. The spatial parameters and the environmental factors analyzed, when used in combination, lead one to choose a sound alternative land re-use type by utilizing a suitability matrix. Further on-site engineering analysis and area-wide market studies would help pin-point the best possible re-use.

Suitability Matrix

The function of the suitability matrix is to demonstrate the degrees of suitability among several variables. As shown in Figure 8, four generalized land use types are considered; urban, recreational, agricultural, and other unrelated uses. These land use types have been placed in a matrix with seven main categories of the major physical and spatial characteristics of the area. They are: water intake areas, water run-off areas, soils, scenic areas, steep areas, areas in juxtaposition to urban areas, and areas that are accessible to an interstate highway. The degree of suitability, ranked as

Physical and functional constraints

Legend
X Possible use
+ Moderate limitations
□ Severe limitations

Alternative uses	Acc. 0-5 mi from interstate	Acc. 5-10	Acc. 10-15	Acc. 15 or more miles	Jux. 0-15 mi from SMSA	Jux. 15-25	Jux. 25-40	Jux. 40 or more miles	Topography over 20% slope	Soil classe C	Soil classe D	Soil classe E	Soil classe F	Water intake areas	Scenic areas	Water run off areas
Urban related uses																
Residential	X	X	+		X	X	+		+	+	+	+		+	+	+
Commercial	X	+			X	+				+	+				+	+
Educational	X	X	+		X	X	+			+	+			+	+	
Industrial	X	+			X	+				+	+				+	
Transportation	X	+	+		X	+			+	+	+			+	+	+
Agricultural uses																
Intensive		+	+	X		+	X	X	X	+				X	+	
Extensive		+	+	X		+	X	X	+	X	+	+		X	+	X
Recreational uses																
Intensive	X	+			X				+	X	+	+	+	+	+	
golf courses	X	+			X				+	X	+	+	+	X	+	
camp grounds	X	+			X	X	X	+	X	X	+	+	+	X	+	
Extensive	+	+	+	X	+	+	+	X	X	+	X	X	X	X	X	X
forest		+	+	X	+	X	X	X	X	X	X	X	X	X	X	X
camping		+	+	X	+	+	X	+	X	X	X	X	X	X	X	X
hunting		+	+	X	+	X	X	X	X	X	X	X	X	X	X	X
fishing		+	+	X	+	X	X	X	X	X	X	X	X	X	X	X
timber		+	+	X	+	X	X	X	X	X	X	X	X	X	X	X
Other uses																
Refuse disposal		+	+		+	+				X	+					
Other non-specified		+	+		+	+				X	+				+	+

Figure 8. Chart showing reclaimed strip mined site alternative land use suitability matrix.

possible, moderately limited, and severely limited (each of the opposing variables were determined by considering both the real and potential economic and social costs involved in an area's use). These three categories of suitability are as follows.

Possible use. Includes areas with slight limitations. All or nearly all the component areas are favorable for the intended use. An unfavorable factor is relatively easy to overcome.

Moderate limitations. Areas of this classification are only reasonably favorable for the intended use. The unfavorable sites may be overcome through correct planning, careful design, good management, or moderate expense.

Severe limitations. Areas of this classification are unfavorable for the intended use. Adapting the environmental and spatial criteria components in question to the intended use may be accomplished only with difficulty and considerable expense.

Where feasible, strip mined sites should be placed in either single or multiple land uses which have the greatest degree of compatibility with the area's predominant physical characteristics. Generally, these would be those of either the possible or moderately limited ratings.

Application of Process

In order to determine the best alternative land use for a reclaimed strip mine site, from a physical standpoint, each of the predominant physical, environmental and spatial features of the study area were identified and depicted on maps. Based on the features shown on the overlays, a map of the area was prepared which showed the possible type of land use and its suitability to the area in generalized terms. Certain areas which are expected to continue in their present uses were identified and not considered from a suitability standpoint. These areas consisted of land in public ownership and areas presently in urban-type uses. The land use suitabilities map suggests what major groups of land use could be implemented rather than its specific use. For example, low density residential development could (and perhaps should) be permitted in an area shown in "agriculture," with the exception of areas of high agricultural value. The generalized land suitability map for the study area was depicted in Figure 7.

Land capability analysis can be defined as a methodology for evaluating the physical components of the environment in terms of their ability to support, withstand, or tolerate various land uses. The analysis is based on an inventory of the physical variables in an area and an assessment of how their inherent properties can be affected by different types of land use.

Four steps are involved in executing the procedure developed for the study area:

1. The inventory and mapping of physical and spatial planning data.
2. The delineation of all hazard areas, areas of critical environmental concerns and areas of environmental significance.
3. Future plan—objectives plus planned public facilities by state, region, or locality. (Consider the future plans of the area.)

4. The analysis of the data to determine the suitability of the land for uses of the four major use groups.

Input from the suitability analysis process can, at this point, be combined with a suitability analysis based on socio-economic and cultural data to determine the overall ability of a site to support various kinds of proposed re-uses.

Subsequent to the analysis of the data, the process was carried two steps further:

5. The environmental impact of proposed major development, such as subdivisions and industrial parks, on the physical environment was evaluated.
6. A detailed analysis of several proposed land use types on selected strip mined sites was conducted.

The physical land capability analysis was modified to include the statistical parameters developed to measure the land use suitability of a site in relationship to its distance from an interstate highway or an urban area in the form of a suitability matrix. The physical data were collected and mapped on transparent overlays which indicated that many sites were more suitable than others for a variety of land uses. Specific sites proved more suitable for various re-uses when studied in detail.

Values were assigned for each factor considered. These were separated into three classes: possible use, moderate limitations, severe limitations. For example, depth to the water table is one component of the residential function. Based on criteria developed by the Soil Conservation Service, if the depth is greater than 5 feet, it is considered a good possibility for residential use; if it is between 2.5 and 5 feet, it is considered to have moderate limitations for residential purposes; and if it is less than 2.5 feet, it is considered to have severe limitations.

The suitability for any strip mined site for the various proposed re-uses can be ascertained from the matrix. The matrix helps to portray the interactions between the various types of land uses and to model the interactions more precisely. After the analysis is complete, an attempt should be made to verify the results to determine if the model has correctly identified problem areas. For example, after mapping an area as severely limited for residential development, it would be valuable to check with people living in the area to see whether they, in fact, have severe problems with their homes.

Summary

The Central Appalachian area contains vast amounts of low-sulfur bituminous coal. Much of this coal can most economically be extracted by surface mining; however, this process severely disrupts the area. This paper has developed a process for determining the most suitable land re-use for the disturbed areas. Physical land capability analysis was modified to include spatial parameters for the determination of the most suitable re-use of the land.

The study area of five Cumberland Plateau counties of Tennessee was chosen and analyzed due to the similarities and homogeneity of the area to the Central Appalachian region. The Tennessee counties are relatively underdeveloped and contain an expanding strip mining industry. The important physical factors were mapped and studied to produce a generalized land suitability map. This map and the spatial parameters developed by the author were combined to produce a reclaimed strip mined site alternative land use suitability matrix.

Four general land use groups were formed—urban, recreational, agricultural, and other unrelated uses—for the purpose of testing the matrix and mapping techniques; however, the process only suggests possible post-mining land uses.

The final product of the modified suitability analysis is a set of data maps and an alternative land use suitability matrix for reclaimed strip mined sites. The suitability of a strip mined site for the various proposed re-uses can be ascertained from the matrix. The matrix helps to illustrate the interactions between the various types of land uses and to model the interactions more precisely. Once this generalized process has been applied to a strip mined site, the economic realities of the situation must be considered. For example, if the site is ideally suited for an industrial park but there is already an industrial park in the area, another proposed use within the urban classification might be chosen.

The data presented for the five counties indicated that most of the strip mined sites within the area would best be returned to their original condition and enjoyed for open space qualities. This process indicated possible land uses that meet the suitability criteria established but did not intend to pinpoint the best re-use. Regional and environmental planners can avoid problems of re-use by taking physical and spatial information into consideration, and economically justifiable land uses can be determined for reclaimed strip mined sites. This methodology is presently being tested for its practical applications in Western Maryland by the Tri-County Council for Western Maryland.

Notes

1. House of Representatives, Sec. 508(a)(3), Surface Mining Control and Reclamation Act of 1975. (Similar language has appeared in most of the 1974-1977 bills.)

2. The actual locations of the various reclaimed strip mined sites are based upon Rowe (1975) and are available on request.

References

Buckner, E. R. and Kring, J. S., 1967. A crop for mine spoils? Keep Tenn. Green J., 7: 1-7.

Fisher, A. C., Krutilla, J. V. and Cicchetti, C. J., 1972. Alternative uses of natural environments: the economics of environmental modifications. In: J. V. Krutilla (Editor), Natural Environment: Studies in Theoretical and Applied Analysis. Johns Hopkins Press, Baltimore, Md., pp. 1-13.

Johnson, J., 1972. Econometric Methods. McGraw-Hill, New York, N.Y., 300 pp.

Kmenta, J., 1971. Elements of Econometrics. Macmillan, New York, N.Y., 655 pp.

Limstrom, G. A. and Merz, R. W., 1949. Rehabilitation of Lands Stripped for Coal in Ohio. Ohio Reclamation Association, Columbus, Ohio, 17 pp.

McHarg, I., 1971. Design with Nature. Natural History Press, Garden City, N.Y., 197 pp.

Newsweek, 1975. Prize for a stripper. Newsweek, 86 (December 1): 86-87.

Paone, J., 1974. Land Utilization and Reclamation in the Mining Industry, 1930-71. Information Circular 8642, U.S. Government Printing Office, Washington, D.C., p. 11.

Rowe, J. E., 1975. An Inventory of the Unique Uses for Reclaimed Strip Mined Land in the Appalachian Region. ARP-33, University of Tennessee, Environment Center, Knoxville, Tenn., 36 pp.

Rowe, J. E., 1976. The Strip Mining Dilemma: The Case of Virginia. Land: Issues and Problems, No. 22, Virginia Polytechnic Institute and State University, Cooperative Extension Service, Blacksburg, Va., 4 pp.

Postscript

The Surface Mining Control and Reclamation Act of 1977 was signed into law by President Carter on August 3, 1977. The law requires "back-to-contour" reclamation except in situations where alternative post-mining uses are warranted. For an examination of the law and its impact on Appalachia, see: J. E. Rowe, 1977. Surface mine reclamation: The 'back-to-contour' constraint. J. Soil Water Conserv. (March-April 1977): 74-75.

24

Land Use Planning
in Surface Mine Areas

James R. LaFevers
Edgar A. Imhoff

Introduction

Pre-mine reclamation planning has long been recognized as an essential element in any cost-effective extraction/reclamation program. There is an additional need, however, to carry the planning process a step further to include local and regional land use planning elements in the reclamation planning process. In that way, mined areas could be reclaimed to help satisfy specific local land use needs and comply with land capabilities, rather than merely revegetated to satisfy legal requirements. Such a planning program could be beneficial to both the mining company and the public. The public would benefit by having land use suitability, whereas the company would benefit from making their reclaimed land more marketable. This type of integrated reclamation/land use planning program will generally require input from both industry personnel and public sector planners. A land use needs inventory for the community, and other data necessary for the development of marketing strategies should be provided to the mining company by public planning agencies in whose jurisdiction the mine will be operating. In reciprocation, the mining company could provide to planning agencies information valuable in the generation of future development scenarios. Water and other mineral availability data, analyses of the overburden and the likely effect of disturbance and baseline data relating to archeological, social, historical, biological and other resources could be provided to planning agencies by the mining company and used for their mutual benefit.

An ongoing Argonne National Laboratory program,[1] funded jointly by the Energy Research and Development Administration and the U.S. Geological Survey, has examined the potential and effective procedures for achieving integrated reclamation and land use planning in several

From the *Proceedings of the Fifth Symposium on Surface Mining and Reclamation* at Louisville, Kentucky, on October 18-20, 1977, pp. 311-319. Reprinted by permission.

extractive industries, including the coal industry. As a part of this program, a series of workshops is being conducted to familiarize local planners with the unique problems and opportunities associated with land use planning in surface mine areas. At the time of this writing, a Federal reclamation bill has just been signed into law by President Carter. Since this law contains at least ten distinct references to the implementation of land use planning as related to mined area reclamation, one might expect that a number of RD&D projects would have been conducted to facilitate this process. An examination of the "Inventory of Federal Energy-Related Environment and Safety Research" by the authors disclosed, however, that of the 2536 programs ($453 million in funding) in the Inventory, only the Argonne program is directed specifically at reclamation/land use planning issues.

Such a low priority has clearly resulted in a need for more data concerning the land use impacts of surface mining. Mining company personnel and public sector planners are required by the Federal law to cooperatively confront a number of land use issues which may never have been addressed jointly before.

Land Use Planning Issues of the Federal Reclamation Law

After more than 15 years of debate, a President and Congress have finally agreed upon the substance of federal regulation of surface coal mine reclamation. According to the preamble of the House version, H.R.2 (the compromise bill not being available at the time of this writing), the new federal law establishes a minimum level of performance for the many activities associated with effective conversion of coal-mined land for beneficial use. Forty states have previously enacted mined-area reclamation laws. These laws, whatever their excellence, will likely have to be modified before equivalency will be attained with the new federal legislation, which is both comprehensive and complex.

A comparison of H.R.2 with the 40 state reclamation laws indicates that another critical issue exists in addition to the much-discussed issues of "approximate original contour" and "protection of alluvial valley floors." It will certainly draw much attention when the coal-producing states approach the matter of attaining equivalency with federal legislation. This "sleeper" issue is: governmental land use planning as related to mined-area reclamation. Public land use planning or plan implementation is cited in ten separate places in H.R.2, as shown in Table 1.

A significance of the ten statements in H.R.2 pertaining to land use planning is that mining companies are now required to have their reclamation plans reviewed by whatever local planning agency is available before a permit will be issued. The authors have discerned that in some cases,

TABLE 1

PUBLIC LAND USE PLANNING OR PLAN IMPLEMENTATION ISSUES
IN FEDERAL BILL H.R.2

Reference	Requirement or Guide (Paraphrased)
201(c) (8)	Technical information on mining and reclamation will be provided to local land use planning agencies.
505(b)	States can enact land use controls more stringent than those required by the Federal law.
508(a) (3)	In reviewing and acting on reclamation plans, the regulatory authority will evaluate the selected land use objective in terms of existing land use policies and plans, including comments of any authorizing local planning bodies.
508(a) (9)	Consideration must be given to making the surface mining and reclamation consistent with applicable state and local land use plans and zoning requirements.
513(a)	Planning agencies shall be notified of applications for mine permits and related opportunities for hearings.
515(b) (2)	The proposed postmining land use shall be consistent with applicable land use policies and plans.
515(c) (3) (A) 515(c) (3) (C)	Certain variances from performance standards may be allowed in mountain-top removal, provided there is consultation with the appropriate land use planning agencies (if any) and certification that the proposed land use is consistent with local land use plans and programs.
522(a) (3) (A)	A specific land area may be designated unsuitable for certain types of coal mining operations if such operations are found to be incompatible with existing public land use plans or programs.
522(a) (5)	Determinations of the unsuitability of land for surface mining shall be integrated with present and future land use planning and regulation processes at the Federal, state and local levels.

as in Fulton County, Illinois, there is a history of cooperation between the mining industry and local planning agencies. In Fulton County, for example, coal mining companies routinely submit their reclamation plans to the County Planning Administrator, who frequently suggests constructive changes. For the most part, however, this type of cooperation has been found lacking in many mining areas. During the more than two year tenure of the Argonne program previously mentioned, which examined coal and non-coal mining, examples of reclamation programs which had been designed with active input from local planners were found to be almost nonexistent. Numerous examples are available of mined areas that have been converted to residential, commercial, recreational, or other uses.

On close examination, however, one finds that in almost every case the land use planning was done after the mining had been completed. One study shows that mined areas can be as much as ten times as attractive for housing sites as adjacent unmined areas (LaFevers, 1974). In every case examined, however, it was found that development was occurring on sites previously reclaimed for other uses (primarily forestry, range, and agriculture).

After-the-fact (of mining) planning and re-reclamation—generally referred to as development—are costly in relation to pre-mine planning. A major midwestern city, in dire need of a landfill site, recently acquired a mined area that had been reclaimed for forestry several years previously. About seven hundred dollars per acre had been spent grading, backfilling and revegetating this site. After acquisition, the city then had to spend about two hundred dollars per acre removing the trees and preparing the site for landfill; a process which will completely obliterate all vestiges of the previous reclamation effort, including the grading and backfilling. Not the entire nine hundred dollars per acre was wasted, since some initial expense would have been required to prevent erosion. It is obvious, however, that a significant savings could have been realized with a little foresight and a little interaction between the coal mining company and the local planners. It is most likely that city planners were aware at the time of mining that in a few years a new landfill site would have to be acquired. Had they contacted the mining company, a jointly drafted long-term reclamation/land use plan could have been generated that would actually have saved both parties a considerable amount of money.

While inspection of the federal mandates and advisements in Table 1 confirms that the new Federal law is not a national land use law in disguise, it is obvious that local and state land use planning has been afforded a position of potential importance in surface mining and reclamation. The planning community has an opportunity through this law to influence such decisions as where not to mine (Section 522) and to what land use the affected areas should be reclaimed (Sections 508, 513, 515).

State Reclamation Laws and Planning Issues

In most states reclamation laws are not particularly conducive to long-term land use planning. Although they may encourage reclamation, in many cases these laws should be considered deterrents to planning, including reclamation planning. The philosophy of planning inherently is based on the ability to exercise options. It is the nature of most reclamation laws, however, to close a large number of what should be available options. By dictating reclamation practices, the laws limit the planning potential. The development of reclamation laws has gradually affected both reclamation planning and the type of reclamation that is actually achieved. Since they do not generally address the issue of land use planning, however, these laws have had a lesser impact on planners than on mining personnel; a situation that will be considerably equalized by the new Federal law.

The reflection of state mined area reclamation laws in the surface expression of site specific reclamation programs is most apparent in coal mining, as compared with other extractive industries, primarily because it is this industry that has been regulated for the longest period of time. At the end of 1975 four states still had reclamation laws covering only land affected for coal, and five other states had at least two reclamation laws so that coal lands could be regulated separately from other minerals. When the first reclamation laws were enacted in the U.S. in 1939 and 1940, only coal mine lands were covered and little was required of the mine operators other than some attempt at revegetation. Early laws were not really designed to improve the long-term economic use of the land, as non-commercial tree species were often planted because of their rapid growth characteristics. In some instances commercially valuable tree species were planted but because of poor accessibility, were never harvested. Early attempts at creating pasture and range lands were often more successful, with some coal companies reporting successes as early as the 1930s (pre-law). Although these early reclamation laws reflected the state-of-the-art of reclamation technology at the time, they were probably inordinately industry oriented because coal company employees actually drafted the requirements of the first bills. Coal companies have lagged behind the sand and gravel industry in developing the concept of integrated reclamation and land use planning. Because most sand and gravel, and other construction aggregate mining, occurs in close proximity to urbanized areas (because of relatively high transportation costs), these mining operations have a long history of encountering close public scrutiny of mining and reclamation procedures. They also encounter intense competition for land, high land values, and an excellent market potential for land that is reclaimed for public use or commercial or industrial development. Thus, the opportunities and obligation existed for the aggregate industry to develop techniques for planning and initiating cooperative programs with local planners for the development of mined out areas. The first authoritative discussion of public planning imperatives in surface mining is given by Ahern (1964) in a report on the sand and gravel industry.

Investigations by Strauss and Kusler (1976), Doyle (1977), and Imhoff (1976) establish the existence of a variety of state statutory constraints against the interference of local planning with mining. Alabama's Surface Mine Reclamation Act of 1975, for example, prohibits local regulation of surface mining. Judicial review has not yet tested this prohibition in terms of its effect on land use planning and plan implementation. Other states with statutory preclusions against local governmental zoning ordinances that interfere with surface mining activities are listed in Table 2.

Rather than statutory prohibitions, state reclamation laws are generally characterized by omissions with respect to land use planning. Ten states with reclamation laws do not require the declaration of the proposed land use to which reclamation will be directed, and twelve states' (Arkansas, Indiana, Iowa, Kansas, Missouri, Montana, North Dakota, Ohio, Oklahoma, Pennsylvania, Tennessee, and West Virginia) reclamation laws are totally silent on the subject of governmental land use planning or plan implementation. In contrast, positive roles for land use planning are set forth in mined-area reclamation laws in those states listed in Table 3.

Constraints to the Attainment of Land Use Planning Goals

A large number of factors can interfere with the ability of the local planner to achieve the land use planning objectives alluded to in H.R.2 and in some state reclamation laws. The most significant of these factors are probably those already discussed (lack of cooperative interaction between local planners and mining industry personnel, and state statutory constraints) and two others: an absence of public planning agencies in the locale of the mining operation; and inadequate information and analytical capability at the planning agency level. A review of the status of county

TABLE 2

STATES EXEMPTING MINING FROM
SPECIFIC LAND–USE CONTROLS OF LOCAL GOVERNMENT[1]

State	Control Exempted
Arizona	Subdivision regulations and zoning
Idaho	Building codes and zoning
Missouri	Subdivision regulations and zoning
Montana	Zoning
West Virginia	Zoning (outside of incorporated areas)
Wyoming	Zoning

[1] Additional exemptions apply in some states in which mining is classified as an agricultural activity under zoning laws.

TABLE 3

BRIEFS OF STATE ASSERTIONS OF THE ROLE OF
LOCAL PUBLIC PLANNING IN REGULATION OF MINE RECLAMATION
(Adapted from USGS Circular 731, Table 1)

State	Role Stipulated
California	Act on mining permits and reclamation plans and mining policy in general plans (of state).
Colorado	Review for conformity with local land-use controls.
Florida	Local government may impose stricter standards (silent on local planning).
Illinois	County board may recommend land use, and may request hearings.
Kentucky	State permits must comply with local zoning laws.
Maryland	State, in acting on applications, takes cognizance of county planning, zoning, and grading permits.
Minnesota	"Rules . . . shall conform with any state and local land-use planning."
New Mexico	Consultation required with soil and water conservation districts. (No word about other types of planning organizations).
Oregon	Department may approve local governmental permitting or reviewing in lieu of state (local planning involved).
South Carolina	Local soil and water conservation districts review and comment.
South Dakota	Incompatibility with local land plans can be basis for state rejection of mining permit application.
Texas	Local governments are notified of mining-reclamation and their comments comprise input to the decision-making process.
Utah	Local governments are notified and their comments are taken under advisement.
Vermont	State action must accord with local plans.
Virginia	Local soil and water conservation districts advise.
Washington	Applicant must show legality of action with regard to local zoning.
Wisconsin	Mining, reclamation and comprehensive plan (for site development) shall conform to local zoning.
Wyoming	County involved in administration of act.

comprehensive planning in coal mine areas, based on information in the files of the National Association of Counties (Washington, D.C.), shows that only half of the coal-mine counties have developed a comprehensive plan, and only part of these counties have developed a land use plan element of such plans. Although there are cases in which planning agencies other than county level may have input to local planning, the absence of county plans, including the absence of even a county planning agency in half the cases, is a severe constraint on the realization of the concept of integration of land use planning and mine reclamation planning.

If there is one condition that seems especially to constrain the ability of a planning agency to fulfill the requirements of H.R.2 it is the fact that local and state planning agencies, with rare exception, don't have the information or analytical capability to perform the complex analyses required. A sampling of ten county level planning/reclamation meetings (Sept. 1976) revealed an amazing paucity of information on and analyses of governmental planning issues, and especially land use planning. In short, it was found that planners were neither generating nor receiving (from the mine operator applicant) answers to the six recurring questions:

1. What is the source, reliability and frequency of collection of the earth sciences information used to support the post-mining land use plan submitted by the mine operator applicant?
2. What analyses (if any) have been applied to compare the applicant's plan with local governmental plans or policies?
3. As mining proceeds to its completion—say at intervals of 5 years— what will be the land cover and land use in the mining affected area, by percent of total and by acreage?
4. What are the estimated impacts on local employment and public income (taxes)?
5. What food and fiber production capability will be lost or gained by the land use conversion?
6. Can the mining and reclamation plan be modified reasonably to reduce impacts adverse to the local government?

It is recognized that some of the above questions are partially addressed in federal environmental impact statements, but EISs apply only to federal lands and do not focus on land use planning considerations. Considering the voluminous amount of data required of state and local planners by the Federal and State reclamation laws, and the sensitive nature of the questions that must be answered if land use planning is to be effective, it appears that the only way an integrated planning/reclamation concept that will be beneficial to the public and private industry can be implemented is through cooperative interaction among all concerned parties. The task is too large and too important to assume that either planners or industry can accomplish the overall goals without the complete cooperation of the other. Although Federal law now mandates that interaction will occur, in the near term the cooperative attitude of industry and planners will determine the success or failure of planning/reclamation programs rather than legal requirements.

Note

1. The Integrated Reclamation and Land Use Planning Program, Energy and Environmental Systems Division, Argonne National Laboratory, funded by the Resources and Land Investigations (RALI) Office of the U.S. Department of the Interior and the Biological and Environmental Research Division of ERDA.

References

Ahern, Vincent p., Jr. 1964. "Land Use Planning and the Sand and Gravel Producer," National Sand and Gravel Association, Washington, D.C., 30 pp.

Doyle, John C., Jr. 1977. *State Strip Mining Laws: An Inventory and Analysis of Key Statutory Provisions in 28 Coal-Producing States.* Environmental Policy Institute, Washington, D.C.

Imhoff, E. A., T. O. Friz, and J. R. LaFevers. 1976. *A Guide to State Programs for the Reclamation of Surface Mined Areas.* U.S. Geological Survey Circular 731. Reston, Virginia.

Imhoff, Edgar A. Sept. 1976. "Planners Can Improve Responsiveness to Surface Mining Reclamation Issues," *Practicing Planner.* American Institute of Planners, Washington, D.C.

LaFevers, James R. 1974. "A Cost and Benefit Analysis of the Reclamation of Land Surface Mined for Coal in Vigo County, Indiana." Ph.D. dissertation, Indiana State University, Terre Haute, Indiana. 149 pp. Available through Argonne National Laboratory, Energy and Environmental Systems Division, Argonne, Illinois.

Strauss, Eric and Jon Kusler. 1976. *Statutory Land Use Control Enabling Authority in the Fifty States.* U.S. Department of Housing and Urban Development, Flood Insurance Administration. 304 pp.

25
Allocation of United States Coal Production to Meet Future Energy Needs

Michael R. LeBlanc
Robert J. Kalter
Richard N. Boisvert

Introduction

As part of the current review of energy policy, increased attention is being focused on coal as an important source of domestic supply. This interest is explained, in part, by its substitutability for oil and natural gas in electricity generation, the abundance of domestic coal reserves, and the growing opposition to nuclear power.

Increased use of coal in the future raises a number of public policy issues, ranging from environmental impacts to leasing of the public domain. Evaluation of these issues requires improved information on present and future coal shipment and development patterns and resource costs. The impact of changing demand, sulfur dioxide emission regulations and transportation alternatives on these factors is critical to an understanding of coal's potential as a future energy source.

The empirical results from a study designed to investigate these policy issues are reported in this paper. The multiperiod model includes a spatial allocation framework, linking coal transportation systems to various supply regions and electrical utility demand centers. Quality differences among supply regions, in terms of sulfur and Btu. content, are recognized, as are the different types of coal reserves (surface or deep mine). Alternative levels of sulfur emission and exogenously specified levels of coal consumption are investigated. Although these features are treated by others [Henderson 1958; Libbin and Boehlje 1976; Nagarvala, Ferrel and Olver 1976; Schlottman and Spore 1976; and Schlottman and Abrams 1977], the present model contains several new features.

From *Land Economics*, vol. 54, no. 3, August 1978, pp. 316-336. Reprinted by permission.

The authors are, respectively, Graduate Research Assistant in Resource Economics, Professor of Resource Economics, and Associate Professor of Agricultural Economics at Cornell University.

The impact of the contract-spot market aspects of coal sales on delivery and development patterns over time is considered specifically. Several transportation modes are considered, including rail, barge and the possibility of mine-mouth electricity generation and the subsequent transportation of electrical energy. Coal shipments through a combination of modes and modal capacity limitations resulting in possible transportation bottlenecks are also incorporated. Optimal delivery patterns for the 1973 through 1990 time horizon are evaluated.

Analytical Framework

To understand the issues involved in moving coal resources from existing or potential production areas to meet growing demands, the 48 contiguous states and the District of Columbia were divided into 33 potential supply regions and 44 demand regions (aggregate regions used in this article are outlined in Figure 1). In each supply region, two types of coal were identified—deep mine and surface. The boundaries of each region were delineated to minimize the differences in quality characteristics of the coal, both in terms of Btu. and sulfur content, within a region. A unique point of origin, or centroid, in each region was derived from a weighting procedure based on the distribution of existing production across the region [U.S. Department of the Interior 1975a].

Because many air pollution and other policy issues are defined around political jurisdictions, states, excluding those with no current or forecast consumption, were chosen as demand regions. Locations of coal-fired electric power plants were used as a proxy for all coal demand [National Coal Association 1974]. The location data were weighted to determine the existing central demand node for each region [LeBlanc 1977].

While there are future plans for other forms of coal transportation, such as coal slurry pipelines, the two predominant modes used in this model are rail and barge. Truck transportation was not considered because its primary use is short-distance hauling. Two shipping alternatives using combined modes are included: rail-to-barge activities using the interior river system, and rail-to-barge activities using the Great Lakes. Finally, mine-mouth generation, transporting electricity rather than coal by converting the electricity to its Btu equivalent, was allowed. Generating facilities for this mode are assumed to exist in the corresponding supply regions.

To account for many of the long-run decisions involved in the allocation of coal resources, the model was also modified to enable solution through time, in a recursive fashion. A primary reason for using a recursive rather than a polyperiod approach was to incorporate long-term contracts.

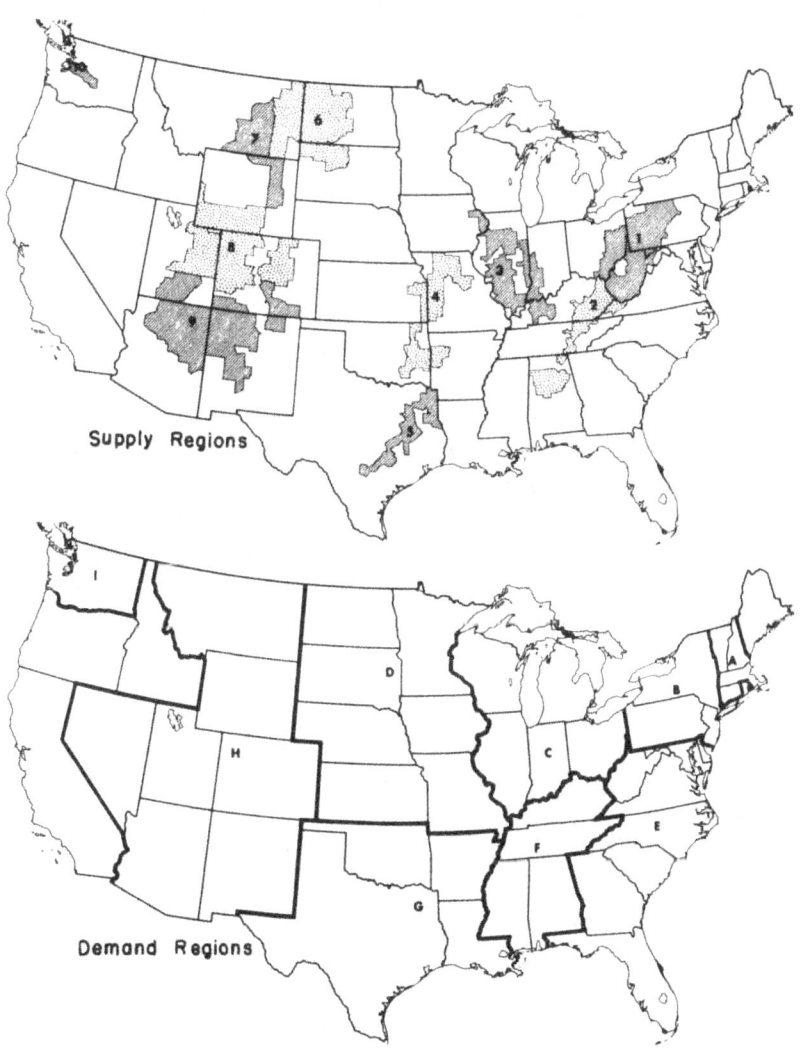

FIGURE 1
AGGREGATE SUPPLY AND DEMAND REGIONS

Historically, nearly 75 percent of all coal deliveries have been made under contracts lasting from 15 to 25 years. The other 25 percent is supplied by the spot market, characterized by marginal mines moving in and out of the market depending upon the price of coal and the supply constraints on mines having contract obligations [Federal Power Commission 1974]. By incorporating this market dichotomy a more realistic characterization of institutional inflexibility on the competitive decision process is rendered. In any given year the reallocation of coal resources may be constrained by decisions made as much as twenty years earlier.

It was assumed that development of new mines resulted only when a 20-year contract was negotiated. Since disaggregate data were not available, the initial contract-spot coal allocations were obtained for each origin and destination by specifying the model in the base period, t_0, absent of any contract restrictions. From this optimal solution, it is assumed that 75 percent of each shipment from origin i to destination j is constrained by long-term contracts. The remaining 25 percent is spot market coal. (The sensitivity of this assumption is discussed later.)

In period $t + 1$, past shipment patterns may be altered by changes in demand from destination j as well as expiration of previously existing contracts. Unexpired contracts continue to lock in supply-demand linkages established to t_0. After an optimal solution is attained, new investments that may have entered the solution are identified and shipments of this new coal are locked in by contract. The procedure is then repeated for all the subsequent time periods ($t = 2, \ldots , T$). If no new investment takes place in a time period, no new contracts are established. While the initial solution assumed that 75 percent of coal shipments were comprised of contract coal, subsequent solutions allow this percentage to vary as a function of contract decay, changing demand and the rate of new mine development.

Objective Function

Within this framework, the objective explicit in the allocation model is to find the shipping and production patterns that minimize the combined transportation and production costs of moving coal resources from the supply regions to the numerous demand points. Defining $X^{gs}_{ijp}(t)$ as the shipment of a physical ton of coal (type g) from region i to region j, using transportation mode s, the objective function for a given year t can be written as:

Minimize

$$Z(t) = \sum_{g=1}^{4} \sum_{i=1}^{33} \sum_{j=1}^{44} \sum_{s=1}^{5} \left(C^{gs}_{ij}(t) + \alpha^{g}_{i}(t) \right) X^{gs}_{ijp}(t) \qquad [1]$$

where $C_{ij}^{gs}(t)$ is the cost per ton of shipping coal of the gth type from region i to region j using transportation mode s; and $\alpha_{i}^{g}(t)$ is the extraction cost per ton of type g coal in region i. In the short run, contract coal shipments cannot be altered; only spot shipments or new production, designated by subscript p, are the decision variables.

The size of the model made the collection of actual rate data on transportation cost difficult; for each of the five transportation alternatives, transportation costs were estimated indirectly. A somewhat different procedure was employed for each transportation mode.

Rail transport of coal is regulated by the Interstate Commerce Commission. Since tariffs are set as maximum levels, the increasing reliance on unit trains (with substantially lower transport costs) has led to rate determination through bargaining between the railroads and coal consumers. Tariff data are therefore inappropriate sources of rail costs.

To circumvent this problem, Zimmerman [1975] has developed a model to simulate this bargaining process. Rates, bounded by the delivered price of an alternative fuel to the utility and the resource cost of delivered coal, are assumed to be a function of the degrees of monopoly power possessed by the railroad and utility. To operationalize the procedure, Zimmerman estimated a linear equation implying that the actual tariff is equal to the resource cost plus some fraction of the difference between the delivered price of the alternative fuel and the delivered resource cost of coal. Based on these calculations, rail costs varied considerably from region to region. For the East and Midwest, costs averaged 0.88 cents per ton-mile and ranged from a high of 2.40 cents to a low of 0.30 cents. For the West, the range was much narrower, 0.80 cents to 0.60 cents per ton-mile. The average was 0.68 cents. Rates calculated in this manner compared favorably with actual rates available from industry sources [1975 Keystone Coal Manual].

Mine-mouth generation of electrical energy currently provides one of the most inexpensive means of effectively transporting coal short distances. To calculate transmission cost, a conversion rate between electric power and coal had to be established. Using perfect conversion of coal to electricity (3,413 Btu. per kilowatt-hour) and a 35 percent average efficiency for current techniques, a conversion factor of 9,750 Btu per kilowatt-hour was estimated [Federal Energy Administration 1975]. Electrical transmission costs also vary inversely with the megawatt capacity of a generating unit and the kilovolt capacity of overhead lines, and directly with distance. Transmission costs were approximated by linking distance and National Power Survey information on costs [Icerman 1974; Federal Power Commission 1971]. Costs per ton-mile averaged 0.14 cents. Transport of electricity rather than coal was limited to a 350-mile radius of a supply center because of rapidly increasing transmission costs [Federal Power Commission 1971].

The cost of barging coal is difficult to ascertain because many shipments are made on a nonregulated contract basis. However, secondary data indicate that barge transportation of coal is less expensive (per ton-mile) than rail transportation. Downstream rates calculated from Project Independence Blueprint [Batelle-Columbus Laboratories 1974] were estimated at an average of 0.29 cents per ton-mile, about half the 0.51 cents upstream rate.

The combined modes use barge and rail transport. These activities are limited to coal originating in the western United States and allow for transport through the Great Lakes system, with all routes interfacing the water portion in Duluth, and the Mississippi-Ohio inland waterways. Since contract rates for Great Lakes activity are generally not announced, transport costs are also estimated from work done by Zimmerman [1975].

Production costs were derived through a complex process to account for differences in mine size, terrain, method of extraction, coal depth, seam thickness and various other factors [LeBlanc 1977]. First, the procedure required the recognition of two groups of extractive activities: existing mines and new development. For existing mines, only variable costs are relevant, since fixed costs, consisting of land, most of the mine structure and much of the equipment used in extraction, cannot be avoided.[1] Investment costs must be considered in the long run; amortized capital costs were included in the extraction costs for new mines.[2]

Second, an extraction function for deep, as well as surface, mining was established for each origin *i*. Surface extraction costs varied with regional differences in stripping ratios and topographically allowable mine sizes. While mine size played an important role in underground extraction, coal seam thickness, rather than stripping ratio, was the other variable considered.[3]

Added to the per unit extraction costs of existing mines, for the recursive portion of the model, is a per unit depletion cost, i.e., an increased cost of extraction over time, representing the increased difficulty of extracting coal through time. Depletion costs were assumed to increase at a rate of 2 percent annually and were assumed to be constant across regions [Tyner, Kalter and Wold 1977].[4]

Constants

To complete the description of the analytical model, one must describe the physical and institutional constraints imposed on the minimization of equation [1].

Supply capacity contraints are given by:

$$\sum_{j=1}^{m} \sum_{s=1}^{6} X_{ijp}^{gs}(t) \le k_i^g(t)$$

$$- \sum_{j=1}^{m} \sum_{s=1}^{6} X_{ijf}^{gs}(t-1)$$

$$(i=1,\ldots,n; g=1,\ldots,4) \qquad\qquad [2]$$

That is, in any time period, the amount of spot market coal of type g shipped from origin i to all destinations must be less than or equal to the production potential, $k_i^g(t)$, less the amount committed for shipment under contract at the start of the period (the subscript f denotes contract shipments). Because X_{ijf}^{gs} for any time period is determined by the net change in contract shipments in the previous period, these variables do not enter the objective function.[5] For the base year, it is assumed that production limits were set at 1973 levels [U.S. Department of the Interior 1974a]. In the recursive solutions, post 1973, regional supplies could be augmented by developing new mines.

The demand constraints require deliveries of coal from all origins to be greater than or equal to the demand at each destination j. However, coal shipped to and demanded at each destination is characterized not only by amount, but also by such characteristics as moisture content, heat value, and fly ash and sulfur content. Because of their importance in recent policy debates, heating value and sulfur content are incorporated into the analysis [U.S. Department of the Interior 1974a].

Since a ton of coal from different supply regions may vary in heat content (i.e., Btu. per ton), one must normalize coal from all origins i $(i = 1, \ldots, n)$ into equivalent units. This is accomplished by defining demand at each destination in normal tons (i.e., a ton averaging 24,000,000 Btu.). Thus, the demand constraints become:

$$\sum_{g=1}^{4} \sum_{i=1}^{n} \sum_{s=1}^{5} a_i \, X_{ijp}^{gs}(t) \ge \Delta_j(t)$$

$$- \sum_{g=1}^{4} \sum_{i=1}^{n} \sum_{s=1}^{5} a_i \, X_{ijf}^{gs}(t-1)$$

$$(j=1,\ldots,m) \qquad\qquad [3]$$

where a_i^* equals Btu. per ton of coal at i divided by Btu. per normal ton and $\Delta_j(t)$ is the demand in region j, measured in normal tons.

T Estimates of utility coal demand for the base year, 1973, were obtained from data compiled by the National Coal Association [1974]. Forecast consumption through 1980 was taken from estimates by Johnson [Gordon

1975]. He uses commitments of planned electrical utilities and estimates coal's share of new capacity as a function of price.[6] By assuming proliferation of nuclear electricity generators, one set of 1990 forecasts shows a reduction in total coal demand from 1980. Since the extent of future nuclear power generation is uncertain, another set of post-1980 forecasts was used—assuming demand growth to continue at pre-1980 rates.

National sulfur dioxide regulations prohibiting emissions of more than 1.2 pounds of SO_2 per million Btu. is the second factor affecting the demand for coal (Public Law 88-206).[7] Coal from each supply region contains, on average, a certain amount (by weight) of sulfur per ton [U.S. Department of the Interior 1975b]. Since the sulfur restriction applies to burning (and not to delivery), one can allow the mixing of high- and low-sulfur coal that when burned, on average, meets the emission standards. Demand is already converted to normal ton equivalents. The adjustments are facilitated by the fact that available data indicate a linear function approximates the relationship between Btu. and sulfur content quite well (see LeBlanc [1977] for detailed calculations).[8] The sulfur content per ton is normalized by using a factor a^*_i defined as Btu. per ton of normal coal divided by Btu. per ton of coal from region i. The constraint becomes:

$$\sum_{g=1}^{4} \sum_{i=1}^{n} \sum_{s=1}^{6} a^*_i \delta_i X_{ijp}^{gs}(t) \le \delta_j \Delta_j(t)$$

$$- \sum_{g=1}^{4} \sum_{i=1}^{n} \sum_{s=1}^{5} a^*_i \delta_i X_{ijf}^{gs}(t-1)$$

$$(j=1, \ldots, m) \qquad\qquad\qquad [4]$$

where δ_i equals the decimal fraction by weight of sulfur per physical ton of coal at i and δ_j equals the allowable sulfur emission (decimal fraction by weight) per normal ton of coal.[9]

Supply from any region is ultimately constrained by total reserves. Separate constraints for underground and surface deposits for each region are specified and take the form:

$$\sum_{s=1}^{5} \sum_{j=1}^{m} X_{ij}^{gs}(t) \le TR_i^g(t)$$

$$(i=1, 2, \ldots, n; g=1, 2, 3, 4) \qquad\qquad [5]$$

where the left-hand side of the inequality equals the total amount of coal extracted in region i in time t and the right-hand side is the associated coal reserve remaining in time period t and for extraction method g.[10]

In addition to reserve constraints, restrictions of five million tons per year or 10 percent of 1973 production, whichever is greater, are placed on underground and surface development in regions east of the Mississippi River.[11] In 28 out of 38 instances, the constraints were 5 million tons per year. Such a constraint characterized development consistent with current regional socioeconomic concerns and land ownership patterns. With such small supply centers, a five-million-ton increase in coal development would probably disrupt the regional economies. Analysis of 1973 eastern mine sizes articulates this point. Average mine sizes for eastern underground and surface deposits are 171.1 and 77.8 thousand tons, respectively [U.S. Department of the Interior 1975a] and a five-million-ton constraint would imply that 29 underground and 64 surface mines may be opened in a given region. For all eastern regions, a total of 522 new underground and 1,152 new surface mines could be opened, representing a 30.9 and 39.6 percent increase per period over the current number.[12]

Supply regions in the West are constrained only by estimated reserves. The heavy reliance on surface mining and increased government and private interests in developing western coal are likely to lead to public and private decisions designed to reduce the short-term production problems currently characteristic of the East. By restricting production only by estimated reserves, solutions to the model also provide an indication of the desirability of western coal development and the importance of not allowing short-term production problems to develop.

The remaining set of constraints deals with possible capacity limitations. These constraints could be of two types: route constraints and rolling or physical stock constraints. Only route constraints are dealt with in this analysis, since physical stocks in the initial time period were adequate and could be expanded to meet needs.

A mode constraint is not specific to any origin or destination. By this specification many different routes (i, j) may be restricted by one constraint. The constraints are:

$$\sum_{sek} \sum_{g=1}^{4} \sum_{i \epsilon z} \sum_{j \epsilon z} X_{ijp}^{gs}(t) \leq M_z^k(t)$$

$$- \sum_{sek} \sum_{g=1}^{4} \sum_{i \epsilon z} \sum_{j \epsilon z} X_{ijf}^{gs}(t-1)$$

$$(k = 1, \ldots, 4; z = 1, \ldots, Z) \qquad\qquad [6]$$

For transport alternatives s, shipping deep or surface mined coal g, using mode k between origin i and destination j, and passing through the area where this mode restriction applied (i.e., i and j), the amount of coal shipped must be less than or equal to the capacity of the constraint,

$M_z^k(t)$, measured in physical tons, less the amount used by shipments locked in by contract.

Barge constraint levels are a function of 1973 lock capacities on the Mississippi and Ohio River systems [Bhutani et al. 1975]. Great Lakes shipments are restricted by the availability of coal transport ships [1975 Keystone Coal Manual]. Mine-mouth generation constraints are based on the capacity of mine-mouth facilities [Federal Power Commission 1971]. It was assumed that new mine-mouth capacity could be added for post-1973 solutions.

Empirical Results

To begin the analysis, a solution to the linear programming model described above was obtained for production and demand conditions prevailing in 1973. Since this solution was to provide the starting point for the multi-period analysis, no initial contract or sulfur constraints were assumed. The proportions of spot and contract coal were allowed to vary from the historical levels (in subsequent multi-period solutions) by permitting contract decay and the opportunity for new investment. It was assumed that 20-year contracts on new development had been signed at a constant rate prior to 1973; thus, contracts in effect as a result of the 1973 solution were decaying at a constant rate (5 percent per year).[13] Demand not bound by a contract can move on the spot market or can elicit new development by negotiating a new contract.

Two alternatives, one with and one without sulfur regulations, are used for the 1975 through 1990 models. In addition, each of these alternatives was run assuming a high and a low demand level for the years 1985 and 1990. It was assumed contracts signed prior to or in 1973 are not subject to the sulfur regulations. Shipments from new development and the spot market are forced to conform to emission standards. Consequently, for the 1990 demand alternatives, only 5 out of 44 regions meet the actual per ton air quality standards; but with all original contracts due to expire by 1993, compliance is almost complete in all regions.

Base Year Solution, 1973

A careful examination of the base year solution is important because a comparison of it with actual production and shipment patterns helps to validate the model. A general overview can be obtained from Tables 1 and 2 (LeBlanc [1977] contains a full report). In 1973, demand and supply are concentrated east of the Mississippi River (82.2 and 83.6 percent, respectively). Established coal fields in the northeast and southeast meet eastern demand. Major interregional shipments move from east to west, with eastern regions shipping 34.3 percent of their production to industrial states from Ohio westward to Wisconsin. Virtually no coal moves across the Mississippi River in either direction.

Not surprisingly, rail is the predominant transportation mode. Only 8.6 percent of the coal produced was shipped by other modes. Mine-mouth generation accounted for 6.8 percent, while 1.8 percent was shipped by barge.

When compared with actual 1973 production and shipment patterns, the performance of the model, on an aggregated regional basis, was quite good. Few discrepancies of any consequence emerged. For example, the model predicts that 106 million tons of coal will be shipped from the Northeast Region to the industrial areas consisting of the East North Central and Middle Atlantic regions. However, actual data indicate the flows to be 95 million tons, nearly a 10 percent differential. The accuracy of aggregated western flows is similar. The model predicts 27 million tons emanating from the Northern Great Plains to the North Central Region. Actual flows are approximately 23 million tons, representing a discrepancy of 15 percent. The major reason for such discrepancies appears to be the failure to account for contractual patterns impacting 1973 shipments. Lack of data on these previously negotiated contracts forced use of the procedure outlined above to determine the contract-spot pattern.

No Sulfur Restrictions

In the situation where no sulfur constraints are imposed, demands in 1975 and 1980 were estimated to rise 22.2 and 52.7 percent above 1973 levels. To meet this increased demand, production rose in most supply regions. With the exception of large shipments (approximately 11 million tons in both years) between the North Central and South Atlantic Regions, distribution patterns are consistent with 1973 patterns. Thus, the eastern regions retain their comparative cost advantages, primarily due to their proximity to large markets.

The production of most existing mines supplying coal in the spot market rose, but 48.8 percent of the increased supply in 1975 was estimated to come from new development (Table 2).[14] By 1980, over 200 million tons per year of coal are produced from new developments, accounting for 88.3 percent of the increased demand since 1973. Because of relatively high Btu. content, 70.1 percent of 1975 new development occurs east of the Mississippi.[15] This proportion drops to 60 percent by 1980. Lower extraction costs result in new development coming largely from surface mines (Table 3).

Beyond 1980, the analysis becomes more complex. Uncertainties with respect to the future of nuclear power led to the specification of a high

446

TABLE 1
AGGREGATED COAL SHIPMENTS
(1000 Physical Tons)

Supply*	Demand*									TOTAL
	New England (A)	Middle Atlantic (B)	East North Central (C)	West North Central (D)	South Atlantic (E)	East South Central (F)	West South Central (G)	Mountain (H)	Washington (I)	(I)
North-East (1)										
1973	204	26,904	78,662		22,325					128,095
1975 WOS**	195	28,011	80,614		34,213					143,033
1975 WS***	835	25,726	59,711		29,967	980				117,219
Southeast (2)										
1973	933	18,692	9,764		59,863	40,306				129,558
1975 WOS	3,886	20,905	8,003		84,582	27,456				144,832
1975 WS	633	22,707	6,591		92,717	41,672				164,320
Northcentral (3)										
1973			68,968	13,229		21,365				103,562
1975 WOS			79,308	19,851	14,990	38,798				152,947
1975 WS			51,316	9,023		14,422				74,761
Central (4)										
1973			1,320	4,423		2,618				8,361
1975 WOS			891	10,312		1,993	1,163			14,359
1975 WS			891	2,994		4,543	65			8,493
Texas (5)										
1973				1,335		105	5,505			6,945
1975 WOS				901		71	7,972			8,944
1975 WS				1,521		71	5,352			6,944

Eastern Northern Great Plains (6)									
1973		3,157	4,062						7,219
1975 WOS		2,131	7,178						9,309
1975 WS		2,131	6,769						8,900
Western Northern Great Plains (7)									
1973	3,448		53,152	18,715	1,948	1,298		1,357	284	20,356
1975 WOS				18,089				1,981	284	20,354
1975 WS				36,347				1,981	284	98,458
Mountain (8)									
1973				1,195	8,567	3,789		16,138		16,138
1975 WOS								18,919		18,919
1975 WS								18,963		35,357
Southwest (9)									
1973							2,843	8,593		8,593
1975 WOS				388			341	11,306		12,035
1975 WS				131			857	11,361		12,349
Washington (10)									
1973									3,255	3,255
1975 WOS									3,254	3,254
1975 WS									3,254	3,254
TOTAL									
1973	1,137	45,596	161,871	41,764	82,188	64,394	5,505	26,088	3,539	432,082
1975 WOS	4,081	48,916	170,947	56,719	133,785	68,318	9,476	32,206	3,538	527,986
1975 WS	4,916	48,433	173,792	57,980	133,199	66,776	9,117	32,305	3,538	530,055

*These regions correspond to those in Figure 1.
**WOS refers to solution without sulfur constraints.
***WS refers to solution with sulfur constraints.

TABLE 2
REGIONAL PRODUCTION, NEW DEVELOPMENT AND CONSUMPTION
(1000 Physical Tons)

Alternatives*

Region	1973	1975		1980		1985 Low Demand		1985 High Demand		1990 Low Demand		1990 High Demand	
	WOS	WOS	WS	WOS	WS	WOS	WS	WOS	WS	WOS	WS	WOS	WS
Total Production (tons/year)	432,082	527,986	530,056	659,574	657,290	656,249	650,475	800,164	801,069	652,281	642,967	999,749	996,241
Eastern** (as % of total)	83.6	83.5	67.2	76.9	66.6	74.9	62.6	71.8	54.7	73.1	59.0	67.3	42.2
Western*** (as % of total)	16.4	16.5	32.8	23.1	33.4	25.1	37.4	28.2	45.3	26.9	41.0	32.7	57.8
Total Cumulative New Development (tons/year)	—	46,819	126,235	200,844	236,709	291,917	319,127	362,007	453,057	377,793	399,789	559,400	730,970
Eastern (as % of total)	—	70.1	19.0	60.2	38.3	67.1	44.1	57.4	34.7	70.4	47.7	54.1	29.5
Western (as % of total)	—	29.9	81.0	39.8	61.7	32.9	55.9	42.6	65.3	29.6	52.3	45.9	70.5
Demand (tons/year)	432,082	527,986	530,056	659,574	657,290	656,249	650,475	800,164	801,069	652,281	642,967	999,749	996,241
Eastern† (as % of total)	82.2	80.7	80.6	76.7	76.8	76.6	76.7	75.2	75.5	76.7	76.6	73.2	73.3
Western†† (as % of total)	17.8	19.3	19.4	23.3	23.2	23.4	23.3	24.8	24.5	23.3	23.4	26.8	26.7

* The alternatives refer to solutions for the corresponding years under condition of high or low demand and with (WS) or without (WOS) sulfur constraints on sulfur emissions.
** Includes supply regions 1 through 3 from Figure 1.
*** Includes supply regions 4 through 10 from Figure 1.
† Includes demand regions A, B, C, E, and F from Figure 1.
†† Includes demand regions D, G, H, and I from Figure 1.

TABLE 3
CUMULATIVE COAL DEVELOPMENT
(1000 Physical Tons/Year)

Alternatives**

Supply*		1975		1980		1985 Low Demand		1985 High Demand		1990 Low Demand		1990 High Demand	
		WOS	WS	WOS	WS	WOS	WS	WOS	WS	WOS	WS	WOS	WS
North-East	(U)†	2,000	2,000	12,000	3,950	12,000	8,950	23,986	11,461	12,000	13,950	38,985	16,461
	(S)††	8,000	4,001	43,000	39,003	78,001	61,049	78,000	62,451	107,999	82,109	112,999	83,610
Southeast	(U)	2,000	10,001	7,000	15,001	12,000	23,034	12,000	30,443	17,001	28,034	22,000	45,443
	(S)	8,000	8,000	21,042	20,542	31,043	30,541	31,044	30,542	41,041	40,543	41,043	41,541
Northcentral	(U)	0	0	0	0	0	0	0	0	0	0	0	0
	(S)	12,801	0	37,800	12,246	62,800	17,246	62,800	22,156	87,800	25,969	87,800	28,656
Central	(U)	0	0	0	0	0	0	0	0	0	0	0	0
	(S)	6,000	2,000	21,000	12,196	36,000	17,196	36,000	17,197	51,000	22,765	51,001	22,198
Texas	(U)	2,000	0	16,685	0	16,685	0	45,543	0	16,685	0	78,590	0
	(S)	0	0	0	491	0	491	0	491	0	491	0	491
Eastern Northern Great Plains	(U)	0	0	0	0	0	0	0	0	0	0	0	0
	(S)	2,091	1,682	5,084	1,682	5,084	1,875	8,579	5,442	5,084	4,686	22,409	19,175
Western Northern Great Plains	(U)	0	0	0	0	0	0	0	0	0	0	0	0
	(S)	0	0	0	0	0	0	0	0	0	0	0	0
Mountain	(U)	0	0	0	0	0	0	0	0	0	0	0	0
	(S)	0	78,104	19,577	94,566	19,577	118,321	29,084	209,975	19,658	137,412	45,269	374,494
Southwest	(U)	0	0	0	0	0	0	0	0	0	0	0	0
	(S)	3,927	20,363	14,630	31,297	15,701	32,447	24,950	41,643	16,499	32,938	38,783	55,341
Washington	(U)	0	0	0	0	0	0	0	0	0	0	0	0
	(S)	0	84	3,026	5,735	3,026	7,977	10,021	21,256	3,026	10,892	20,521	43,560
TOTAL		46,819	126,235	200,844	236,709	291,917	319,127	362,007	453,057	377,793	399,789	559,400	730,970

* Supply regions correspond to those in Figure 1.
** The alternatives refer to solutions for the corresponding years under condition of high or low demand and with (WS) or without (WOS) sulfur constraints on sulfur emissions.
†(U) refers to underground mines.
††(S) refers to surface mines.

and a low level of coal demand in 1985 and 1990. The low demand alternative assumes the nuclear generation of electricity becomes increasingly important; coal demand falls by 0.5 percent between 1980 and 1985. Demand in 1990 is below that in 1985 and 1.1 percent below 1980 levels.

Despite the small reduction in demand, production and distribution patterns change little from those established in the 1973-1980 period. The fact that the proportion of production originating in the East is down slightly is accounted for almost entirely by the slight increase in demand in the West.

One of the most interesting aspects of these solutions is the continued growth in new mine development. Although demand is about the same as in 1980, cumulative new development increased from 201 million tons per year in 1980 to 292 million tons per year in 1985 and to 378 million tons per year by 1990. Approximately 70 percent of the new development is estimated to come from east of the Mississippi River. Depletion of old mines and the increased cost of producing remaining reserves from older mines makes new development attractive. Thus, from a public policy perspective, the question of where to encourage new coal development will arise in the near future even if demand stabilizes in the face of increased adoption of nuclear power generation.

The other demand alternative examined is based on the assumption that nuclear power generation is not adopted at a rapid rate. Demand is assumed to rise steadily after 1980, increasing 21.3 and 51.6 percent by 1985 and 1990, respectively. In the face of these increased demands, major distribution patterns still follow the ones established in 1975. However, one finally sees the first signs of erosion in the dominant position of eastern supply regions. In 1985, eastern regions still supply 74.9 percent of the coal, while by 1990, this figure drops to 67.3 percent. Under assumptions of high demand, only 54.1 percent of new development by 1990 originates east of the Mississippi. Added production costs of recovering remaining reserves in eastern mines begin to shrink the East's cost advantage. Despite longer distances, lower per unit production and transportation costs begin to make western coal more competitive. Increases in production west of the Mississippi are concentrated in the Western Northern Great Plains Region (45 million tons in 1990 under high demand assumptions); but shipments still reach no further east than the North Central Region.

Throughout the analysis when no sulfur constraints are imposed, rail remains the predominant transportation mode. This result obtains, in part, from the fact that no constraints are placed on the movement of coal along existing rail lines. Ex post examination of the solutions suggests that rail capacity was not a limiting factor, a conclusion consistent with a recent study estimating that current track utilization is approximately 30 percent of full capacity [Bhutani et al. 1975].

In post-1975 solutions, however, there is an upward trend in barge traffic. By 1985 and 1990, under high demand assumptions, increases of 14 and 26 million tons over the 1980 level are established. In general, the flow of coal via barge follows patterns which are established in the 1973 and 1975 model solutions. The increased amounts shipped indicate intensification of established patterns rather than initiation of new supply and demand linkages. Demand destinations of barge shipments focus on Kentucky, Pennsylvania, Ohio and Mississippi.

Solutions assuming no sulfur constraints show no coal being transshipped, but indicate increased growth of mine-mouth generation. Use of this latter mode rises from 30 million tons in 1973 to a high of 59 million tons in 1990, under conditions of high demand. Greatest increases in mine-mouth generation occur in Maryland, Illinois, and Arizona due to proximity of these demand nodes to supply regions.

National Sulfur Restrictions

Perhaps the most effective way of deriving the policy implications of imposing restrictions on the amount of sulfur emissions is to compare the model's results with those assuming no sulfur restrictions. One comparison of interest is certainly the added production and transportation costs associated with imposition of emission restrictions of 1.2 pounds of SO_2 per million Btu. To meet 1975 demand levels, the combined cost of production and transportation would be 11.0 percent higher when sulfur restrictions are imposed. The costs are only 8.3 percent higher in 1980. For 1985 (1990) the costs are 13.0 percent (16.3 percent) and 14.1 percent (18.2 percent) higher, assuming low and high demand, respectively, as a result of the added restrictions.[16]

The other significant implications (and the reasons for higher costs) of implementing sulfur restrictions begin with the production and new development patterns in 1975 (Table 2). As a result of the restrictions, only 67.2 percent of the coal production originates in eastern supply regions, compared with 83.5 percent without the restrictions. Western regions dominated new development from the outset. Only 19.0 percent of the new development occurs east of the Mississippi River in 1975.

In comparison with the 1973 and 1975 solutions without sulfur constraints, one begins to see a significant reversal in the east to west shipping patterns (Table 1). The Northern Great Plains and the western regions begin shipments of 74.3 million tons to areas east of the Mississippi, including Illinois, Indiana, Michigan and Ohio. Closer analysis reveals that many of the new west to east linkages (approximately 81 percent or 60 million tons) originate in southern Montana and northern Wyoming. Large differences also take place in tonnage shipped from other supply regions, except Texas and the Eastern Northern Great Plains.[17] The lower sulfur content and per mile transport costs of these western coal reserves more than compensate for the lower Btu. content and longer distances.[18]

These reversed trends (when compared to the situations with no sulfur constraints) are accentuated in future years. Increased demand in the high demand alternative for 1985 and 1990 leads to further development of western reserves, until by 1990 57.8 percent of all coal production originates in the West. Even at lower levels of demand, western supply regions produce 41.0 percent of the coal by 1990. While demand has not increased as substantially in this situation, much of the high-sulfur coal locked in by contract in 1973 (i.e., not subject to restriction) is replaced with western coal as the contracts decay. The patterns of development and shipment remain approximately the same.

The most heavily developed area when sulfur constraints are imposed and high demand conditions are assumed is the Northern Great Plains. Nearly 400 million tons per year are extracted from this area by 1990. If it is assumed that the average mine produces between 5 and 10 million tons per year, this translates into from 40 to 80 new mines developed in this area by 1990. Under these assumptions, the Southwest Region also shows important levels of development, with 44 million tons per year extracted in 1990.

Analysis of extraction methods still reveals a heavy preference for surface techniques (see Table 3). Eastern coal regions show the only evidence of underground mining while western coal regions are devoid of any new underground development. The low extraction cost of surface techniques in conjunction with ample reserves eliminates any need for the development of underground mining in western areas.

In the alternatives where sulfur restrictions are incorporated, rail is also the dominant transportation mode. Ex post examination of the solutions again revealed that rail capacities were not exhausted. Under the high demand assumption, almost the entire production from the Western Northern Great Plains (210 and 375 million tons in 1985 and 1990) is shipped to the East or Southeast. With this concentrated movement, serious consideration of future capacity expansion along these routes is likely to be needed if sulfur constraints are enforced.

The use of barges is quite consistent with the patterns formulated in the alternatives where sulfur constraints are not imposed. However, beginning in 1975, with the implementation of sulfur restrictions, there is an immediate use of combined modes, with nearly 25 million tons being transshipped. Great Lakes transshipment (17 million tons) to

Ohio accounts for nearly 70 percent of the total.

Destinations of transshipped coal in later years are centered on the North Central Region, though many areas received such shipments. Shipments extend as far east as New York, and as far south as Mississippi and Louisiana. The origin of nearly all transshipments is the Northern Great Plains.

Although mine-mouth generation of electricity is a fairly attractive alternative with no sulfur restrictions, the absolute level of mine-mouth generation climbs in the presence of sulfur regulations to a peak of only 39 million tons per year in 1980 and falls to a low of 26 million tons in the 1990 low demand situation and 33 million tons in the 1990 high demand situation. The lack of proximity between low-sulfur coal deposits and major consumption centers appears to be the major reason for this result.

Supplementing the information given above, dual variable analysis provides interesting insights into the changes in the objective function to marginal changes in constraint levels. (See LeBlanc [1977] and Hadley [1962] for an exact interpretation of the dual programming problem.) If one assumes each factor receives compensation equal to its marginal contribution, the interpretation of the dual variables associated with the model's demand constraints is that of delivered coal prices at destination points.

Regional shadow prices for solutions containing sulfur constraints indicate New England, Middle Atlantic, South Atlantic, and East South Central regions are the areas most severely affected by sulfur regulation. The average price increase for these areas from 1973 to 1975 was $7.65 per ton. Comparisons with the solution not constrained by sulfur restrictions show the average price increase for these areas to be $.20. Analysis through time shows that the greatest differences between delivered prices for solutions with and without sulfur constraints occur in 1975. For example, the average difference in 1975 was $7.87 while a 1985 solution shows a difference of only $5.11 per ton.

In addition to information concerning delivered prices, preliminary insights into the impact of stack-gas scrubbers can be examined by inspecting the shadow prices associated with the sulfur constraints. That is, the effect of installing scrubbers could be approximated by allowing higher sulfur emission levels (relaxing sulfur constraints). Regional variation in sulfur constraint shadow values followed a pattern similar to one found for demand constraints. They are generally higher in the East and Northeast than the West (i.e., a pattern generated in large part by the location of low-sulfur coal supplies). Thus it appears that the use of gas-stack scrubbers should be considered most seriously in the East. Moreover, comparison of sulfur constraint shadow price magnitudes with the amortized cost of scrubbers between 1980 and 1985 is a viable alternative to the use of low-sulfur coal for meeting emission standards.[19]

Summary and Policy Implications

The purpose of this paper has been to describe a programming model of the coal industry and study the implications of changing demand and the possible imposition of limits on sulfur emissions from fossil fuel electrical generating plants on coal production, new mine development and coal distribution patterns. The model incorporates, explicitly, alternative transportation modes and generates production, new development and distribution patterns minimizing the combined production and transportation cost of meeting coal demand over the 1973-1990 time horizon. The existence of a long-term contract market, as well as a spot market, for coal is also considered. While a limited number of alternatives are examined, the results have implications for future energy costs, federal leasing policy, land use, and transportation policy.

One important implication is that even with rising demand and the imposition of restrictions on sulfur emissions, demand for low-sulfur coal can be met through a combination of increased production of existing mines and development of new reserves. Total costs of production and transportation increase significantly in the face of rising demand, the imposition of sulfur restrictions, and the utilization of lower Btu. coal. Implementation of sulfur regulations reverses a significant proportion of the east to west flow of coal which has historically dominated the U.S. market. Even without the imposition of sulfur constraints, the west to east distribution pattern would begin to emerge in the future if demand for coal increases substantially.

Examination of the shadow prices (dual variables) associated with the demand constraints also provides some indication of expected changes in coal prices. Because of the inability of eastern supply regions to adequately supply low-sulfur coal, large regional price differentials among the supply regions arise. The long distances over which western coal is shipped and the rising demand for low-sulfur coal would precipitate coal price increases in the East significantly higher than those in the West. In the absence of sulfur restrictions, even in the face of high future demand, price differentials would remain much lower. Undoubtedly, much of these price increases would be passed through to eastern consumers. The trade-off between environmental quality and energy costs must be faced squarely in the near future.

Regardless of the assumptions made concerning future levels of demand or the imposition of sulfur constraints, a significant amount of the demand by 1990 will be met from new mine development. In the event that demand rises slowly and sulfur restrictions are not imposed, much of the new development would occur in the East. Under all other conditions, high future demand and/or restricted sulfur emissions, development is concentrated in the West. While not considered explicitly in this analysis, the locations of these potential new developments have important implications for leasing public lands in the West. Strong indications are given that a western leasing strategy should focus on the Northern Great Plains Region, particularly southeast Montana and northeast Wyoming. Other areas of importance, though to a lesser degree, are southwest Wyoming, northwest Colorado, and the Four Corners Region of New Mexico and Arizona.

One intent of leasing procedure is to allow for efficient coal utilization, while extracting as much economic rent from the lessee as possible. Although this may sound attractive from a national perspective, the impacts on the economies of local areas and land use patterns in the West complicate the issue and may conflict with national environmental objectives. The recent moratorium on the leasing of federal coal lands is indicative of increased concern.

The strict enforcement of national ambient air standards for sulfur dioxide emissions may cause extremely large flows of investment into the Northern Great Plains. Social impacts and the concomitant problems for municipal and state governments are likely to arise from the creation of "boom town" conditions. Severance taxes imposed by western states in an attempt to slow development or add to state revenues could increase costs and conflict with national policy objectives. These local fiscal implications and local development impacts were beyond the scope of this study, but without a detailed analysis of these problems, the total impact on society from future coal development cannot be ascertained.

The imposition of sulfur regulations has implications for the environment in addition to the obvious impact on air quality. It is possible, for example, that increased use of low-sulfur, surface mined western coal combined with that region's scarcity of water may have repercussions on the future productive potential and use of nearby land resulting from changes in regional hydrology.[20] A thorough analysis of individual sites would be required to accurately determine the potential damage resulting from surface mining.

Specific alternatives to limiting sulfur emissions to maintain air quality standards have not been examined in this study. For example, one could certainly extend the model to include the option of taxing sulfur emissions to encourage the use of low-sulfur coal.[21] From a technological perspective, one could also examine in much greater detail the feasibility of stack-gas scrubbers which would permit the more extensive use of high-sulfur coal.

From the standpoint of national transportation policy, the study's results are encouraging. Rail is the dominant mode of transportation under all specifications of the model. The rail capacity was adequate in most instances. However, rail facilities in the Northern Great Plains were approaching capacity limitations under the assumption of both sulfur restrictions and high demand late in the time period analyzed. Although there is an indication that other modes (such as mine-mouth generation and barge) will become more important in the future, there was excess capacity in these modes along major distribution routes. Thus, it is unlikely that other modes will threaten rail's dominant position. A transportation policy designed to encourage the maintenance and expansion of rail capacity along major distribution routes is likely to contribute to efficient use of our energy resources.

In conclusion, the results provided by this study generate information to allow decision makers greater insight into coal development policy alternatives. The analysis is focused on the dynamic effects of change in sulfur constraints and demand intensity. As with any large programming model, the validity of the results is contingent on the validity of the data and model specification. Of particular importance in this study are the constraints on eastern coal development and the simplifying assumption regarding increased future depletion costs. Although these assumptions were based on the best available data and familiarity with current production and cost patterns, the policy implications of this analysis may not change in the directions indicated above if these conditions fail to obtain in the future. In addition, a more in-depth analysis of sulfur removal devices, reclamation costs, experimental transportation modes and the impact of state and local policies is needed to completely understand the role of coal resources in meeting future energy needs.

Notes

1. For the base period, coal extraction costs are based on estimates from Project Independence [U.S. Department of Interior 1974a]. These estimates are shutdown prices of the production activities observed in the depressed 1972 coal market, and serve as useful proxies for actual production costs for existing mines.

2. Cost data for new underground and surface mines were obtained from model mine estimates developed for the Project Independence Blue-

print [TRW Systems Group 1974]. In the case of new mines, extraction costs αg_i are increased to reflect a 15 percent return on investment. Capital costs are a function of mine size and method of extraction. Holding other things constant, for example, average yearly capital costs for a mine producing one million tons per year is $1.96 per ton higher than a mine producing three million tons per year. Average yearly capital costs for new surface mines also vary directly with mine size. For example, annual capital costs of mines producing one million tons per year average $1.99 per ton more than surface mines producing five million tons per year.

3. Extraction costs vary directly with stripping ratios (cubic tons of overburden per ton of coal); for example, extraction costs for mines with stripping ratios of 15:1 average $2.17 per ton more than mines with stripping ratios of 5:1. As seam thickness falls from 6 to 4 feet, extraction costs of underground mines increase by an average of $1.38 per ton.

4. It is recognized that these depletion costs are probably an increasing function of the age of an existing mine. Therefore, to the extent that the average age of existing mines in the East is higher than the age of existing mines in the West, this assumption of constant depletion costs may underestimate slightly the rate of increase in costs of eastern low-sulfur coal production and make western sources of coal somewhat more attractive. Data needed to refine these cost estimates were not available.

5. Shipments under new contracts in year t can, for that year only, be treated in the same way as spot shipments (i.e., their levels are chosen in that year). Before the model is rerun for $t + 1$, new contract coal in year t is allocated to the right-hand side of equation 2 in a side calculation. This eliminates the need for additional variables in the model and is made possible by the assumption that only new developments can move into the contract market.

6. Demand figures used for the model include all coal uses except export and coking coal. The increase in demand for uses other than utility demand was assumed negligible over the time horizon of the study.

7. In addition to this national standard, the amendments to the Clean Air Act require that states submit air pollution control plans called State Implementation Plans (SIPs) which may be more restrictive than the minimum standards adopted by the Environmental Protection Agency. State standards were not, however, dealt with in this analysis.

8. Calculations also show this to be a unique linear relationship for each supply region. Data are obtained from U.S. Department of the Interior [1975*b*].

9. The national ambient air standard currently specifies δ_j to be .0075.

10. Reserve constraints for each supply region were developed from U.S. Department of the Interior known reserve estimates obtained from their

Data Processing Division in Denver [1975*b*]. These constraints proved to be of minimal importance, impacting only three of the possible new development activities.

11. The recursive logic of the model involved solving the allocation problem for 1973, 1975 and then every five years through 1990. Solutions for a given time period were assumed to be constant over the five-year period. For example, development constraints restrict new mine openings to be no greater than five million tons over five years for each type of mine.

12. Sensitivity analysis on new development constraints indicates a broad range over which activities remain in the optimal solution with changes in the development restrictions. For example, two 1985 solutions, one with sulfur constraints and the other without sulfur constraints, allow increases in new development of 23.3 and 31.9 percent, respectively, before the solution would change. Decreases of 22.1 and 8.0 percent, respectively, would be allowed. Investigating the number of new development constraints which are binding shows that the importance of the development constraints is minor in many supply regions. The 1985 cases show 19 of 36 constraints binding for the solution without sulfur constraint while the solution with sulfur constraints has only 11 of 36 new development constraints binding.

13. Sensitivity of results to assuming an initial 75-25 contract-spot split can be ascertained by investigating the range over which supply constraints can vary while activities would remain in the optimal solution. Such ranges indicate a broad area of relative stability. For example, 1985 solutions show allowable decreases of 23.4 and 11.8 percent for the with and without sulfur constraint solutions, respectively. Increases indicated are 14.4 and 40.9 percent, respectively.

14. Spot coal supply constraints for 1975 and subsequent years are augmented by 49 million tons to allow for the precipitous increase in spot allocations due to recent energy shortages [National Coal Association 1976]. All further increases in supply through time, however, result from new development.

15. The Btu. content of an average ton of coal from supply regions east of the Mississippi ranges from 12,171 to 14,115 Btu. per pound; in western coal, the range is from 10,556 to 13,345 Btu. per pound [U.S. Department of the Interior 1975*a*].

16. Comparing the 1990 low and high demand solutions without sulfur constraints, the high demand solution implies production and transportation costs 37.5 percent higher. For the same comparison with sulfur constraints, the cost differential was 39.0 percent. Thus, from a policy perspective, efforts to reduce reliance on fossil fuel generated electricity

could possibly pay large dividends in terms of cost. Such a curtailment can only be assessed on the basis of the extent to which it resulted from a reduction in demand and/or in the substitution of a less expensive source of electrical energy. This in no way implies that alternative energy sources are free of socioeconomic and environmental problems.

17. Compared with other models [Libbin and Boehlje 1976; and Nagarvala, Ferrel and Olver 1976] this analysis shows shipments of western coal to demand areas further east. These differences are due primarily to the spot-contract dichotomy considered in this model, and to a lesser extent to the inclusion of development constraints on eastern supply areas.

18. The sulfur content of estimated coal reserves averages 2.7 percent and ranges from 0.8 percent to 5.2 percent, by weight, in the East. The western average is 0.7 percent and it ranges from 0.3 to 1.4 percent [U.S. Department of the Interior 1975a].

19. Current estimates of limestone and lime scrubbing systems, presently the most feasible means of sulfur removal, place the annualized cost of investment and operation between $2.64 to $2.88 per ton of coal burned for 85 percent sulfur removal [Abelson 1974]. The 1985 high demand solution indicates the average eastern and western shadow prices for sulfur constraints to be $3.66 per ton and $1.91 per ton, respectively, thus limiting reasonable use of stack scrubbers to eastern areas.

20. Production cost estimates for new development included ten cents per ton for reclamation. A more accurate approach would include regional differences in reclamation costs following the work of Schlottman and Spore [1976]. They argue that the highest reclamation costs occur in eastern fields and the lowest in the West. To the extent that our estimates understate the actual reclamation cost it would appear that our model has a bias towards the production of eastern coal. In addition to reclamation costs, further work is needed in the area of regional physical restrictions on reclamation as well as state environmental and tax policies relating to coal development.

21. The work of Schlottman and Abrams [1977] is one example of a programming model application to analyze sulfur taxes.

References

Abelson, Philip, ed. 1974. *Energy: Use, Conservation, and Supply, A Science Compendium.* Washington, D.C.: American Association for the Advancement of Sciences.

Battelle-Columbus Laboratories. 1974. "Modal Transportation Costs for Coal in the United States." Columbus.

Bhutani, J., et al. 1975. *An Analysis of Constraints on Increased Coal Production.* McLean, Virginia: Mitre Corporation.

Federal Energy Administration. 1975. "Monthly Energy Review." Washington, D.C.: FEA, May.

Federal Power Commission. 1971. *The 1970 National Power Survey: Part I.* Washington, D.C.: Government Printing Office.

———. 1974. *FPC News* (Nov.).

———. 1975. *Power Generation: Conservation, Health and Fuel Supply.* Washington, D.C.

Gordon, Richard. 1975. *U.S. Coal and the Electric Power Industry.* Baltimore: Johns Hopkins University Press for Resources for the Future.

Hadley, G. 1962. *Linear Programming.* Reading, Mass.: Addison-Wesley Publishing Company, Inc.

Henderson, James. 1958. *The Efficiency of the Coal Industry.* Cambridge: Harvard University Press.

Icerman, Larry. 1974. "Relative Costs of Energy Transmission for Hydrogen, Natural Gas, and Electricity." *Energy Sources* 1:435-46.

LeBlanc, Michael. 1977. "A Transportation Model for the United States Coal Industry." Masters thesis, Cornell University.

Libbin, James, and Boehlje, Michael. 1976. "Interregional Structure of the United States Coal Economy." Paper presented at Agricultural Economic Association's Annual Meetings, Pennsylvania State University, State College, Pennsylvania, August.

Nagarvala, P.; Ferrel, George; and Olver, Leon. 1976. *Regional Energy System for the Planning and Optimization of National Scenarios, Final Report: Clean Coal Energy, Source to Use Economics Project.* San Francisco: Bechtel Corporation.

National Coal Association. 1976. *Coal Traffic Annual.* 1976 ed. Washington, D.C.

———. 1974. *Steam-Electric Plant Factors.* 1974 ed. Washington, D.C.

1975 Keystone Coal Manual. 1975. New York: McGraw-Hill.

Public Law 88-206, Clean Air Act of 1963.

Schlottman, Alan, and Spore, Robert. 1976. "Economic Impacts of Surface Mine Reclamation." *Land Economics* 52 (Aug.):265-77.

Schlottman, Alan, and Abrams, Lawrence. 1977. "Sulfur Emission Taxes and Coal Resources." *Review of Economics and Statistics* 59 (Feb.): 50-55.

TRW Systems Group. 1974. *Coal Program Support Report.* Redondo Beach, California, June 28.

Tyner, Wallace E.; Kalter, Robert J.; and Wold, John. 1977. *Western Coal: Promise or Problem?* A. E. Research 77-13, Department of Agricultural Economics, Cornell University, Ithaca, N.Y., August.

U.S. Department of the Interior, Interagency Coal Task Force, Project Independence Blueprint. 1974*a*. *Final Task Force Report*. Washington, D.C.: U.S. Department of the Interior.

———. 1974*b*. "Description of Quantitative Data and Assumptions Used in Their Derivation." Washington, D.C.

U.S. Department of the Interior, Bureau of Mines. 1975*a*. *Mineral Industry Surveys: Coal—Bituminous and Lignite in 1973*. Washington, D.C.: Government Printing Office.

———. 1975*b*. Data Processing Division. "Computer Tape of Estimated Coal Reserves in the United States." Denver: August 8.

Zimmerman, Martin. 1975. "Long Run Mineral Supply: The Case of Coal in the United States." Ph.D. dissertation, Massachusetts Institute of Technology.

Glossary

Glossary

Glossary of Surface Mining and Reclamation Technology

Bituminous Coal Research, Inc.

Abandoned Mine: A mining operation where coal is no longer being produced and it is the intent of the operator not to continue production from the mine.

Abatement (mine drainage usage): The reduction of the pollution effects of mine drainage.

Abutment: The point of contact between the ends of an embankment and the natural ground material.

Access Road: Any haul road or other road constructed, improved, maintained, or used by the operator which ends at the pit or bench, and is located within the area of land affected.

Acid Producing Materials (acid forming): Rock strata containing significant pyrite which if exposed by coal mining will, when acted upon by air and water, cause acids to form.

Acid Mine Drainage: Any acid water draining or flowing on, or having drained or flowed off, any area of land affected by mining.

Acid Soil: Generally, a soil that is acid throughout most or all of the parts of it that plant roots occupy. Commonly applied to only the surface-plowed layer or to some other specific layer or horizon of a soil. Practically, this means a soil more acid than pH 6.6; precisely, a soil with a pH value less than 7.0. A soil having a preponderance of hydrogen over hydroxyl ions in the soil solution.

Acid Spoil (see also **Spoil** and **Toxic Spoil**): Usually spoil material containing sufficient pyrite so that weathering produces acid water and the pH of the soil determined by standard methods of soil analysis is between 4.0 and 6.9.

Acre-Foot: A term used in measuring the volume of water, equal to the quantity required to cover 1 acre 1 ft in depth, or 43,560 cu ft.

From *Glossary of Surface Mining and Reclamation Technology*, 1974, pp. 1-23. Reprinted by permission.

Active Surface Mine Operation: A mining operation where land is being disturbed in preparation for and during the removal of a mineral.

Aeration: The act of exposing to the action of air, such as, to mix or charge with air.

Afforestation: The artificial establishment of forest crops by planting or sowing on land that has not previously, or not recently, grown tree crops.

Agricultural Limestone: Contains sufficient calcium and magnesium carbonate to be equivalent to not less than 80 percent calcium carbonate and must be fine enough so that not less than 90 percent shall pass through a U.S. Standard No. 10 sieve and not less than 35 percent shall pass through a U.S. Standard No. 50 sieve.

Alkaline Soil: Generally, a soil that is alkaline throughout most or all of the parts of it occupied by plant roots; although the term is commonly applied to only a specific layer or horizon of a soil. Precisely, any soil horizon having a pH value greater than 7.0. Practically, a soil having a pH above 7.3.

Alluvial: Describes material such as earth, sand, gravel, or other rock or mineral materials transported by and laid down by flowing water.

Amendment: Any material, such as lime, gypsum, sawdust, or synthetic conditioners, that is worked into the soil to make it more productive. Technically, a fertilizer is also an amendment but the term, "amendment" is used most commonly for added materials other than fertilizer.

Angle of Dip: The angle an inclined stratum makes with the horizontal.

Angle of Repose: The greatest angle to the horizontal that any loose or fragmented solid material will stand without sliding or come to rest when poured or dumped in a pile or on a slope.

Angle of Slide: The slope measured in degrees of deviation from the horizontal, on which loose or fragmented solid materials will start to slide. It is a slightly greater angle than the angle of repose.

Annual Plant (annuals): A plant that completes its life cycle and dies in 1 year or less.

Anticline: A fold of rock beds that is convex upward.

Aquifer: A water bearing formation through which water moves more readily than in adjacent formations with lower permeability.

Aquitard: A rock unit with relatively low permeability that retards the flow of water.

Area Mining: Surface mining that is carried on in level to gentle rolling topography on relatively large tracts.

Argillaceous: An adjective describing a rock unit having a high clay content.

Artesian: Ground water under sufficient pressure to cause the water level in a drilled hole to rise above the top of the rock unit.

Aspect: The direction toward which a slope faces. Exposure.

Attitude of Bedrock: A general term describing the relation of some directional features to a rock in a horizontal surface.

Auger Mining: Generally practiced but not restricted to hilly coal bearing regions of the country. Utilizes a machine designed on the principle of the auger, which bores in to an exposed coal seam, conveying the coal to a storage pile or bin for loading and transporting. May be used alone or in combination with conventional surface mining. When used alone, a single cut is made sufficient to expose the coal seam and provide operating space for the machine. When used in combination with surface mining the last cut pit provides the operating space.

Available Nutrient: The part of the supply of a plant nutrient in the soil that can be taken up by plants at rates and in amounts significant to plant growth.

Available Water: The part of the water in the soil that can be taken up by plants at rates significant to their growth. Usable: obtainable.

Back Blade: In regrading, to drag the blade of a bulldozer or grader as the machine moves backward, as opposed to pusing the blade forward.

Backfill: The operation of refilling an excavation. Also the material placed in an excavation in the process of backfilling.

Background Level (water quality usage): The natural or normal concentration of a chemical or other constituents in water.

Band: Slate or rock interstratified with coal; also called "slate band" or "sulphur band."

Bank: A mound of mine refuse. Synonyms include tip, culm, gob, refuse pile, slate dump, stack, and heap.

Base of Highwall: The point of intersection between the highwall and the plane formed at the base of the excavated material.

Basin: A natural depression of strata containing a coal bed or other stratified deposit.

Basin, Stilling: A structure or excavation which reduces velocity or turbulence of flowing or falling water.

Bed: A stratum of coal or other sedimentary deposit.

Bedrock: Any solid rock underlying soil, sand, clay, silt, and any other earthy materials.

Bench: The surface of an excavated area at some point between the material being mined and the original surface of the ground on which equipment can set, move or operate. A working road or base below a highwall as in contour stripping for coal.

Bench Method, Extended: The use of a large capacity walking dragline in deep overburden operating from a machine-supporting bench formed by filling the V between the bank and the spoil. This V-section is formed from material that falls from the bank or is placed by the dragline and it must be rehandled. Rehandling averages approximately 10 to 25 percent of the solid bank, depending on the height of the bench.

Bench Terrace (see **Terrace Types**).

Berm: A strip of coal left in place temporarily for use in hauling or stripping. A layer of large rock or other relatively heavy stable material placed at the outside bottom of the spoil pile to help hold the pile in position (a toe wall). Also used similarly, higher in the spoils for the same purpose.

Berm Interval: Vertical distance from crest of berm to its underlying toe as in a bank or bench.

Binder: A streak of impurity in a coal seam.

Blind Drain: A trench filled with stones selected so as to fill a trench, yet to allow the flow of water through it.

Block Coal: A bituminous coal which breaks into large cubical blocks.

Block Cut Method: This method of surface mining removes overburden and places it around the periphery of a box-shaped cut. After coal is removed the spoil is pushed back into the cut and the surface is blended into the topography.

Blossom: The decomposed outcrop of a coal bed.

Bone: Slaty coal or carbonaceous shale found in the coal seam. Also called "bone coal" or "bony coal."

Box Cut: The initial cut driven in a property, where no open side exists; this results in a highwall on both sides of the cut.

Breaker Refuse: Slate, bone, or rock as removed by hand or mechanical cleaners in all classes of breakers.

Breakthrough: Interception of an underground mine by surface or auger mining.

Broadcast Seeding: Scattering seed on the surface of the soil. Contrast with drill seeding which places the seed in rows in the soil.

Buffer: Substances in soil or water that act chemically to resist changes in reaction or pH.

Buffer Strips: (1) Unaffected areas between the mining operation and areas designated for other public and private use. (2) Strips of grass or other erosion-resisting vegetation between or below surface or auger mining disturbances.

Calcareous Soil: Soil containing sufficient calcium carbonate (often with magnesium carbonate) to effervesce visibly when treated with cold 0.1 normal hydrochloric acid.

Capillary Water: The water held in the "capillary" or small pores of a soil, usually with tension greater than 60 centimeters of water. Much of this water is considered to be readily available to plants.

CFS: Cubic feet per second—measurement of water flow.

Channel Stabilization: Erosion prevention and stabilization of velocity distribution in a channel, using jetties, drops, revetments, vegetation, and other measures.

Check Dam: Small dam constructed in a gully or other small watercourse to decrease the streamflow velocity, minimize channel scour, and promote deposition of sediment.

Clay (soils): (1) A mineral soil separate consisting of particles less than 0.002 millimeter in equivalent diameter. (2) A soil textural class. (3) (engineering) A fine-grained soil that has a high plasticity index in relation to the liquid limits.

Coal Horizon: The stratigraphic position where a coal should occur—generally used to signify the position that a coal should occupy but is absent.

Coal Seam: A layer, vein, or deposit of coal. A stratigraphic part of the earth's surface containing coal.

Coefficient, Permeability: The rate of flow of a fluid through a unit cross section of a porous mass under a unit hydraulic gradient, at a temperature of 60 degrees Fahrenheit. The standard coefficient of permeability used in the hydrologic work of the United States Geological Survey is defined as the rate of flow of water at 60 degrees Fahrenheit, in gallons or million gallons a day, through a cross section of 1 sq ft, under a hydraulic gradient of 100 percent.

Compaction: The closing of the pore spaces among the particles of soil and rock, generally caused by running heavy equipment over the area, as in the process of leveling the overburden material of strip mine banks.

Companion Crop (see **Nurse Crop**).

Concentration (water quality usage): The amount of a substance occurring in a given amount of water—common unit is parts per million (ppm) or milligrams per liter (mg/l).

Condensation Trap: A condensation trap is a deep planting basin containing a stock seedling. The basin is covered with a plastic sheet secured around the edge to hold a large amount of air. A center hole allows the plant to grow and the sheeting collects moisture which trickles down, irrigating the plant.

Conifer: A tree belonging to the order Coniferae, usually evergreen with cones and needle-shaped or scale-like leaves and producing wood known commercially as "softwood."

Contour: An imaginary line connecting points of equal height above sea level as they follow the relief of the terrain.

Contouring: (a) Sometimes used to mean contour mining. (2) Returning spoil banks to even sloped terrain features.

Contour Stripping or Surface Mining: The removal of overburden and mining from a coal seam that outcrops or approaches the surface at approximately the same elevation, in steep or mountainous areas.

Cool-Season Plant: A plant that makes its major growth during the cool portion of the year, primarily in the spring but in some localities in the winter.

Copperas Water: Water containing by suspension or in solution ferrous sulfate formed by oxidation of sulfur and sulfide of iron, usually green in color later turning reddish-brown.

Core Drilling: The process by which a cylindrical sample of rock and other strata is obtained through the use of a hollow drilling bit which cuts and retains a section of the rock or other strata penetrated.

Core, Impervious: The interior portion of an embankment so designed and constructed as to prevent the passage of water.

Crest: The top of a dam or spillway to which water must rise before passing over the structure. It is frequently restricted to the overflow portion.

Crop Coal: Coal at the outcrop of the seam. It is usually considered of inferior quality due to partial oxidation, although this is not always true.

Crust: A thin layer of hard soil that forms on the surface of many soils when they are dry.

Culm: The waste or slack of the Pennsylvania anthracite mines.

Culvert: A closed conduit for the free passage of surface drainage water under a roadway or other embankment.

Cut: Longitudinal excavation made by a strip-mining machine to remove overburden in a single progressive line from one side or end of the property.

Daylighting: A term to define the surface mining procedure for exposing an entire underground mined area to remove all of the remaining mineral underlying the surface.

Deciduous: Refers to a tree that sheds all its leaves every year at a certain season.

Deep Chiseling: Deep chiseling is a surface treatment that loosens compacted spoils. The process creates a series of parallel slots on the contour in the spoils surface which impedes water flow and markedly increases infiltration.

Deep Mine: An underground mine.

Density, Forage: The percent of ground surface which appears to be completely covered by vegetation when viewed directly from above.

Density, Stand: Density of stocking expressed in number of trees per acre.

Dibble: A planting tool used for planting seedlings.

Dip (see Angle of Dip).

Direct Seeding: A method of establishing a stand of vegetation by sowing seed on the ground surface.

Dissolved Solids: The difference between the total and suspended solids in water.

Disturbed Land: Land on which excavation has occurred or upon which overburden has been deposited, or both.

Diversion Ditch: A man-made waterway used for collecting surface runoff on the uphill side of a mine in order to keep it out of the workings; a ditch designed to change the normal or actual course of water.

Dozer or Bulldozer: Tractor with a steel plate or blade mounted on the front end in such a manner that it can be used to cut into earth or other material and move said material primarily forward by pushing.

Dragline: An excavating machine that utilizes a bucket operated and suspended by means of lines or cables, one of which hoists or lowers the bucket from a boom; the other, from which the name is derived, allows the bucket to swing out from the machine or to be dragged toward the machine for loading. Mobility of draglines is by crawler mounting or by a walking device for propelling, featuring pontoon-like feet and a circular base or tub. The swing of the machine is based on rollers and rail. The machine usually operates from the highwall.

Drainage Basin: The area from which water is carried off by a drainage system, a watershed or a catchment area.

Drainage Plan: The proposed methods of collection, treatment and discharge of all waters within the affected drainage area as defined in the pre-mining plan.

Draw Slate: A soft slate of shale immediately above certain coal seams which falls quite easily when coal support is withdrawn.

Drift: A deep mine entry driven directly into a horizontal or near horizontal mineral seam or vein when it outcrops or is exposed at the ground surface.

Ecology: The science that deals with the mutual relation of plants and animals to one another and to their environment.

Ecosystem: A total organic community in a defined area or time frame.

Effective Precipitation: That portion of total precipitation that becomes available for plant growth. It does not include precipitation lost to deep percolation below the root zone or to surface runoff.

Effluent: Any water flowing out of the ground or from an enclosure to the surface flow network.

Emergency Spillway: A spillway designed to convey water in excess of that impounded for floor control or other beneficial purposes.

Environment: All external conditions that may act upon an organism or soil to influence its development, including sunlight, temperature, moisture and other organisms.

Erodibility: The relative ease with which one soil erodes under specified conditions of slope as compared with other soils under the same conditions; this applies to both sheet and gully erosion.

Erosion: The wearing away of the land surface by running water, wind, ice, or other geological agents, including such processes as gravitational creep. Detachment and movement of soil or rock fragments by water, wind or ice, or gravity.

Essential Element (plant nutrition): A chemical element required for the normal growth of plants.

Evapo-transpiration: A collective term meaning the loss of water to the atmosphere from both evaporation and transpiration by vegetation.

Excavation: The act of removing overburden material.

Ferric Iron: An oxidized or high-valence form of iron (Fe^{+3}) responsible for red, yellow, and brown colors in soils and water.

Ferrous Iron: A reduced or low-valence form of iron (Fe^{+2}), imparting a blue-gray appearance to water and some wet subsoils on long standing.

Fertilizer: Any natural or manufactured material added to the soil in order to supply one or more plant nutrients.

Fertilizer Grade: The guaranteed minimum analysis in whole numbers, in percent, of the major plant nutrient elements contained in a fertilizer material or in a mixed fertilizer. For example, a fertilizer with a grade of 20-10-5 contains 20 percent nitrogen (N), 10 percent available phosphoric acid (P_2O_5), and 5 percent water-soluble potash (K_2O). Minor elements may also be included. Recent trends are to express the percentages in terms of the elemental fertilizer [nitrogen (N), phosphorus (P), and potassium (K)].

Fill: Depth to which material is to be placed (filled) to bring the surface to a predetermined grade. Also, the material itself.

Fill Bench: That portion of the bench which is formed by depositing overburden beyond the cut section.

Filter: A device or structure for removing solid or colloidal material, usually of a type that cannot be removed by sedimentation, from water, sewage, or other liquid.

Final Cut: Last cut or line of excavation made on a specific property or area.

Fire Clay: Soft, unbedded, gray or white clay, high in silica and hydrated aluminum silicates, and low in iron and alkalies. Fire clay forms the seat earth of many coalbeds and may have value of refractory clay.

Flood, 10-year: The flow of a stream which has been equaled or exceeded, on the average once in 10 years (or other designated period).

Flood, Annual: Annual—a flood equal to the mean of the discharges of all of the maximum annual floods during the period of record. Minimum—the smallest of the annual floods during the period of record.

Flood, Possible, Maximum: The largest flood that theoretically can occur at a given site during present geologic and climatic era, assuming simultaneous occurrence of all possible flood producing factors in the area.

Flood, Probable: The maximum flood for which there is a reasonable chance that it will occur on a given stream at a selected site. It is often assumed to be equal to the maximum flood observed in areas having the same or similar physiographic and meteorological characteristics. Such a flood would very likely be less than the maximum possible flood.

Flora: The sum total of the kinds of plants in an area at one time.

Flume: An open channel on a prepared grade.

Flushout: An accumulation of water that is removed suddenly from surface-mined lands by heavy precipitation. Such water may contain contaminants in sufficient concentration to result in pollution of the receiving stream.

Fold Axis: A line drawn on a rock stratum at the point of intersection of the axial plane (which is a plane passing through the trough of a syncline or the peak of an anticline).

Forage: Unharvested plant material which can be used as feed by domestic animals. Forage may be grazed or cut for hay.

Forest Land: Land bearing a stand of trees at any age or stature, including seedlings and of species attaining a minimum of 6 feet average height at maturity or land from which such a stand has been removed but on which no other use has been substituted. The term is commonly limited to land not in farms, forests on farms are commonly called woodland or farm forests.

Fracture: The character or appearance of a freshly broken surface of any rock or mineral.

Freeboard: The vertical distance between the normal maximum level of the surface of the liquid in a reservoir and the top of a dam or levee, etc., which is provided so that waves and other movements of the liquid will not overtop the confining structure.

French Drain: A covered ditch containing a layer of fitted or loose stone or other pervious material.

Furrow Grading: A method of land surface grading which uses narrow trenches or furrows longitudinally on a sloping area for the purpose of retaining rainfall, preventing erosion, and enhancing plant growth.

Gabion: A mesh container used to confine rocks or stones and used to construct dams and groins or lining stream channels.

Game Food: Numerous grasses, legumes, shrubs, and trees that are planted on reclaimed mined-land to provide sustenance and cover for wildlife whose habitat is natural or adaptable to that particular site.

Georgia V-Ditch: Grading is performed to create positively draining swales midpoint between and parallel to the highwall and lowwall to convey water runoff to drains established to carry the water away from the spoil area.

Germination: Sprouting; beginning of growth.

Gob: Waste coal, rock pyrites, slate, or other unmerchantable material of relatively large size which is separated from coal and other mined material in the cleaning process.

Gouging: Gouging is a surface configuration intended to trap precipitation, increase infiltration and reduce erosion.

Gradation: A term used to describe the series of sizes into which a soil sample can be divided.

Grade: (1) The inclination or slope of a stream channel or ground surface, usually expressed in terms of the ratio or percentage of number of units of vertical rise or fall per unit of horizontal distance. (2) The finished surface of a road bed, top of an embankment or bottom of an excavation. (3) To establish a profile by backfilling.

Gradient (see Texture): The rate of change of any characteristic per unit of length, or slope. The term is usually applied to such things as elevation, velocity, pressure, etc.

Grain Size: Physical size of soil particle, usually determined by either sieve or hydrometer analysis.

Ground Cover: Any living or dead vegetative material producing a protecting mat on or just above the soil surface.

Ground Water: Subsurface water occupying the *saturation zone,* from which wells and springs are fed. In a strict sense the term applies only to water below the water table. Also called *plerotic water; phreatic water.*

Growing Season: The season which in general, is warm enough for the growth of plants, the extreme average limits of duration being from the average date of the last killing frost in spring to that of the first killing frost in autumn. On the whole, however, the growing season is confined to that period of the year when the daily means are above 42 F.

Grubbing: The operation of removing stumps and roots.

Gully Erosion: Removal of soil by running water, with formation of deep channels that cannot be smoothed out completely by normal cultivation.

Habitat: The environment in which the life needs of a plant or animal are supplied.

Haulback Mining Method (see Lateral Movement).

Haul Road: Road from pit to loading dock, tipple, ramp, or preparation plant used for transporting mined material by truck.

Head of the Hollow (also Valley Fill Method): Basically, overburden material from adjacent contour or mountain top mines is placed in compacted layers in narrow, steep-sided hollows so that surface drainage is possible.

Head (of water): Water pressure expressed as the feet of elevation difference between the top of the water and the point at which the pressure is exerted.

Highwall: The unexcavated face of exposed overburden and coal in a surface mine or the face or bank on the uphill side of a contour strip mine excavation.

Hot: Refers to material in the overburden or gob piles either capable of spontaneous ignition or containing highly acid producing material.

Hp: The difference in elevation of the emergency spillway crest and the water surface elevation in the reservoir when required to pass the design emergency spillway discharge.

Hydrology: The science that relates to the water systems of the earth.

Hydroseeding: Dissemination of seed hydraulically in a water medium. Mulch, lime, and fertilizer can be incorporated into the sprayed mixture.

Impervious: Prohibits fluid flow.

Impoundment: A reservoir for collection of water. Collection of water by damming a stream or the like. Used in connection with the storage of tailings from a mine.

Infiltration: Water entering the ground water system through the land surface.

Interstice: A pore or small open space in rock or granular material, not occupied by solid mineral matter. It may be occupied by air, water, or other gaseous or liquid material. Also called void or void space.

Inundate: To flood an entire mine with water or a large section of the workings.

Intermittent Stream: A stream or portion of a stream that flows only in direct response to precipitation. It receives little or no water from springs and is dry for a large part of the year.

Land Classification: Classification of specific bodies of land according to their characteristics or to their capabilities for use. A use capability classification may be defined as one based on both physical and economic considerations according to their capabilities for man's use, with sufficient detail of categorical definition and cartographic (mapping) expression to indicate those differences significant to men.

Land Use Planning: The development of plans for the uses of land that, over long periods, will best serve the general welfare, together with the formulation of ways and means for achieving such uses.

Lateral Movement: This method removes coal by stripping and augering with no material being placed on the downslope. Lateral movement reduces disturbed acreage by nearly two thirds when compared with conventional surface mining because the overburden is hauled by truck laterally along the bench and then backfilled against the highwall.

Leaching: The removal of materials in solution by the passage of water through soil.

Leachate: Liquid that has percolated through a medium and has extracted dissolved or suspended materials from it.

Legume: A member of the legume or pulse family, leguminosae. One of the most important and widely distributed plant families. Includes many valuable food and forage species, such as the peas, beans, peanuts, clovers, alfalfas, sweet clovers, lespedezas, vetches and kudzu. Practically all legumes are nitrogen-fixing plants.

Level Terrace: A broad surface channel or embankment constructed across sloping soil on the contour, as contrasted to a graded terrace which is built at a slight angle to the contour.

Lime: Lime, from the strictly chemical standpoint, refers to only one compound, calcium oxide (CaO); however, the term lime is commonly used in agriculture to include a great variety of materials which are usually composed of the oxide, hydroxide, or carbonate of calcium or of calcium and magnesium. The most commonly used forms of agricultural lime are ground limestone, marl, and oyster shells (carbonates), hydrated lime (hydroxides), and burnt lime (oxides).

Quickline	– limestone + heat (calcined) $\rightarrow CaO$
Hydrated lime	– quicklime + $H_2O \rightarrow Ca(OH)_2$
Slaked lime	– same as hydrated but slaking equipment is used for adding water
Milk of lime	– water mixture containing lime in solution + lime in suspension

Lime Requirement: The amount of standard ground limestone required to bring a 6.6 inch layer of an acre (about 2 million pounds of mineral soils) of acid soil to some specific lesser degree of acidity, usually to slightly or very slightly acid. In common practice, lime requirements are given in tons per acre of nearly pure limestone, ground finely enough so that all of it passes a 10-mesh screen and at least half of it passes a 100-mesh screen.

Limestone: A bedded sedimentary deposit consisting chiefly of calcium carbonate ($CaCO_3$) which yields lime (CaO) when burned. A general term for that class of rocks which contain at least 80 percent of the carbonates of calcium or magnesium.

Litter: Freshly fallen or slightly decomposed organic debris.

Load (water quality use): The quantity of material carried by flowing water—generally expressed as pounds per day.

Longwall Stripping: Longwall mining accomplished in areas of shallow cover where surface mining might normally have been conducted. The outby end, where the longwall controls, pumps, and face conveyor discharge end are located, is located in a ditch that is exposed to the surface. Roof chocks are used to protect the mining area and the roof (or overburden) is allowed to settle into the mined out section.

Low Wall: The vertical wall, on the downslope side of the mining operation, consisting of the deposit being mined and some overlying rock and soil strata.

Lysimeter: Structure containing a mass of soil and so designed as to permit the measurement of water draining through the soil.

Mattress: A blanket made up of brush and poles interwoven or otherwise lashed together, placed on the banks of streams to prevent scour from currents, and weighted with rocks, concrete blocks, etc. to hold it in place.

Mg/L: Abbreviation for milligrams per liter, which is a weight to volume ratio commonly used in water quality analysis. It expresses the weight in milligrams of a substance occurring in one liter of liquid.

Micro-climate: A local climatic condition near the ground resulting from modification of relief, exposure, or cover.

Micro-nutrients: Nutrients in only small, trace, or minute amounts.

Mine Drainage: Any water forming on or discharging from a mining operation. May be alkaline or acid in nature.

Mine Dumps: Surface area used for disposal of mine and/or preparation plant waste.

Mined-Land: Land with new surface characteristics due to the removal of mineable commodity by surface mining methods and subsequent surface reclamation.

Modified Block Cut: This method of surface mining adapts the "Block cut" method to steeply sloped areas. The modification essentially is backfilling with spoil from succeeding blocks rather than from the spoil-producing block.

Mountain Top Removal: In this mining method, 100 percent of the overburden covering a coal seam is removed in order to recover 100 percent of the mineral. Excess spoil material is hauled to a nearby hollow to create a valley fill.

Mulch: A natural or artificial layer of plant residue or other materials placed on the soil surface to protect seeds, to prevent blowing, to retain soil moisture, to curtail erosion, and to modify soil temperature.

Multiple Seam Mining: Surface mining in areas where several seams are recovered from the same hillside.

Natural Draining: Any water course which has a clearly defined channel, including intermittent streams.

Natural Revegatation: Natural re-establishment of plants; propagation of new plants over an area by natural processes.

Natural Seeding (volunteer): Natural distribution of seed over an area.

Neutralization: The process of adding an acid or alkaline material to water or soil to adjust its pH to a neutral position.

Neutral Soil: A soil in which the surface layer, at least to normal plow depth, is neither acid nor alkaline in reaction. For most practical purposes soil with a pH ranging from 6.6 through 7.3.

Nitrogen Fixation: The conversion of atmospheric (free) nitrogen to nitrogen compounds. In soils the assimilation of free nitrogen from the air by soil organisms (making the nitrogen eventually available to plants). Nitrogen fixing organisms associated with plants such as the legumes are called symbiotic; those not definitely associated with plants are called nonsymbiotic.

Nurse Crop: A planting or seeding that is used to protect a tender species during its early life. A nurse crop is usually temporary and gives way to the permanent crop. Sometimes referred to as a companion crop.

Nutrients: Any element taken into a plant that is essential to its growth.

Opencut: A method of excavation in which the working area is kept exposed.

Open Pit Mining: Surface mining, a type of mining in which the overburden is removed from the product being mined and is dumped back after mining; or may specifically refer to an area from which the overburden has been removed and has not been filled.

Operation: All of the premises, facilities, railroad loops, roads, and equipment used in the process of extracting and removing a mineral commodity from a designated surface mine or in the determination of the location, quality, and quantity of a natural mineral deposit.

Orphan Banks: Abandoned surface mines, operated prior to the enactment of comprehensive reclamation laws, that require additional reclamation.

Outcrop: Coal which appears at or near the surface; the intersection of a coal seam with the surface.

Outslope: The exposed area sloping away from a bench cut section.

Overburden: The earth, rock, and other materials which lie above the coal.

Over the Shoulder: A method of handling overburden whereby it is moved parallel to the highwall instead of at right angles to the wall as normally done.

Parent Material: The unconsolidated mass of rock material (or peat) from which the soil profile develops.

Peak: The tops of strip mine banks before grading.

Percolation: Downward movement of water through soils.

Permeability: The measure of the capacity for transmitting a fluid through the substance.

pH: The symbol or term refers to a scale commonly used to express the degrees of acidity or alkalinity. On this scale pH of 1 is the strongest acid, pH of 14 is the strongest alkali, pH of 7 is the point of neutrality at which there is neither acidity or alkalinity. pH is not a measure of the weight of acid or alkali contained in or available in a given volume.

Photogrammetrics: The process of creating topographical mapping from stereo aerial photographs.

Piezometer: An instrument for measuring pressure head in a conduit, tank, soil, etc. It usually consists of a small pipe or tube tapped into the side of the container, the inside end being flush with, and normal to, the water face of the container, connected with a manometer pressure gauge, mercury or water column, or other device for indicating pressure head.

Pit: Used in reference to a specifically describable area of open cut mining. May be used to refer to only that part of the open cut mining area from which coal is being actively removed or may refer to the entire contiguous mined area.

Pit, Borrow: A bank or pit from which earth is taken for use in filling or embanking.

Pitch (see Angle of Dip).

Pit, Test: Test pits, dug by hand or machine, are open excavations large enough to permit a man to enter and examine formations in their natural condition. It is by far the most accurate method of determining the character of materials, but it is also the most costly. Samples in the disturbed or undisturbed condition can be readily taken.

Pollution: Environmental degradation resulting from man's activities or natural events.

Pond: A body of water of limited size either naturally or artificially confined and usually smaller than a lake.

Porosity—Effective: The ratio, usually expressed in percentage of (a) the volume of water or other liquid which a given volume of rock or soil, after being saturated with the liquid, will yield under any specified hydraulic conditions, to (b) the total volume of soil or rock.

Portal: The surface entrance to an underground mine.

Power Shovel: A large machine for digging, the digging part of which the bucket is the terminal member of an articulated boom. Power to the bucket is supplied either through hydraulic cylinders or cables.

Pre-law: Term used to refer to strip mine operations conducted previous to the passage of a state's first reclamation act.

Preplanning: Process of foreseeing reclamation problems and determining measures to minimize off-site damages during the mining operation and to provide for quick stabilization of the mining.

Prospecting: The removal of overburden, core drilling, construction of roads, or any other disturbance of the surface for the purpose of determining the location, quantity, or quality of the natural mineral deposit.

Pullback Machine: This is the use of a smaller dragline to assist the prime-mover machine when the cover is beyond the capability of the shovel or dragline to spoil the overburden without filling the area adjacent to the highwall.

Pyrite: A yellowish mineral, iron disulfide, FeS_2, generally metallic appearing; also known in "fool's gold."

Rail Waste Dumps: A dump which should be a train length long or more.

Rain: (1) Heavy—Rain which is falling at the time of observation with an intensity in excess of 0.30 in. per hr (over 0.03 inch in 6 min). (2) Light—Rain which is falling at the time of observation with an intensity of between a trace and 0.10 in. per hr (0.01 inch in 6 min). (3) Moderate—Rain which is falling at the time of observation with an intensity of between 0.11 in. per hr (0.01+ inch in 6 min) and 0.30 in. per hr (0.03 inch in 6 min).

Range Land: Land where the potential natural vegetation is predominantly grasses, grasslike plants, forbs, or shrubs, where natural herbivory was an important influence in its precivilization state, and that is more suitable for management by ecological rather than agronomic principles.

Rate, Percolation: The rate, usually expressed as a velocity, at which water moves through saturated granular material. The term is also applied to quantity per unit of time of such movement, and has been used erroneously to designate infiltration rate or infiltration capacity.

Reclamation: The process of reconverting mined land to its former or other productive uses.

Recreation Land: Land and water used, or usable primarily as sites for outdoor recreation facilities and activities.

Red Dog: A gob pile after it has burned. The material is generally used as a road surfacing material; it has no harmful acid or alkaline reaction.

Reforestation: The natural or artificial restocking of an area with forest trees.

Refuse: All the solid waste from a coal mine, including tailings and slurry. Other synonyms are: dirt, gob, shale, slate, etc.

Regrading: The movement of earth over a depression to change the shape of the land surface. A finer form of backfilling.

Rehabilitation: Implies that the land will be returned to a form and productivity in conformity with a prior land use plan, including a stable ecological state that does not contribute substantially to environmental deterioration and is consistent with surrounding aesthetic values.

Reject: The material extracted from the feed coal during cleaning for retreatment or discard. The stone or dirt discarded from a coal preparation plant, washery or other process, as of no value.

Reservoir: A pond, lake, tank, basin, or other space, either natural in its origin, or created in whole or in part by the building of engineering structures, which is used for storage, regulation, and control of water.

Restoration: The process of restoring site conditions as they were before the land disturbance.

Retention: The amount of precipitation on a drainage area that does not escape as runoff. It is the difference between the total precipitation and total runoff.

Revegetation: Plants or growth which replaces original ground cover following land disturbance.

Reverse Terrace (see **Georgia V-Ditch**).

Rider Coal Seam: A "stray" coal seam usually above and divided from the main coal bed by rock, shale or other strata material. The rider seam is generally thin and seldom merchantable.

Ripping: The act of breaking, with a tractor-drawn ripper or long angled steel tooth, compacted soils or rock into pieces small enough to be economically excavated or moved by other equipment as a scraper or dozer.

Riprap: Broken rock, cobbles, or boulders placed on earth surfaces, such as the face of a dam, bank of a stream or lining drainage channels, for protection against the action of water.

Rise: Same as dip except taken in the upward direction versus the downward.

Rock-Fill Dam: A dam composed of loose rock usually dumped in place, often with the upstream part constructed of handpacked or derrick-placed rock and faced with rolled earth or with an impervious surface of concrete, timber or steel.

Runoff: That portion of the rainfall that is not absorbed by the deep strata: is utilized by vegetation or lost by evaporation or may find its way into streams as surface flow.

Runoff Volume: The total quantity or volume of runoff during a specified time. It may be expressed in acre-feet, in inches depth on the drainage area, or in other units.

Saline-Alkali Soil: A soil having a combination of a harmful quantity of salts and either a high degree of alkalinity or a high amount of exchangeable sodium, or both, so distributed in the soil profile that the growth of most crop plants is less than normal.

Saline Soil: A soil containing enough soluble salts to impair its productivity for plants but not containing an excess of exchangeable sodium.

Sandstone: A cemented or otherwise compacted detrital sediment composed predominantly of quartz grains, the grades of the latter being those of sand.

Saturation: Completely filled; a condition reached by a material, whether it be in solid, gaseous, or liquid state, which holds another material within itself in a given state in an amount such that no more of such material can be held within it in the same state. The material is then said to be saturated or in a condition of saturation.

Scalping: Removal of vegetation before mining.

Scarify: To loosen or stir the surface soil without turning it over. Also, in the case of legume seeds, abrasion of the hard coat to decrease time required for germination.

Scoria: A slaglike clinker deposit characteristic of burned-out coal beds, especially in the western Great Plains.

Seam: A stratum or bed of coal.

Secondary Blasting: Reblasting oversize pieces of rock to reduce them to a size suitable for handling by the excavating, materials transporting and crushing equipment available.

Secondary Coal Recovery: This is analogous to "mining the pillars" in deep mining in that the coal is taken on the retreat or after primary mining has been completed. When the final cut has been made in an open-cut coal mine the coal seam remains exposed in the bank and three methods are used to recover as much of the coal as can be economically won (a) coal augers, (b) push-button miner and (c) punch mining by underground machines.

Sediment: Solid material, both mineral and organic, that is in suspension, is being transported, or has been moved from its site of origin by air, water, gravity, or ice and has come to rest on the earth's surface either above or below sea level.

Sediment Basin: A reservoir for the confinement and retention of silt, gravel, rock, or other debris from a sediment-producing area.

Sediment Structure: A barrier or dam constructed across a waterway or in other suitable locations to form a silt or sediment basin.

Seedbed: The soil prepared by natural or artificial means to promote the germination of seed and the growth of seedlings.

Seep: A more or less poorly defined area where water oozes from the earth in small quantities.

Shaft Mine: One where the coal seam is reached by a vertical shaft which may vary in depth from less than 100 feet to several thousand feet. A mine in which the main entry or access is by means of a shaft.

Shale: Sedimentary or stratified rock structure generally formed by the consolidation of clay or clay-like material.

Shotgun Mixture: Seeding a number of species at random.

Shovel: Excavating or coal-loading machine that utilizes a bucket mounted on and operated by means of a handle or dipper stick that moves longitudinally on gears and which is lifted or lowered by cable. The entire machine is mounted on crawlers for mobility and the upper structure is mounted on rollers and rail for swing or turn.

Shovel Overcasting: The process of overcasting material from the bank into the mined-out pit when digging straight ahead and spoiling at a swinging angle averaging 90 degrees or less.

Side Slopes: The slope of the sides of a canal, dam, or embankment. It is customary to name the horizontal distance first as 1.5 to 1.0 or frequently 1-1/2:1 meaning a horizontal distance of 1.5 feet to 1 foot vertical.

Silt: Small mineral soil grains the particles of which range in diameter from 0.05 to 0.002 mm (or 0.02-0.002 mm in the international system).

Slate: A miner's term for any shale or slate accompanying coal. Geologically it is a dense, fine textured, metamorphic rock, which has excellent parallel clearage so that it breaks into thin plates or pencil-like shapes.

Slip or Slide: A mass of spoil material that moves downward and outward to a lower elevation due to the force of gravity, generally caused by overloading of the downslope, freezing and thawing, or saturation of the fill.

Slope (see **Grade (1)**).

Slope Down: Modification of grade.

Slope Mine: A mine opened by a slope or incline. A mine with an inclined opening used for the same purpose as a shaft or a drift mine.

Slope Reduction Method: Overburden from the first cut is spread over a predetermined area of the outslope in such a manner that the natural angle of slope is reduced.

Slope Stability: The resistance of any inclined surface, as the wall of an open pit or cut, to failure by sliding or collapsing.

Sludge: The precipitate resulting from chemical treatment of water, co-agulation, or sedimentation.

Sludge Density: A measure of the weight of solids contained in the sludge in relation to total weight.

Slurry: Refuse separated from the coal in the coal cleaning process of relatively small size which is readily pumpable in the washing plant effluent. A pulverized coal-liquid mixture transported by pipeline.

Soil (see Acid Soil and Alkaline Soil): Surface layer of the earth, ranging in thickness from a few inches to several feet composed of finely divided rock debris mixed with decomposing vegetative and animal matter which is capable of supporting plant growth.

Soil Condition: Any material added to a soil for the purpose of improving its physical condition.

Soil Conserving Crops: Crops that prevent or retard erosion and maintain or replenish rather than deplete soil organic matter.

Soil Porosity: The degree to which the soil mass is permeated with pores or cavities. It is expressed as the percentage of the whole volume of the soil which is unoccupied by solid particles.

Soil Profile: A vertical section of the soil through all its horizons and extending into the parent material.

Soil Structure: The combination or arrangement of primary soil particles into secondary particles, units, or beds.

Solid Bench: That portion of the bench located on undisturbed material.

Solum: The upper part of a soil profile, above the parent material, in which the processes of soil formation are active. The solum in mature soils includes the A and B horizons. Usually the characteristics of the material in these horizons are quite unlike those of the underlying parent material. The living roots and other plant life and animal life characteristic of the soil are largely confined to the solum.

Spillway: A waterway in or about a dam or other hydraulic structure for the escape of excess water.

Spoil (see Acid Spoil and Toxic Spoil): The overburden or non-coal material removed in gaining access to the coal or mineral material in surface mining.

Spoil Bank (spoil pile): Area created by the deposited spoil or overburden material prior to backfilling. Also called cast overburden.

Stabilize: Settle, fix in place, non-moving, usually accomplished on overburden by planting trees, shrubs, or grasses, or by mechanical compaction or aging.

Stratified: Composed of, or arranged in, strata or layers, as stratified alluvium. The term is applied to geological materials. Those layers in soils that are produced by the processes of soil formation are called horizons, while those inherited from parent material are called strata.

Strike: The course or bearing of the outcrop of an inclined bed or structure on a level surface. Also the direction or bearing of a horizontal line in the plane of an inclined stratum, joint, fault, cleavage plane, or other structural plane. It is always perpendicular to the dip direction.

Strike-off: Mechanically removing the apex of a spoilbank to provide a truncated condition. Spoils are partly graded so that the tops of the ridges are reduced 4 to 6 feet and the depressions are partly filled in.

Strip: To mine a deposit by first taking off the overlying burden.

Strip Mine: Refers to a procedure of mining which entails the complete removal of all material from over the product to be mined in a series of rows or strips; also referred to as "open pit," or "surface mine."

Strip Mining (see Surface Mining).

Stripping: The removal of earth or non-ore rock materials as required to gain access to the ore or mineral materials wanted. The process of removing overburden or waste material in a surface mining operation.

Stripping Ratio: The unit amount of spoil or waste (overburden) that must be removed to gain access to a similar unit amount of ore or mineral material.

Structure Contour: A line shown on a map that is drawn through all points of equal elevation on an inclined strata.

Subdrain: A pervious backfilled trench containing a pipe or stone for the purpose of intercepting ground water or seepage.

Subsidence: The surface depression over an underground mine that has been created by subsurface caving.

Subsidence Break: The fracture in the rocks overlying a coal seam or mineral deposit as a result of its removal by mining operations. The subsidence break usually extends from the face upwards and backwards over the unworked area.

Subsoil: The B horizon of soils with distinct profiles. In soils with weak profile development, the subsoil can be defined as the soil below the plowed soil (or its equivalent of surface soil) in which roots normally grow. Although a common term, it cannot be defined accurately. It has been carried over from early days when "soil" was conceived only as the plowed soil and that under it as "subsoil."

Succession Planting: Seeding and planting to accelerate natural vegetative succession.

Sulfuric Materials: Mineral material or compounds containing sulfur that can be oxidized in the presence of moisture to form acid.

Surface Mining: Mining method whereby the overlying materials are removed to expose the mineral for extraction.

Surface Soil: That part of the upper soil of arable soils commonly stirred by tillage implements or an equivalent depth (5 to 8 inches) in non-arable soils.

Survival Rate: Percentage of plants surviving after a given time period.

Suspended Solids: Sediment which is in suspension in water but which will physically settle out under quiescent conditions (as differentiated from dissolved material).

Swale Grading (see Georgia V-Ditch).

Swallow Tail (see Georgia V-Ditch).

Sweet: Refers to the lime content or calcareous condition of the overburden which indicates a neutral or slightly alkaline material capable of supporting certain calcium-demanding plants; indicates a pH of 7 or above.

Syncline: A fold of rock beds that is convex downward.

Tacking: The process of binding mulch fibers together by the addition of a sprayed chemical compound.

Tailings: Mineral refuse from a milling operation usually deposited from a water medium.

Tailings Dam: One to which slurry is transported, the solids settling while the liquid may be withdrawn.

Terrace: Sloping ground cut into a succession of benches and steep inclines for purposes of cultivation or to control surface runoff and minimize soil erosion.

Terraced Slope: A slope that is intersected by one or more terraces.

Terrace Types: Absorptive—a ridge type of terrace used primarily for moisture conservation. Bench—a terrace approximately on the contour, having a steep or vertical drop to the slope below, and having a horizontal or gentle sloping part. It is adapted to steeper slopes. Drainage—a broad channel-type terrace used primarily to conduct water from the area at a low velocity. It is adapted to less absorptive soil and regions of high rainfall.

Texture: The character, arrangement and mode of aggregation of particles which make up the earth's surface.

Tilth: The physical condition of a soil in respect to its fitness for the growth of a specified plant.

Toe: The point of contact between the base of an embankment or spoil bank and the foundation surface. Usually the outer portion of the spoil bank where it contacts the original ground surface.

Topography: The shape of the ground surface, such as hills, mountains or plains. Steep topography indicates steep slopes or hilly land; flat topography indicates flat land with minor undulations and gentle slopes.

Topsoil: Presumed fertile soil material—used as a top dressing. Distinction has been made among synthetic, weathered, and geologic topsoil. Synthetic topsoil can include sand and stone chips as well as fly ash, sawdust, or manure not usually a part of geological soil and rock. Weathered topsoil is the natural top-dressing material that has been subjected to weathering throughout geologic time.

Toxic Spoil (see also **Acid Spoil**): Includes acid spoil with pH below 4.0. Also refers to spoil having amounts of minerals such as aluminum, manganese, and iron that adversely affect plant growth.

Transpiration: The normal loss of water vapor to the atmosphere from plants.

Tubelings: Tubelings are plant seedlings planted and nursery developed in reinforced paper cores or tubes. When the root system develops, the tubeling is ready for transplanting.

Unconsolidated (soil material): Soil material in a form of loose aggregation.

Underclay: A bed of clay highly siliceous in some cases and highly aluminous in many others, occurring immediately beneath a coal seam, and representing the soil in which the trees of the Carboniferous swamp forests were rooted.

Undergrowth: Seedlings, shoots, and small saplings under an existing stand of trees.

Valley Fill (see **Head of the Hollow**).

Vegetation: General term including grasses, legumes, shrubs, trees naturally occurring and planted intentionally.

Vegetative Cover: The entire vegetative canopy on an area.

Volunteer: Springing up spontaneously or without being planted; a volunteer plant.

Washery Refuse: The refuse removed at preparation plants from newly mined coal.

Waste (see **Refuse**).

Water Bar: Any device or structure placed in or upon a haul or access road for the purpose of channeling or diverting the flow of water off the road.

Water Conservation: The physical control, protection, management, and use of water resources in such a way as to maintain crop grazing and forest lands, vegetal cover, wildlife and wildlife habitat for maximum sustained benefits to people, agriculture, industry, commerce and other segments of the national economy.

Watershed Area: Surface region or area contributing to the supply of a stream or lake, drainage area, drainage basin, catchment area.

Water, Subsurface: Water that occurs beneath the surface of the earth. It may be liquid, solid or gaseous state. It comprises suspended water and ground water.

Weathering: The group of processes, such as chemical action of air and rainwater and of plants and bacteria and the mechanical action of changes in temperature, whereby rocks, on exposure to the weather, change in character, decay, and finally crumble.

Weir: A notch over which liquids flow and which is used to measure the rate of flow. A dam across a stream for diverting or measuring the flow. Note: The essential difference between an orifice and a weir is implicit in the expression: water flows through an orifice but over a weir. Orifice usually refers to comparatively small devices whereas weir includes both small and large structures. Further, the stream of water which is known as jet when issuing from an orifice becomes the nappe, sheet or vein when flowing over a weir.

Wildlife: Undomesticated vertebrate animals, except fish, considered collectively.

Zingg Bench Terrace: A special type of bench terrace designed for dryland moisture conservation. It employs an earthen embankment similar to the ridge terrace: a part of the terrace interval immediately above the ridge is bench levelled. Runoff water from the sloping area is retained on the levelled area and absorbed by the soil.

Definition Sources

"A Dictionary of Mining, Mineral, and Related Terms," U.S. Bureau of Mines, 1968.

Box, T. W., "Land rehabilitation: prompt passage of federal reclamation law recommended by Ford Foundation Study," Coal Age 79 (5) 108-111, 113-115, 117-118 (1974).

Chironis, N. P., "New surface mining methods simplify reclaiming of spoils," Coal Age 78 (4) 80-84 (April 1973).

"Coal Mine Health and Safety Inspection Manual for Coal Waste Deposits," U.S. Department of the Interior Sept. 1973.

"Coal Mine Refuse Disposal in Great Britain," The Pennsylvania State University Special Research Report March 31, 1971.

Cummins, A. B. and Given, I. A., "SME Mining Engineering Handbook— Seeley W. Mudd Series," 1973.

"Forest Terminology: A Glossary of Technical Terms Used in Forestry," 3rd ed., Washington: Society of American Foresters, 1958. 97 pp.

Harris, T. O., "Engineers Handbook on Strip Mining in Eastern Kentucky," Dept. for Natural Resources and Environmental Protection.

Hodder, R. L., "Surface mine, land reclamation research in eastern Montana," Symposium on Mined Land Reclamation, pp. 85-86, 1970.

House Bill No. 36 Gen'l Assembly Commonwealth of Ky.

Jones, Donald C., "Coal Mining," Pennsylvania State College, 1946.

Kentucky Guide for Classification, Use and Vegetative Treatment of Surface Mine Spoil, U.S. Dept. of Agriculture Soil Conservation Service, 1971.

Krause, R. R., "New methods in mined land reclamation," AIME Environmental Conservation Session, St. Louis, Mo., Oct. 22, 1970, Preprint No. 70F-354.

Krause, R. R., "New mining methods being developed," Green Lands 2 (11) 8-9 (1972). 4th Mine Drainage Symposium, p. 427.

May, R. F. (Central States Forest Exp. Sta.) "Predicting outslopes of spoil banks," U.S. Forest Service Research Note CS-15, Nov. 1963.

Mine Drainage Pollution Watershed Survey, Skelly and Loy, Baker-Wibberly & Assoc. Inc., State of Maryland Department of Natural Resources—Publisher.

Mulhern, J. J. and Lusk, B. E., "WVSMRA receives federal grant, new surface mining technology to minimize environmental disturbance." Green Lands 3 (3) 16-18 (1973).

Peele, Robert, "Mining Engineers Handbook," Vol. II, Feb. 1959.

Processes, Procedures and Methods to Control Pollution from Mining Activities EPA-430/9-73-011. Skelly & Loy, Engineers—Consultants and Penn Environmental Consultants, Inc. October 1973.

Riley, C. V., "Chemical alterations of strip mine spoil by furrow grading—revegetation success," Ecology and Reclamation of Devastated Land, Vol. 2, R. J. Hutnik and G. Davis, Eds., New York: Gordon & Breach Paper VI-12, pp. 315-331.

Soil Survey Material U.S.D.A. Agriculture Handbook 18, 1951, p. 185.

Smith, R. M., "Choosing topsoil to fit the needs," Green Lands 3 (2) 30-1 (1973).

"Steep slope mining . . . a new concept," Green Lands 3 (2) 4-6 (1973).

Strip Mining in Kentucky, The Kentucky Dept. of Natural Resources, 1965.

"Surface Mining," Seeley W. Mudd Series by E. P. Pfleider AIME 1972.

"Surface mining coal via longwall method," Coal Mining & Processing 10 (10) 60-1 (1973).

Sutton, P., "Establishment of vegetation on toxic coal mine spoils," Paper presented Research Applied Technology Symp. Mined Land Reclamation 1973, pp. 153-58.

"Thirty member companies attend tour of 'Pennsylvania Box Cut'," Green Lands 4 (1) 27 (1974) Journal.

U.S. Department of Agriculture, "Glossary," in "Soil: the Yearbook of Agriculture, 1957," Washington: Government Printing Office, 1957, pp. 751-770.

U.S. Environmental Protection Agency, "Environmental Protection in Surface Mining of Coal," Elmore C. Grim and Ronald D. Hill, Washington: (to be published November 1974) 291 pp.

U.S. Geological Survey, "General introduction and hydrologic definitions," by W. B. Langbein and K. T. Iseri, in "Manual of Hydrology: Part 1. General Surface-Water Techniques," Water-Supply Paper 1541-A (1960). 29 pp.

Weimer, W. H., "Production engineering in surface coal mines," Surface Mining, E. P. Pfleider, Ed., New York: AIME Seeley W. Mudd Ser. 1968, Reprinted 1972, pp. 224-246.

"West Virginia Surface Mining and Reclamation Association: Special Report," undated.

3 20